Physical and Technical Problems of SOI Structures and Devices

NATO ASI Series

Advanced Science Institutes Series

A Series presenting the results of activities sponsored by the NATO Science Committee, which aims at the dissemination of advanced scientific and technological knowledge, with a view to strengthening links between scientific communities.

The Series is published by an international board of publishers in conjunction with the NATO Scientific Affairs Division

A	Life Sciences	Plenum Publishing Corporation
B	Physics	London and New York
C	Mathematical and Physical Sciences	Kluwer Academic Publishers
D	Behavioural and Social Sciences	Dordrecht, Boston and London
E	Applied Sciences	
F	Computer and Systems Sciences	Springer-Verlag
G	Ecological Sciences	Berlin, Heidelberg, New York, London,
H	Cell Biology	Paris and Tokyo
I	Global Environmental Change	

PARTNERSHIP SUB-SERIES

1. Disarmament Technologies — Kluwer Academic Publishers
2. Environment — Springer-Verlag / Kluwer Academic Publishers
3. High Technology — Kluwer Academic Publishers
4. Science and Technology Policy — Kluwer Academic Publishers
5. Computer Networking — Kluwer Academic Publishers

The Partnership Sub-Series incorporates activities undertaken in collaboration with NATO's Cooperation Partners, the countries of the CIS and Central and Eastern Europe, in Priority Areas of concern to those countries.

NATO-PCO-DATA BASE

The electronic index to the NATO ASI Series provides full bibliographical references (with keywords and/or abstracts) to more than 50000 contributions from international scientists published in all sections of the NATO ASI Series.
Access to the NATO-PCO-DATA BASE is possible in two ways:

– via online FILE 128 (NATO-PCO-DATA BASE) hosted by ESRIN,
Via Galileo Galilei, I-00044 Frascati, Italy.

– via CD-ROM "NATO-PCO-DATA BASE" with user-friendly retrieval software in English, French and German (© WTV GmbH and DATAWARE Technologies Inc. 1989).

The CD-ROM can be ordered through any member of the Board of Publishers or through NATO-PCO, Overijse, Belgium.

Series 3: High Technology – Vol. 4

Physical and Technical Problems of SOI Structures and Devices

edited by

J. P. Colinge
Université Catholique de Louvain,
Maxwell-DICE,
Louvain-la-Neuve, Belgium

V. S. Lysenko
and

A. N. Nazarov
Institute of Semiconductor Physics,
Ukrainian Academy of Sciences,
Kiev, Ukraine

Kluwer Academic Publishers

Dordrecht / Boston / London

Published in cooperation with NATO Scientific Affairs Division

TK
7871.85
P469
1995

Proceedings of the NATO Advanced Research Workshop on
Physical and Technical Problems of SOI Structures and Devices
Gurzuf, Ukraine
November 1–4, 1994

A C.I.P. Catalogue record for this book is available from the Library of Congress.

ISBN 0-7923-3600-3

Published by Kluwer Academic Publishers,
P.O. Box 17, 3300 AA Dordrecht, The Netherlands.

Kluwer Academic Publishers incorporates the publishing programmes of
D. Reidel, Martinus Nijhoff, Dr W. Junk and MTP Press.

Sold and distributed in the U.S.A. and Canada
by Kluwer Academic Publishers,
101 Philip Drive, Norwell, MA 02061, U.S.A.

In all other countries, sold and distributed
by Kluwer Academic Publishers Group,
P.O. Box 322, 3300 AH Dordrecht, The Netherlands.

Printed on acid-free paper

All Rights Reserved
© 1995 Kluwer Academic Publishers
No part of the material protected by this copyright notice may be reproduced or utilized in any form or by any means, electronic or mechanical, including photocopying, recording or by any information storage and retrieval system, without written permission from the copyright owner.

Printed in the Netherlands

CONTENTS

Preface vii
Contributors ix

SOI MATERIALS 1

Low Dose SIMOX for ULSI Applications
A.J. Auberton-Hervé, B. Aspar and J.L. Pelloie 3

Why Porous Silicon for SOI?
V.P. Bondarenko and A.M. Dorofeev 15

Defect Engineering in SOI Films Prepared by Zone-Melting Recrystallization
E.I. Givargizov, V.A. Loukin and A.B. Limanov 27

Ion Beam Processing for Silicon-on-Insulator
W. Skorupa 39

Semi-Insulating Oxygen-Doped Silicon by Low Temperature Chemical Vapor Deposition for SOI Applications
J.C. Sturm, P.V. Schwartz and Z. Liliental-Weber 55

Direct Formation of Thin Film Nitride Structures by High Intensity Ion Implantation of Nitrogen into Silicon
R. Yankov and F. Komarov 67

Stimulated Technology for Implanted SOI Formation
V.G. Litovchenko, B.N. Romanyuk, A.A. Efremov and V.P. Mel'nik 73

Behaviour of Oxygen and Nitrogen Atoms Sequentially Implanted into Silicon
A.B. Danilin 79

SOI Fabrication by Silicon Wafer Bonding with the Help of Glass-Layer Fusion
N.I. Koshelev, A.I. Ermolaeva and V.Z. Petrova 87

Crystallization of a-Si Films on Glasses by Multipulse-Excimer-Laser Technique
A.B. Limanov 93

Microzone Laser Recrystallized Polysilicon Layers on Insulator
A.A. Druzhinin, V.G. Kostur, I.T. Kogut, I.M. Pankevitch and
Y.L. Deschinsky 101

SOI MATERIALS CHARACTERIZATION TECHNIQUES 107

Electrical Characterization Techniques for Silicon on Insulator Materials and Devices
S. Christoloveanu 109

The Defect Structure of Buried Oxide Layers in SIMOX and BESOI Structures
A.G. Revesz and H.L. Hughes 133

IR Study of Buried Layer Structure on Different Stages of Technology
V.G. Litovchenko, I.P. Lisovskii, V.B. Lozinskii, B.N. Romanyuk and
V.P. Melnik 157

Optical Investigation of Silicon Implanted with High Doses of Oxygen and
Hydrogen Ions
P.A. Aleksandrov, E.K. Baranova, I.V. Baranova, V.V. Budaragin and
V.L. Litvinov 163

Electrical Properties of ZMR SOI Structures: Characterization Techniques and
Experimental Results
T.E. Rudenko, A.N. Rudenko and V.S. Lysenko 169

SOI DEVICES 181

Fabrication and Characterisation of Poly-Si TFTs on Glass
S.D. Brotherton, J.R. Ayres, D.J. McCulloch and N.D. Young 183

Hot Carrier Reliability of SOI Structures
D.E. Ioannou 199

Novel TESC Bipolar Transistor Approach for a Thin-Film Silicon-on-Insulator
Substrate
C.J. Patel, N.D. Jankovic and J.-P. Colinge 211

Problems of Radiation Hardness of SOI Structures and Devices
A.N. Nazarov 217

Fabrication of SIMOX Structures and IC's Test Elements
G.G. Voronin, L.V. Degtyarenko, I.G. Lukitsa, V.G. Malinin, V.V. Starkov,
Y.V. Fedorovitch and L.N. Frolov 241

Low-Frequency Noise Characterization of Silicon-on-Insulator Depletion-mode
p-MOSFETS
N.B. Lukyanchikova, M.V. Petrichuk, N.P. Garbar, E. Simoen and C. Claeys .. 247

SOI CIRCUITS 253

SOI Devices and Circuits: An Overview of Potentials and Problems
J.-P. Colinge 255

1.2 µm CMOS/SOI on Porous Silicon
V.P. Bondarenko, Y.V. Bogatirev, L.N. Dolgyi, A.M. Dorfeev, A.K. Panfilenko,
S.V. Shvedov, G.N. Troyanova, N.N. Vorozov and V.A. Yakovtceva 275

SOI Pressure Sensors Based on Laser Recrystallized Polysilicon
V.A. Voronin, I.I. Marymova, A.A. Druzhinin, E.N. Lavitska and
Y.M. Pankov 281

Index 287

PREFACE

This book contains the contributions of the speakers who attended the NATO Advanced Research Workshop on "Physical and Technical Problems of SOI Structures and Devices", which was held in the Central Military Sanatorium of Gurzuf, near Yalta, In Crimea, Ukraine, on November 1-4, 1994.

For over 10 years, scientists of the West and from the East have been working on Silicon-on-Insulator (SOI) technologies. But USSR scientists were publishing in Russian, and virtually no SOI scientist in the West can read that language. Beside the language barrier, security matters were sometimes preventing USSR scientists from publishing their work in western journals. But a third and unfortunately very high barrier has now arisen: the economical barrier. In Gurzuf, the participants from NATO countries have had the chance to meet excellent scientists with remarkable ideas and achievements, but the economic situation is such that Former Soviet Union (FSU) scientists have now very reduced means for research and, of course, for travelling abroad. This is why it was decided to hold the Workshop on their ground.

One of the goals of this Workshop was to break as much as possible the barriers between NATO and FSU researchers. This goal was fully met since many friendly personal contacts and concrete proposals for assistance or collaboration arose during the week of the Workshop.

Another goal of the Workshop was of course to exchange information, experience and visions about Silicon-on-Insulator technology. It is now well admitted that SOI devices offer unique advantages in fields such as of radiation hardness, high-temperature operation, sensors, VLSI and low-power, low-voltage integrated circuits. The problems associated with SOI structures and devices are, however, generally less publicized. These problems are, of course common to NATO and FSU scientists, but unique ways to address and solve them have been devised on both sides. Hence the interest of our meeting in Gurzuf, where all major SOI technologies being developed in Eastern and Western countries were represented, namely the Separation by IMplanted OXygen (SIMOX), wafer bonding, zone melting recrystallization (ZMR); oxidation of porous silicon (FIPOS), and low-temperature recrystallization of poly-Si on glass. Various designs of SOI devices and techniques for characterization of SOI systems were discussed as well. .

All the participants to the Workshop want to express their gratitude to the NATO ARW - International Scientific Exchange Programme, which has made the meeting possible, as well as to local support from the State Committee of Ukraine for Science and Technology, the State Innovation Fund of Ukraine, the Crimean Branch of the State Innovation Fund of Ukraine, the National Academy of Sciences of Ukraine, the Institute of Semiconductor Physics and from Kvazar-IPAN. Our acknowledgements also go to G. Naumovetz, our interpreter and to V.Kilchitskaya, S.Djurenko, M. Lokshin, A. Yurchenko, Yu. Tkachev and I. Barchuk for clerical and organizational help, and to V. Scheuren for her help in organizing this book.

Jean-Pierre Colinge
Louvain-la-Neuve, Belgium

Alexei N. Nazarov
Vladimir S. Lysenko
Kiev, Ukraine

NATO Advanced Research Workshop
on "Physical and technical problems of SOI structures and devices"
Gurzuf, Crimea, Ukraine, Nov. 1-4, 1994

Photograph of participants.

CONTRIBUTORS

Auberton-Hervé, A.J.
SOITEC
Site technologique ASTEC
15 Rue des Martyrs
38054 Grenoble CEDEX 9
France

Baranova, E.
Russian Research Center "Kurchatov Institute"
1, Kurchatov Square
123182 Moscow
Russia

Baranova, I.
Russian Research Center "Kurchatov Institute"
1, Kurchatov Square
123182 Moscow
Russia

Bondarenko, V.
Dep.Microelectronics
University of Informatics and Radioelectronics
220000 Minsk
Petrusya Brovka str., b.6
Belarus

Brotherton, S.
Philips Research Labs
Cross Oak Lane
Redhill, Surrey RH1 5HA
England

Colinge, J.P.
Université Catholique de Louvain
Maxwell-DICE
Place du Levant, 3
1348 Louvain-la-Neuve
Belgium

Cristoloveanu, S.
ENSERG-LPCS
23 Rue des Martyrs
BP 257
38016 Grenoble CEDEX
France

Danilin A.
Centre for Analysis of Substances
Elektrodnaya st. 9,
111524 Moscow
Russia

Druzhinin, A.
"Lviv Politechnika" State university
Kotlyarevskogo st. 1,
290013 Lviv
Ukraine

Givargizov, E.
Institute of crystallography
Russian Academy of sciences
Leninsky pr. 59
Moscow 117333
Russia

Ioannou, D.
George Mason University
Dept. of Electrical and Computer Eng.
Fairfax, VA 22030-4444
USA

Koshelev, N.
Moscow State Institute of Electronic Engineering
103498 Zelenograd
Moscow
Russia

Limanov, A.
Institute of Crystallography
Leninskii Prospect 59
117333 Moscow B-33
Russia

Litovchenko, V.
Institute of Semiconductor Physics
Prospect Nauki 45
252028 Kiev
Ukraine

Lukyanchikova, N.
Institute of Semiconductor Physics
Prospect Nauki 45
252028 Kiev
Ukraine

Maryamova, I.
"Lviv Politechnika" State University
Kotlyarevskogo st. 1,
290013 Lviv
Ukraine

Nazarov, A.
Instit. of Semiconductor Physics
Ukrainian Academy of Sciences
Pr. Nauki 45
252028 Kiev
Ukraine

Patel, Ch.
Microelectronics Centre
Middlesex University
Bounds green Road
London N11 2NQ
England

Revesz, A.
Revesz Associates
7910 Park Overlook Drive
Betseda, MD 20817
USA

Rudenko, T.
Institute of Semiconductor Physics
Prospect Nauki 45
252028 Kiev
Ukraine

Skorupa, W.
Forschungszentrum Rossendorf e.V.
Institut für Ionenstrahlphysik und
Materialforschung
Postfach 510119
01314 Dresden
Germany

Sturm, J.
Dept. of Electrical Engineering
Princeton University
Princeton, NJ 08544
USA

Voronin, V.
"Lviv Politechnika" State University
Kotlyarevskogo st. 1,
290013 Lviv
Ukraine

Yankov, R.
Forschungszentrum Rossendorf e.V.
Institut für Ionenstrahlphysik und
Materialforschung
Postfach 510119
01314 Dresden
Germany

Section 1:
SOI Materials

Low dose SIMOX for ULSI applications

A.J.Auberton-Hervé[1], B.Aspar[2], J.L.Pelloie[2]

[1] SOITEC SA 15 Av des Martyrs, 38054 Grenoble Cedex 9
[2] LETI (CEA Technologies Avancées) DMEL CEN/G 17 Av des Martyr, 38054 Grenoble cedex 9

Introduction:

The most recent developments in SOI material using the SIMOX technique concern the oxyden dose reduction. As the material cost is the key of the SOI developments the use of lower oxygen dose and therefore the implantation time reduction is one of the parameters in the cost reduction. A very unique process window has been found around a dose 5x lower ($4\ 10^{17}$ O$^+$/cm^2) than the standard oxygen dose ($1,8\ 10^{18}$ O$^+$/cm^2) of the SIMOX process. Around this dose a continuous SiO$_2$ film can be formed with a single implantation and the obtained Buried OXide thickness of the Silicon On Insulator structure is of 80nm to be compared with 400nm in case of standard SIMOX.

The material specification in case of the "low dose" process are not yet at the same performances than the standard SIMOX. We will discuss the parameters which need to be improved and the road map for the SOI development. But one question has to be answered : is such a thickness adapted to the ULSI developments ?

The SOI material developments:

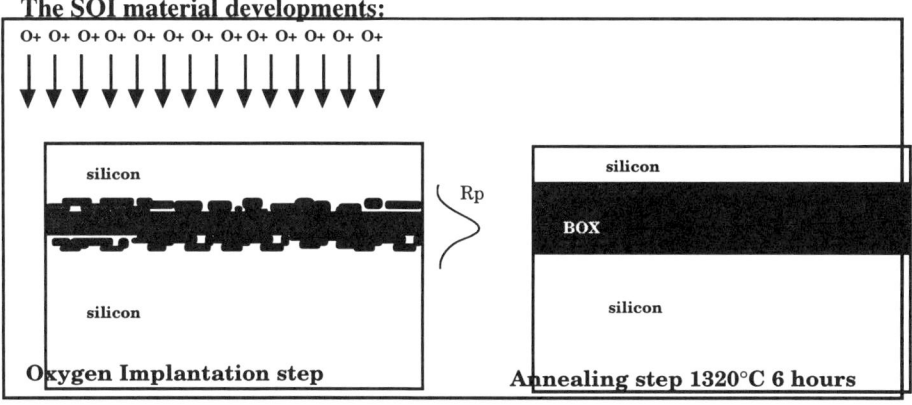

Fig.1 Schematic of the SIMOX technology

Two main technologies are competing today for the SOI wafer market. The SIMOX technology (Fig.1) and the Bonding technology (Fig.2). The SIMOX technology uses two key processes: an oxygen implantation step using a dedicated machine (100mA,

200keV of O+ ions) to locate underneath the initial silicon surface a high concentration of oxygen, and a high temperature anneal to regenerate the crystalline quality of the silicon layer remaining over the oxide; this anneal also drives the chemical reaction which forms the stochiometric oxide buried in the silicon wafer.

Just as SIMOX is best suited for thin film applications, bonding is an inexpensive technique for manufacturing thick film of both oxide and silicon. Starting from two silicon wafers, at least one with an oxide layer on top, these two wafers are bonded together using Van der Walls forces. Subsequent annealing increases the mechanical strength of the bonded interface by the chemical reaction which can occur at this interface. Then one of the substrates is thinned down to 1μm starting from several 100μm; mechanical grinding and polishing can achieve SOI films of 1μm within 10 to 20% uniformity.

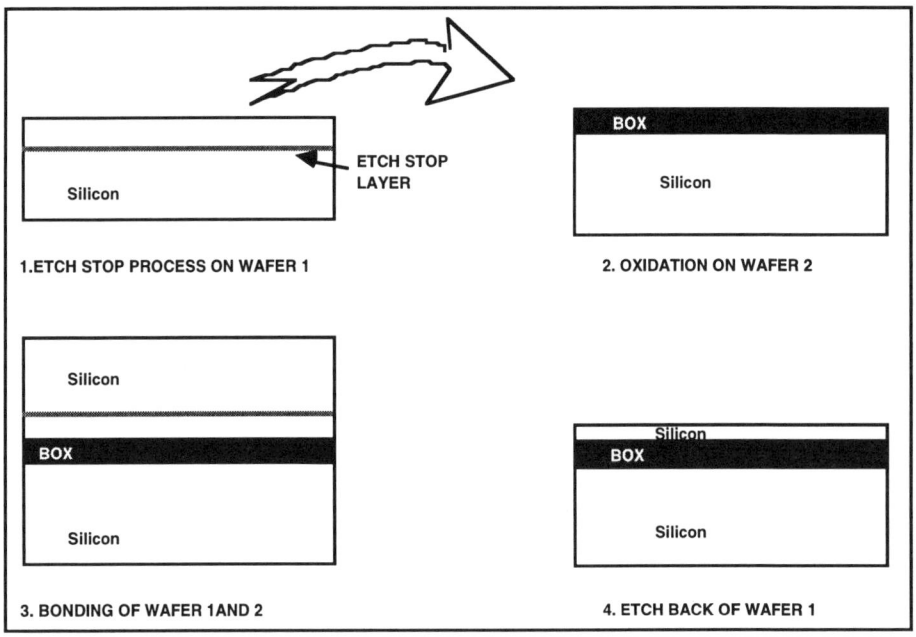

Fig.2 Schematic of the Bonded and Etched back technology BESOI.

However, to compete in the thin film market such a technique cannot be used. New techniques have been developed, most of them using a chemical etch stop technique (Fig.2) to achieve good uniformity. Several etch stops have been reported: boron doped layer, silicon germanium, carbon implanted, porous silicon. However, only a few are compatible with high temperature treatment and the Si/SiO2 bonded interface could be either the upper one or the buried one depending on the etch stop technique. The uniformity depends on the selectivity of the last etching step which varies from 10 to 10^5 depending on the etch stop.

An original technique has been developed using a localized plasma etch. Starting from a non-uniform wafer, an accurate measurement of the top silicon film is performed; then a localised plasma etch is used to reduce the variation of the topography [1].

The crystalline quality of SIMOX and Bonding films are reported to be in the same range for thin film products.[2]. The main limitations are the implantation induced defects in the SIMOX technology and the nanovoids for the bonding. The defect density has been drastically reduced in the SIMOX technology; the same learning curve is expected from the wafer bonding technique.

The SOI material market is mainly segmented by the two main parameters of a SOI wafer : the top silicon film thickness and the buried oxide film thickness. We have summarized in Fig.3 the different markets where SOI has already been used and the associated thickness requirements. The most advanced products will require the thinnest films of both silicon and oxide.As the main market potential is for ULSI applications, the main material requirements will come from this market.

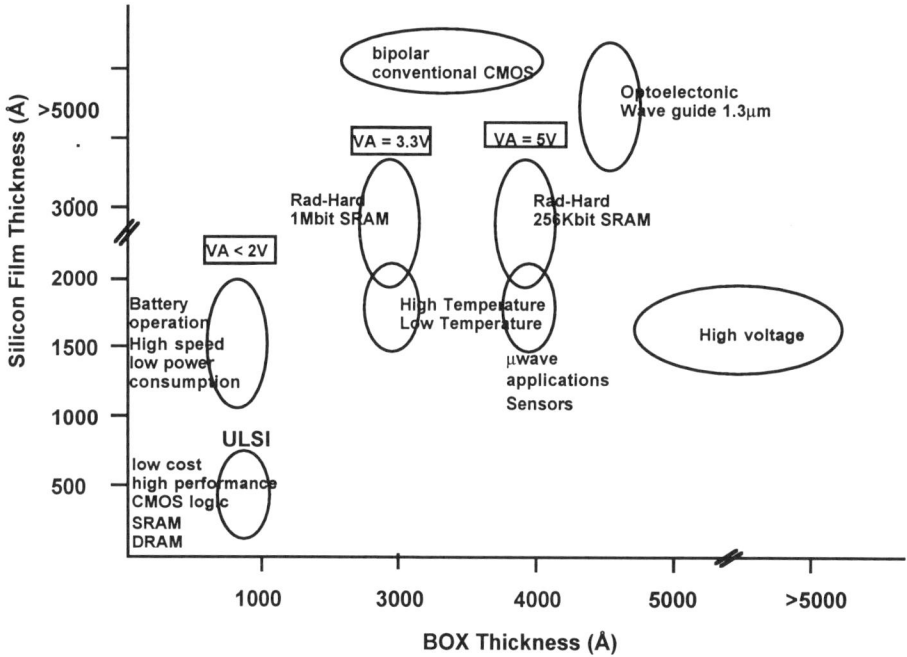

Fig.3 SOI material segmentation versus the SOI film thicknesses.

ULSI developments using SOI.

As the power consumption is $\approx C*V^2*f + V*I_{leak}$ (C total capacitance, V the supply voltage, f the frequency, Ileak the standby current), the supply voltage must be reduced to form low power IC's. The target (Sematech road map) is to reach 0,9V using a single battery, by the end of the century. However, at such a low voltage, performance is also reduced due to a lower transistor drivability.

SOI provides many advantages :
- by reducing the junction capacitance thereby inducing a reduction of the total capacitance by 15% to 30% depending on the circuits design;
- by increasing the switching behaviour of the MOS devices (sharper subthreshold slope), allowing a shrink of the threshold voltage thus increasing the current drivability at low voltage;
- by reducing the junction area at least by two decades, which also decreases the leakage current;
- by lowering the threshold voltage temperature sensitivity.

The main SOI advantage for low voltage, low power, logic circuits arises from the junction capacitance reduction[3]. We have plotted in Fig.4 the different values of the junction capacitance versus the drain voltage. The comparison with standard bulk technology is performed for a substrate doping under the junction of $10^{17}/cm^2$, which is optimistic for ULSI devices. The gain by using SOI is between 5X to 10X. Two conditions are plotted for the buried oxide thickness as 400nm and 80nm. These thicknesses correspond to the two main SIMOX products. The use of a low-doping substrate P-type (resistivity 14-22 $\Omega.cm$) lowers the sensitivity to the BOX thickness reduction due to a depletion region occurring in the substrate. The low dose SIMOX keeps some advantage on the junction capacitance reduction. However, the PMOS drain capacitance is degraded compared to standard SIMOX mainly, at low voltages.

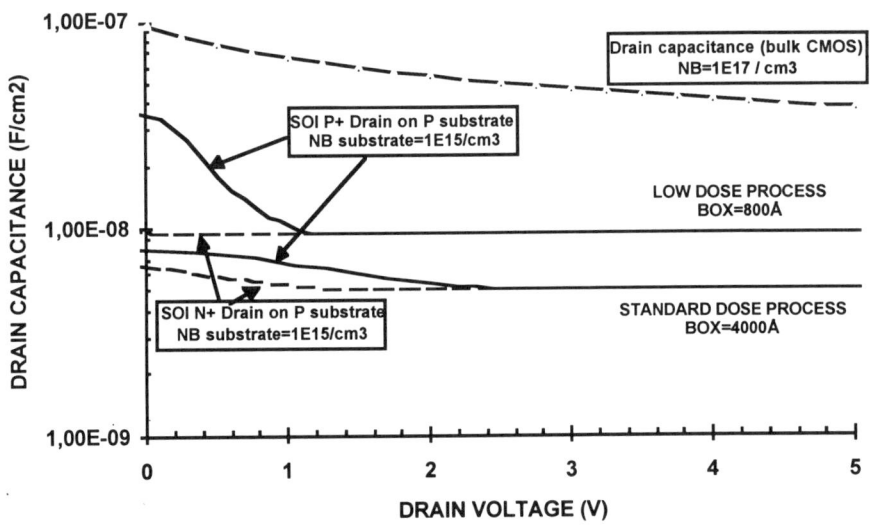

Fig.4 Comparison of drain junction capacitance for standard CMOS on silicon , on standard SIMOX and on "Low dose" SIMOX.

The low power market will also include memories; SRAMs as well as DRAMs have been fabricated using SOI technology. The SOI structure for DRAM provides a better soft error immunity and allows a reduction of the cell capacitor area. More than 5X reduction of the cell capacitor seems possible as reported by MITSUBISHI [4] and data retention is 6X better on SOI as reported by NEC[5]. The operating voltage range is also wider on SOI DRAM[4]. This could allow reduction of the supply voltage below 1,5V without loosing noise margin or performance.

Most of the recent DRAM developments are using a SOI structure obtained by oxygen implantation (SIMOX) [4] [5].] One unique approach is performed by SONY: the SOI-DRAM capacitor is buried under the SOI layer using a bonding technique[6] Fig.5).

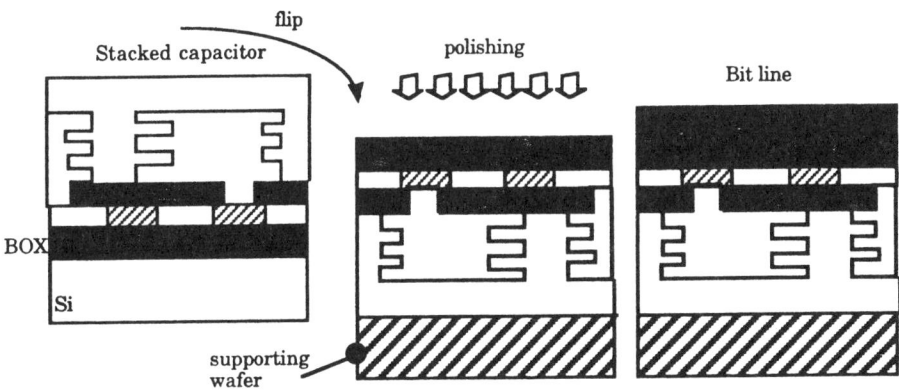

Fig.5 Use of wafer bonding technology to transfer already processed devices. Application to DRAM [6]

Regardless of the SOI technique, the top silicon film thickness used is in the range of 100nm. Such a thin film favours the soft error reduction and the decrease of leakage current. Both the collection area for α-particles and the leakage current are directly proportional to the junction area. The use of thin film SOI reduces this area by at least one decade .

Therefore, SOI-DRAM exhibit very attractive technical performances. However, in DRAM applications, the most important issue is definitely the cost.

Two parameters have to be balanced: the initial SOI wafer cost Fig.6 and the reduced process cost obtained by using SOI. The answer on this second point was already addressed: "Mitsubishi Electric ... expects 256Mbit and 1Gbit SOI substrate DRAMs would require 40%-50% less cost to produce compared to conventional silicon-substrate comparable DRAMs" [7].

The choice of the buried oxide thickness will be defined by the cost aspect but also by the choice of the SOI device best suited to the ULSI applications.

Two different kinds of devices are currently investigated : partially-depleted devices, using a thin silicon film (>1000Å)[9] and fully-depleted devices, using a very thin silicon film (<1000Å)[10]. The advantages and the drawbacks of these different devices will be discussed in the following sections. The choice of the device determines the SOI material specification.

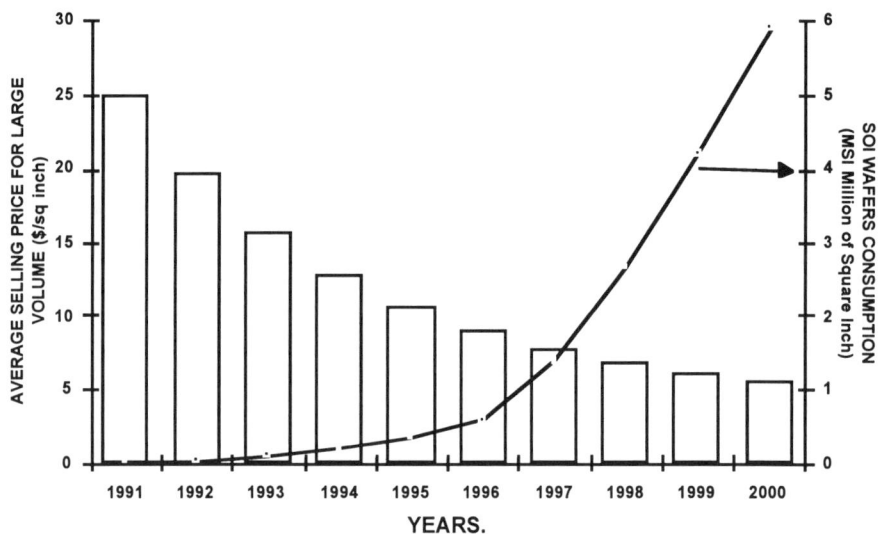

Fig.6 Price and volume forecast for SOI wafers [8]

Partially depleted devices or fully depleted devices?

Partially-depleted devices have been used mainly for space and military applications. Due to the requirements of these specific applications and to minimize the floating-body effect, body ties have been introduced in the device layout leading to a loss in the integration density. For low-voltage applications, the body ties are no longer necessary. On the contrary, the floating-body effect is now considered to be an advantage due to the consequent reduction of the threshold voltage and the subthreshold swing. A first way to design partially-depleted devices is similar to the bulk approach, creating very shallow junctions combined with a halo structure [11]. The source-drain capacitance reduction is still preserved as the shallow junction is only formed below the spacer; thus, the speed advantage of SOI technologies is ensured.

In a partially-depleted SOI MOSFET, one can find the combination of two parallel devices : the main front transistor and the parasitic back transistor formed by the buried oxide (BOX) and the underlying substrate. When optimizing a 0.2 µm gate length device, both front and back transistors have to be controlled. Using a thin gate oxide (50 Å) greatly contributes to the reduction of the short channel effect in the front transistor. As the back transistor gate oxide is the buried oxide, a strong drain induced barrier lowering (DIBL) effect affects the behavior of the device. This is due to the low vertical electric field at the back interface arising from the BOX thickness (usually more than 800 Å for a SOI wafer). Technological solutions must be found to design this parasitic 0.2 µm gate length device with a thick gate buried oxide to avoid any leakage current due to punchthrough between source and drain and the additional component due to floating-body effect.

SOI devices exhibit a self-heating effect due to low thermal conduction of the buried oxide and resulting in a negative resistance effect which shows up on the Id(Vd) characteristic. This effect can be minimized by using a thin buried oxide, thus decreasing the bottom thermal resistance. A further advantage is the reduction of short channel effect for the back transistor. However, the back threshold voltage is reduced if the doping level at the back interface is not increased. This, in combination with the floating-body effect, can lead to a worst case behavior.

The behavior of the back transistor is strongly influenced by the presence of charges at the back interface. If a low doping level is enough to insure a low leakage current without any charge in the buried oxide, this is no longer the case when charges are taken into account; a high doping level at the back interface needs to be achieved in this latter case. Charges in the buried oxide can be present in the starting material or generated during the process.

Sub-0.25 µm SOI technologies making use of partially-depleted devices will offer the advantage of simple process integration by reducing the number of technological steps compared with bulk technologies. Nevertheless, the PMOSFET device will require for a P+ gate process which can be considered as the critical point identical to bulk devices.

SOI devices are fully-depleted when the depletion depth below the gate is more than the silicon film thickness. Several kinds of devices can be defined depending on the body type. Enhancement-mode devices are achieved when the body type is opposite in polarity to the source-drain type : NMOSFET with P-type body and N+ drain, PMOSFET with N-type body and P+ drain. Devices are operating in accumulation-mode when the body type is identical to the source-drain type : NMOSFET with N-type body and N+ drain, PMOSFET with P-type body and P+ drain [12]. In order to avoid a low off-current matched to a suitable threshold voltage, the body-type must be correctly adapted to the gate type. Four different devices can then be used : enhancement-mode NMOSFET and PMOSFET with N+ and P+ gate respectively, accumulation-mode NMOSFET and PMOSFET with P+ and N+ gate respectively. Other possible combinations of the gate and body types generally show poor electrical characteristics.

Fully-depleted devices exhibit a different electrical behavior from partially-depleted devices. The threshold voltage varies with the back gate bias for enhancement-mode and accumulation-mode devices due to the coupling effect between the front and the back gates when the silicon film is fully depleted. This variation does not occur in partially-depleted devices where the depletion depth is less than the silicon film thickness. As a result of the coupling effect, the threshold voltage of fully-depleted devices becomes a function of the silicon and buried oxide thicknesses; accurate analytical models of the threshold voltage for long-channel devices have been developed for both kinds of devices [13,14].

Two-dimensional numerical simulations show that a very thin silicon layer (about 300 Å) is needed to achieve a good design for fully-depleted devices and control of two-dimensional effects (punchthrough, short channel effect). A 0.5 V threshold voltage target is reached in such thin silicon film with a high doping level (8×10^{17} cm^{-3}) for enhancement-mode devices. When the impact ionization mechanism is taken into account, the behavior of the fully-depleted transistor is found to be similar to a partially-depleted transistor. Although the threshold voltage varies with the back gate bias, indicating the fully-depleted nature of the device at low drain voltage, the generated holes due to impact ionization at high drain voltage are accumulated near the source side. Thus, impact ionization is also responsible for an increase of the current in the subthreshold range for the accumulation-mode device.

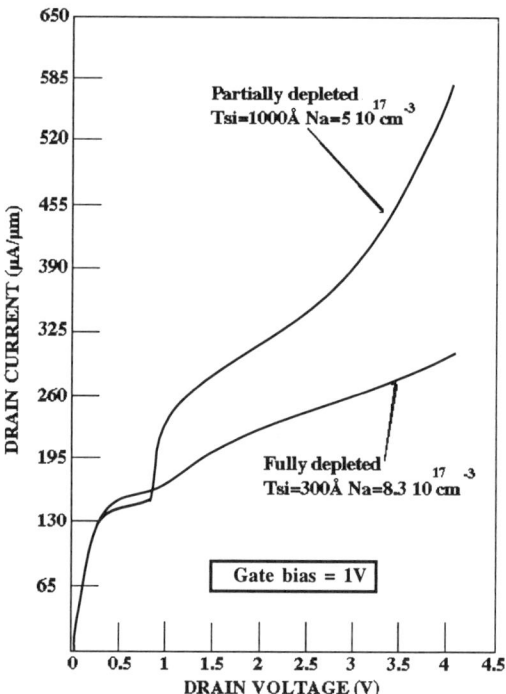

Fig.7 Enhancement mode NMOS Transistor. Comparison of partially and fully depleted SOI MOSFET. Lg=0,2μm, Gate oxide=50Å, BOX=4000Å (Simulation results obtained with FIELDAY using impact ionization)

Long-channel fully-depleted devices exhibit low subthreshold slope close to the ideal value of 60 mV/decade at room temperature. However, two-dimensional effects lead to a degradation of the slope 65 mV/decade for enhancement-mode devices and 76 mV/decade for accumulation-mode devices (0.2 µm gate length). Let us point out that the accumulation-mode device is more sensitive to two-dimensional effects than the enhancement-mode device, since in this case, the barrier height between source and body is reduced as these two regions have the same type (N-type body and N+ source for NMOSFET). The lower subthreshold slope allows devices to be designed with a lower threshold voltage without degrading the off-current. Although a kink effect still exists for enhancement-mode fully-depleted devices, this effect is strongly reduced compared with partially-depleted devices (Fig. 7).

From a circuit point of view, fully-depleted devices are generally preferred to partially-depleted devices for low voltage applications, different combinations of devices can be used depending on the desired function; for instance, accumulation-mode devices are more convenient for analog parts because they eliminate any undesirable floating-body effects. From a technological point of view, fully-depleted devices have several advantages : a simple N+ gate process can be conserved by mixing an enhancement-mode NMOSFET with an accumulation-mode PMOSFET; a simple lateral isolation can be implemented due to the very thin silicon layer used.

However, fully-depleted SOI technologies will emerge if the technological key points are overcome. As the threshold voltage depends on the silicon thickness, one must guarantee a good uniformity of a 300 Å silicon film (±20%); a high series resistance due to the very thin source-drain regions must, therefore, be reduced using a selective silicon epitaxy or a selective deposition thus decreasing the corresponding sheet resistance.[15]

The self-heating effect is more pronounced in fully-depleted structures due to the thinner silicon film; a thinner buried oxide will, therefore, be necessary to minimize it. The limitation for thinning the buried oxide is imposed by the variation of the threshold voltage with the back gate bias, equivalent to a variation of the source potential which is found, for instance, in a NAND gate. For a very thin buried oxide the variation of the back gate bias can be such that the fully-depleted device operates in a partially-depleted way with an accumulation layer created at the back interface. The buried oxide thickness will probably be limited to about 800Å for a 2 V supply voltage.

In case of enhancement-mode device, future use of mid-gap gate materials will make fully-depleted devices less sensitive to the variation of the silicon thickness as a lower doping level is required to fix the threshold voltage. This threshold is therefore essentially determined by the work function difference between the gate and the transistor channel.

What is the impact of the device choice on the material characteristics? Fully depleted devices require both a reduction of the silicon film and of the buried oxide. For the partially depleted devices, both the top silicon layer and the buried oxide will remain thicker.

A consensus to use partially depleted as a first step towards SOI development has been reached mainly due to the process complexity of fully depleted devices. A possible road map is the following:

SOI MATERIAL ROAD MAP		
	1995	1998
Design rule	0,35μm	0,2μm
Device type	Partially depleted	Fully depleted
Wafer size	150mm-200mm	200mm
SOI thickness	100 nm	50 nm
SOI uniformity	10 nm	5 nm
BOX thickness	400 nm	80nm
BOX uniformity	20nm	4 nm
Dislocations density	$< 10^3 /cm^2$	$<100/cm^2$
Pipes density	$< 0,2 /cm^2$	$< 0,2 /cm^2$
Roughness	3 Å (max-min 20Å)	2 Å (max-min 10Å)
Metal contamination	$< 10^{11} /cm^2$	$< 5\ 10^{10} /cm^2$

Low dose SIMOX as a long term solution for ULSI:

The so called "low dose" SIMOX seems best suited for the ULSI market as it offers two advantages :
    - from the material side, the oxygen implantation dose is reduced from 1,8 $10^{18}O^+/cm^2$ (standard SIMOX) to $4\ 10^{17}O^+/cm^2$. The cost of oxygen implantation decreases in the same ratio and the top silicon quality is improved,
    - from the device side, the Buried OXide (BOX) thickness reduction improves both heat dissipation and the short channel effects.

The "low dose" SIMOX is obtained at a specific dose window around $4\ 10^{17}/cm^2$ [16]. With a single implantation and a 6 hour anneal at 1320°C, a continuous BOX of 80nm thick is formed. Fig.8 shows the structure of the buried oxide versus the dose around the process window for an energy of 120keV [17]. The choice of the implantation energy is a critical parameter. At 190kev, which is the standard SIMOX energy, some SiO2 islands exist in the top silicon film Fig.9. As the peak of defects and the peak of oxygen mean range are separate, two sites of precipitation can occur. A reduction of the implantation energy down to 120kev is sufficient to merge the two sites of precipitation.

The dose reduction induces a better top silicon quality. The dislocation density obtained is of $300/cm^2$.

Electrical characterization demonstrates a reduction of the breakdown field of the BOX for large capacitors. This can be attributed to small silicon island inclusions inside the BOX as shown in Fig.10. However, the breakdown voltage is over 20V for $17mm^2$ capacitors which is sufficient for 1V operation.

The thickness of the buried oxide is fixed to 80nm. Any adjustment to thicker films is possible by a second step of implantation and annealing. Any kind of thicknesses can be reached between 80nm and 400nm. However, the cost increases as the buried oxide thickness increases.

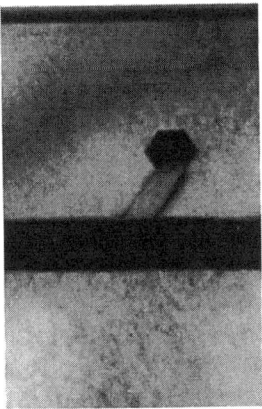

Fig.8 Silicon dioxide precipitates in case of high energy (190keV) "low dose" SIMOX

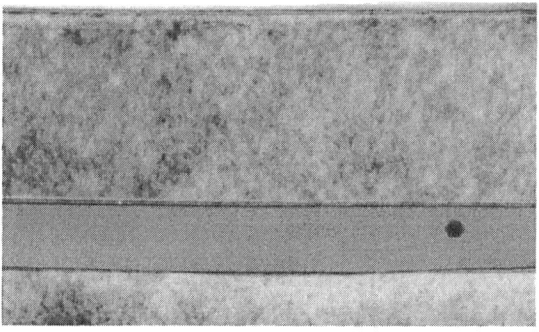

Fig.9 Silicon island in "low dose" SIMOX

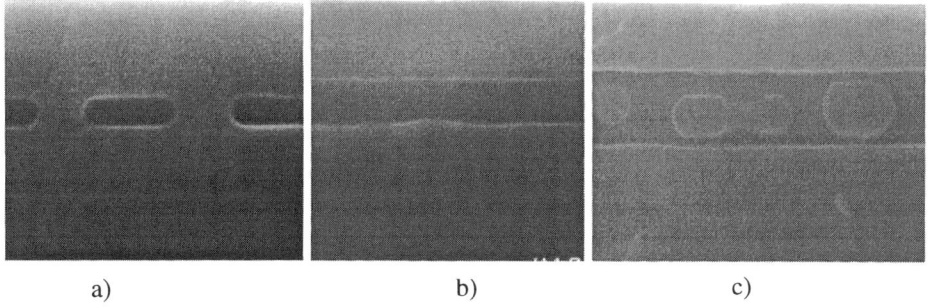

a)   b)   c)

Fig.10 Evolution of the buried oxide structure for a dose of
a) $3 \cdot 10^{17}$ O^+/cm^2, b) $4 \cdot 10^{17}$ O^+/cm^2, c) $5 \cdot 10^{17}$ O^+/cm^2 and an energy of 120keV. The optimum energy and dose for a low dose SIMOX are 120keV and $4 \cdot 10^{17}$ O^+/cm^2.

CONCLUSION

Silicon On Insulator technologies are experiencing increased interest due to the "power crisis" and portable system boom. The low power, low voltage market was 4% of the total IC market in 1993, but is expected to reach 40% in 1998 (source ICE 1994) due to the increase of portable communication systems and laptop computers. However, portable system performances are jeopardised by the need for higher power consumption caused by increased IC's performances and complexity. SOI provides a gain of 3X in term of merit factor at low voltage[18]. Therefore , SOI is the ideal solution to the increasing demand for low power, low voltage volume production. A rapid increase in world demand for the SOI wafer (50% annual growth) is anticipated with large volume production expected in 97-98 for mainstream applications [19]. The greatest challenge for SOI in competition with standard silicon will result from the DRAM developments now being performed in Japan.

References :

[1] P.BMumola,G.J.Gardopee, Extended Abstracts of the 1994 Solid State Devices and Materials (SSDM) p.256
[2] D.K.Sadana, Y.J.Hovel,K.Petrillo , 1994 IEEE International SOI Conference Proceedings, p.111.
[3] A.Yoshino,K.Kumagai,N.Hamatake,S.Kurosawa, K.Okumura, 1994 IEEE International SOI Conference Proceedings, p107
[4] T.Tanigawa,A.Yoshino,H.Koga,S.Ohya , Proceedings of 1994 IEEE Symposium on VLSI Technology p.37
[5] K.Suma,T.Tsuruda,H.Hidaka. & al.Proceedings of ISSCC 1994 p138
[6]T.Nishihara,H.Moriya,N.Ikeda,&al, Proceedings of 1994 IEEE Symposium on VLSI Technology p.39
[7] 02-23-94 Japan Industrial P.7
[8] O'MARA ASSOCIATES :"Silicon-On-Insulator Materials, Technology and Markets" 1992. O'MARA & ASSOCIATES, 2443 Ash streets, Palo Alto, CA, 94306 USA.
[9] G. G. Shahidi et al., Tech. Dig. of IEDM, p. 813, 1993.
[10] Y. Kado et al., Tech. Dig. of IEDM, p. 243, 1993.
[11] G. G. Shahidi et al., Tech. Dig. of Symp. on VLSI Tech., p. 53, 1993.
[12] J. L. Pelloie, Y. C. Sun, Proc. 6th Symp. on SOI Tech. Dev., Elec. Chem. Soc., p. 263, 1994.
[13] H-K. Lim, J. G. Fossum, IEEE Trans. Elec. Dev., vol. ED-30, p. 1244, 1983.
[14] D. Flandre, A. Terao, Solid-State Elevtron., vol. 35, p. 1085, 1992.
[15] J.M.Hwang,R.Wise,E.Yee,T.Houston, G.P.Pollack, 1994 Symposium on VLSI Technology Digest of technical papers p33.
[16] S.Nakashima,Y.Omura,K.Izumi, PROc.5th International On Siliocn On Insulator Technologies, ECS, Vol92-13 St Louis 1992 p 358.
[17] B.Aspar, C.Pudda, A.M.Papon, A.J.Auberton-Herve, J.M.Lamure, The Electrochemical Society ; proceedings volume 94-11 page 62, abstract 541 Silicon On Insulator Technology and Devices edited by S. Cristoloveanu.
[18] G.G.Shahidi, T.H.Ning,R.H.Dennard,B.Davari, Extended Abstracts of the 1994 Solid State Devices and Materials (SSDM) p.265
[19] S. Kawamura 1993 IEEE International SOI Conference Proceedings, p6

WHY POROUS SILICON FOR SOI ?

V.P.BONDARENKO and A.M.DOROFEEV
Belarusian State University of Informatics and Radioelectronics
P.Brovka 6, 220027 Minsk, BELARUS

1. Introduction

Most operations of microelectronic technology are based on heterogeneous reactions taking place at "working media (solution, gas, plasma etc) / semiconductor surface" interface. Activation of these reactions is usually performed by changing the "working media" parameters (temperature, pressure, electromagnetic stimulation etc).

There is another way to activate these heterogeneous reactions: changing the properties of the semiconductor material. In the case of silicon very interesting transformation of crystal properties can be done by electrochemical anodic treatment in electrolytes containing hydrofluoric acid (HF). Under certain conditions of electrochemical anodization, a localized dissolution of the monocrystal occurs leading to the formation of pore network within the bulk of the silicon monocrystal [1].

This so-called porous silicon (PS) layer retains the monocrystalline structure of initial silicon but has very large internal surface area. There is no other semiconductor material than porous silicon that has such unique combination of crystalline structure and large internal surface area. As a new morphological form of silicon, porous silicon, is of great scientific and practical interest.

Since 1982 authors of the present contribution have dealt with SOI technology based on porous silicon but the results in this field are practically unknown to the scientific community. Our results concerning SOI technology based on selective anodization of n^+-layer in $n/n^+/n$-epitaxial structure are given in the present book [2]. The advantages and limitations of this technology are discussed there as well.

The present contribution is devoted to the results that have been obtained recently. The main idea of these investigations is the integration of SOI structures with optoelectronic elements based on porous silicon. We will discuss the perspective concerning the new properties and effects recently discovered in PS and its application for optoelectronic devices compatible with SOI structures.

2. Porous Silicon Processing

There is a saying in the silicon community that the bulk of the crystal was created by God, while the surface was made by the devil [3]. In the case of porous silicon the properties of the surface are introduced into monocrystal up to the depth of pores, so it is obvious with whom we deal. Due to its very large internal surface area (200-800 m^2/cm^3) PS exhibits very high activity in a lot of physical-chemical reactions. It provides significant stimulation of thermal processes (oxidation, diffusion etc) and synthesis of thick semiconducting or dielectric layers on the basis of porous silicon. Besides, due to its monocrystalline structure PS is a promising material for deposition of homo- and heteroepitaxial layers and different metal films.

2.1. THERMAL SINTERING

When heated at temperatures higher than 400-450°C, the original microstructure of PS coalesces leading to the formation of large cavities and thick silicon blocks. Moreover, a pore-free layer ("crust") is formed at the surface of porous silicon as a result of sintering at high temperatures [4,5]. The sintering prevents any thermal oxidation and therefore limits the usefulness of porous silicon. Fortunately, PS sintering can be easily avoided by a mild "pre-oxidation" at 300-350°C in dry oxygen before any restructuring takes place [6]. One can obtain thick doped silicon layers and silicide layers when sintering process is performed after PS doping or metal incorporation into porous silicon [7,8].

2.2. THERMAL OXIDATION

Due to the large surface/volume ratio porous silicon has a very high rate of oxidation. This allows oxidation of thick porous layers in a short time. There are three temperature regions for PS oxidation: (i) low (less than 400°C), (ii) moderate (400-900°C), (iii) high (900-1200°C). Low temperature oxidation is used for PS structure stabilization by passivating the pore inner walls and for thinning interporous regions to obtain a nanoscale Si skeleton. Pre-oxidation step is an integral part of PS oxidation process in SOI technology [9].

Preoxidized porous silicon can be completely oxidized at moderate temperatures (800-900°C), however the quality of the material obtained is still very different from that of standard thermal dioxide. A densification step at 1050-1150°C in wet oxygen followed by a nitrogen annealing is necessary to form an oxidized porous silicon equivalent to thermal SiO_2 [9]. Very good results can be obtained under high pressure oxidation at 800-950°C in oxygen followed by a nitrogen annealing at 1150-1200°C [10].

Both electrical and optical characteristics of oxidized porous silicon strongly depend on oxidation regimes and PS porosity. Low temperature oxidation results in only partial transformation of porous silicon to oxide. Optical properties of partially oxidized PS are in intermediate range between silicon characteristics and that of silicon oxide. Electrical and optical properties of PS oxidized optimally at high temperature were very close to that of thermal dioxide (resistivity $\rho = 1 \cdot 10^{16} - 4 \cdot 10^{16}$ Ohm·cm, dielectrical constant $\varepsilon = 3.66-4.2$, fixed charge $Q = 7 \cdot 10^{10} - 3 \cdot 10^{11}$ cm^{-2}, refraction index n = 1.4- 1.45).

2.3. ELECTROCHEMICAL OXIDATION

This process is carried out at room temperature and, in contrast to thermal processes, is not accompanied by sintering. Electrochemical oxidation of porous silicon can be done in traditional electrolytes for Si oxidation. The remarkable feature of the process is the possibility to introduce doping impurities from the electrolyte into the depth of porous layer. Electrical properties of anodically oxidized PS are poor. Subsequent thermal oxidation and nitrogen annealing improve these properties sufficiently. The impurities having been introduced into PS during anodic oxidation provide change of its electrical and optical properties.

2.4. NITRIDIZATION AND CARBIDIZATION

The PS ability to be converted to silicon carbide or oxynitride is one of its most remarkable features. Direct interaction of silicon/nitrogen (or ammonia) and silicon/carbon results in silicon oxynitride and silicon carbide formation only at very high (more than 1200°C) temperatures. However, even rather long duration of such treatment results in the formation of very thin films.

We studied interaction between PS and carbon- or nitrogen-containing gases in temperature range of 1000-1300°C. The process was performed in epitaxial reactor with the induction coil heating. Nitrogen, ammonia, methane, trichloroethylene, etc were used as reaction gases. Formation of thick silicon oxynitride [11] and silicon carbide layers as a result of interaction between PS and the reaction gases was established. SiC films consisted of two layers: polycrystalline subsurface SiC layer and inner SiC layer of mosaic structure. SiC films had an adsorption band at 12.6 μm region and exhibited luminescence in a blue range of the spectrum.

2.5. CHEMICAL/ELECTROCHEMICAL DEPOSITION OF METALS

Different metals can be easily deposited chemically or electrochemically on porous silicon. These metal films deposited on/into porous silicon could be used as ohmic or injection contacts in PS-based devices. The presence of pores provided sufficient increase of metallic film adhesion to PS surface. During chemical deposition metals are deposited only in pore entrances. In case of electrochemical deposition one has a possibility to put metals deep into porous silicon [12,13]. Deposition of silicide metals, for instance Ni or Co, is the most interesting from the practical point of view. PS layers filled with these metals could be converted to nickel or cobalt silicides by subsequent heat treatment [14].

The established recently [15,16] possibility of rare-earth elements introducing in PS seems to be especially promising. In particular, erbium concentration in PS of about $3 \cdot 10^{19}$ cm^{-3} at a depth of 5-10 μm has been achieved. After rapid thermal anneal by incoherent light PS:Er layers exhibited intense 1.54 μm luminescence even at room temperature.

2.6. EPITAXIAL GROWTH

Formation of high quality epitaxial layers of compound semiconductors on Si substrates is a serious technical problem because of the differences in lattice constants of epitaxial layers and silicon.

Porous silicon can be used as a buffer layer for epitaxy of semiconductors on silicon wafer [17]. Beneficial influence of porous silicon buffer layer on the structure and properties of GaAs [18] and PbS [19] films epitaxially grown on PS was explained by lower mechanical stress level in the epitaxial structures obtained. The similar results have been obtained for diamond films as well [20]. It is reasonable to forsee such effects for other films and to use PS as a universal buffer layer for deposition of different materials on Si substrate, especially for integration of electronic silicon components and optoelectronic compound semiconductor devices.

3. Optoelectronic Application of Porous Silicon

It is common knowledge that only GaAs, or other III-V compound semiconductors can support all the aspects of integrated optical systems including light emission, waveguiding, and detection due to their direct bandgap properties. Silicon provides an interesting low-cost alternative to the III-V semiconductors for the realization of fully integrated optoelectronic components and circuits. The base silicon technology, that is very well established, is already used for the fabrication of visible range photodetectors and high-speed electronic devices. However, the development of Si-based optoelectronics has been limited by the indirect nature of Si bandgap. Radiative recombination does exist in Si, but the quantum efficiency is very low (10^{-5} to 10^{-4}) at room temperature. Continuous efforts have been made to overcome this hurdle over the years. Studies with Ge_mSi_n ordered superlattices, and impurity/defect centers were undertaken in an attempt to increase the radiative recombination efficiency. The recent observations [19] that porous silicon emits visible light has generated the great interest due to the high intensity of emission.

Practical importance of this effect is clear because of perspectives to create optoelectronic devices on the basis of well developed Si technology. In order to exploit fully the potential of silicon integrated optoelecronics, emitter-detector architectures are likely to be optically connected via waveguides. An integration of all these componenets in SOI structures seems to be very attractive to use all the benefits of electronic SOI circuits together with optoelectronic circuits.

3.1. PROBLEMS OF SOI BASED OPTOELECTRONIC CIRCUITS

There are at least two ways for organizing the architecture of optoelectronic circuits in SOI structures. The light from light emitting diode (LED) has to go through the silicon film and the dielectric layer and has to be adsorbed in the photodetector (PD) active region, resulting in a change of its electrical characteristics (Figure 1).

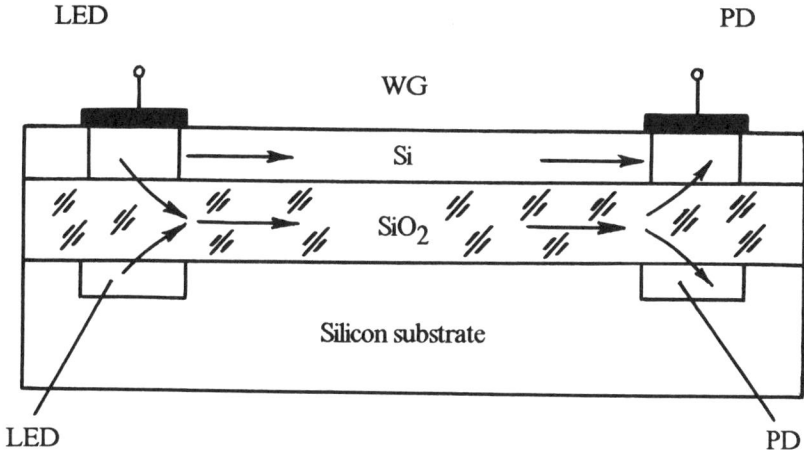

Figure 1. Schematic view of SOI based micro- optoelectronic circuits.

The first way proposes the organization of optical inter-connections between LEDs and PDs directly in silicon layer of an SOI structure. For the realization of this idea the IR range LEDs and PDs are necessary because Si film can exhibit the waveguiding properties in the IR range only. The absence of effective IR range Si based LEDs and PDs is a key question of this approach. IR range LEDs can be made by silicon doping with rare-earth elements in particular, erbium. IR range PDs can be formed by epitaxial deposition of narrow-band semiconductors.

The second way proposes the use of the dielectric layer of the SOI structure for optical interconnections between LEDs and PDs disposed in Si film or in Si substrate. This technique is still a dream because a light goes out from SiO_2 layer to silicon due to unfavourable ratio of refraction indexes ($n_{Si} > n_{ox}$). A silicon dioxide-based optical waveguide (WG) has significantly wider bandwidth (IR, visible and UV) than a silicon-

3.2. LIGHT EMITTING DIODES

For the last two years great advances have been made in development of LEDs based on porous silicon. Although the exact nature of emission is still unclear and highly controversial, the proposed mechanism prevalent in the literature centers around quantum size effects in c-Si quantum wires or mesh structures formed in high porosity PS [21]. Nevertheless, LEDs emitting the light in red, orange, and yellow bands of visible spectrum have been already made. Observation of blue light emission have appeared as well.

Quantum efficiency of these devices is still rather low. Another problem consists of very short working period (no more than several hours) of LEDs based on porous silicon. High chemical activity of porous silicon seems to be the reason of its degradation during operating as a part of LED.

Lazarouk et al [22] have developed LED characterized by increased stability and durability. As seen from Figure 2, the LED consists of Si substrate, PS layer, and Al electrodes built-in in Al_2O_3. The Al_2O_3 layer has been formed by selective electrochemical anodization of initial 0.5 µm thick Al film. Visible light emission from PS edges was observed as soon as voltage between Al electrode and Si substrate exceeded 5 V.

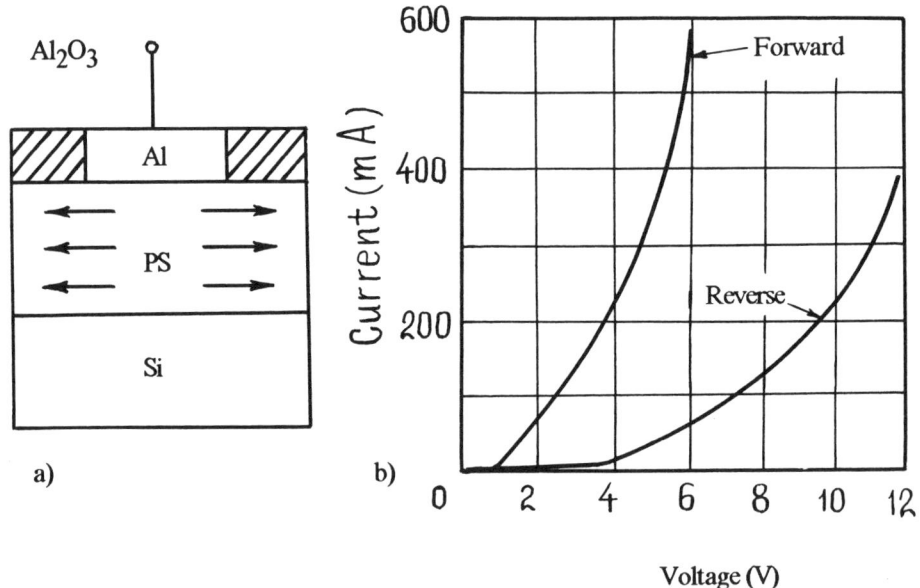

Figure 2. Construction (a) and voltage-current characteristics (b) of PS based LED [22].

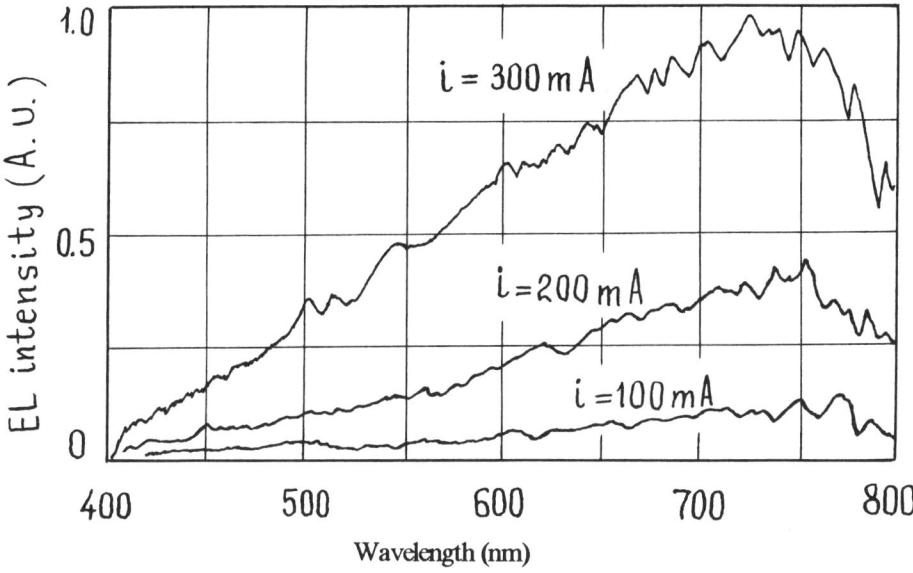

Figure 3. Electroluminescence spectra of PS based LED [22].

As seen from Figure 3, the electroluminescence spectra had a maximum at 750 nm and a prolonged shoulder in the IR range. The main feature of the LED is very high stability of Schottky contact characteristics. It has been achieved due to the passivation of diode periphery by anodic Al_2O_3. As a result, the high quality protection of active PS layer provided a durable and stable operation of this LED, in contrast to other known constructions. According to [22], the current density of 500 A/cm^2 resulted in sufficient porous silicon heating due to Joule heating. But no any measurable shift of LED parameters was observed during device operation for several hundred hours.

3.3 OPTICAL WAVEGUIDES

Optical waveguide is a structure which is used for light concentration and guiding in integrated optical circuits. Optical fiber with a round cross section is an example of optical waveguide. Planar structure such as films and trenches are more promising for integrated optics.

The PS refraction index is known [23] to vary from 1.3 till 3.0 depending on porosity. PS oxidation results in an increase of its refraction index value close to that of SiO_2. We attempted, varying the oxidation conditions, to make a SOI structure suitable for light waveguiding in oxidized PS layer provided that thermal SiO_2 layer was formed at the oxidized PS/Si interface and $n_{ox} < n_{ops}$. Figure 4 presents microphotograph of such SOI cross section and a scheme of lightguiding in the structure. Optical properties of oxidized porous silicon (OPS) and SiO_2 have to be very similar, but n_{ops} has to be higher than n_{ox}.

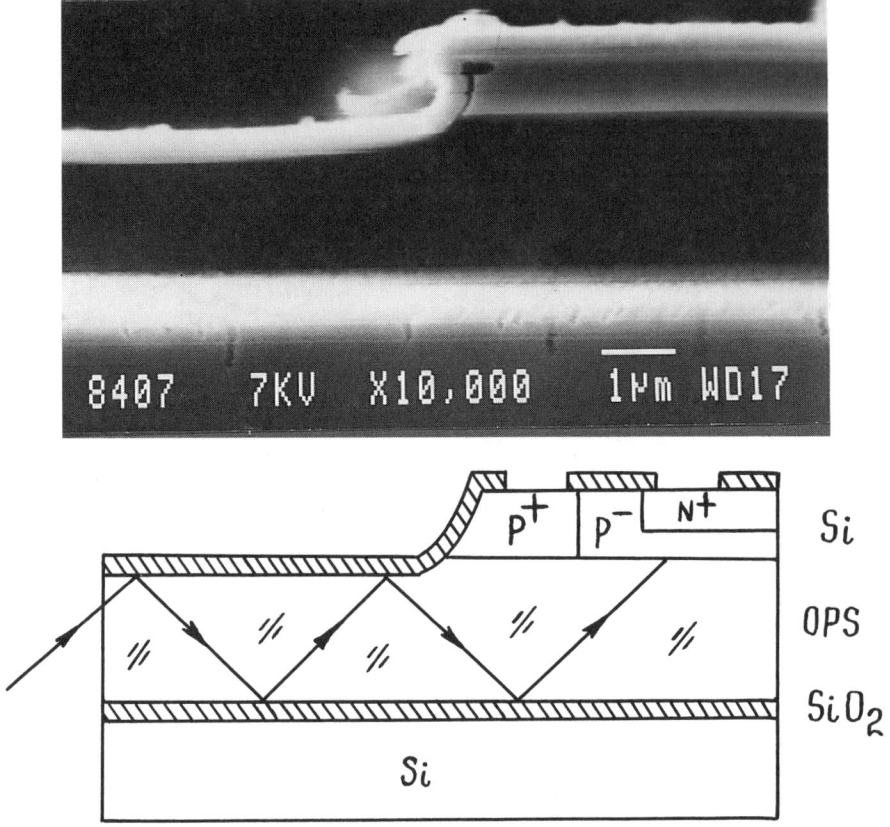

Figure 4. Microphotograph (a) and scheme (b) of lightguiding in OPS based SOI structure.

The SOI structure of entire OPS layer is an example of planar optical waveguide without lightguiding restriction in the OPS plane. In a channel OPS waveguide construction the light is guided along the walls like in optical fiber.

The process sequence for fabrication of an optical waveguide consists of three main steps [24]: (i) silicon anodization in hydrofluoric acid electrolyte through the silicon nitride mask to form local PS layers; (ii) thermal oxidation of PS; (iii) densification of oxidized PS.

P-type Si wafers of (100) orientation and 0.01 Ohm-cm resistivity were used as initial substrates. Selective anodization of the wafers through a 0.2 μm thick silicon nitride mask was performed in HF/alcohol electrolyte under an anodic current density of 10-30 mA/cm^2. As a result, local porous silicon layers of single fiber or Y-shape waveguide topologies were formed. The waveguides had a length of 1.5-3.0 cm, a width and thickness of 5-15 μm.

After mask film removal and cleaning the wafers have been oxidized in a diffusion furnace by a three- step process. First, the structure of porous material was stabilized and prevented from sintering by low-temperature oxidation at 300°C for 1 h in dry oxygen.

Porous silicon was fully oxidized using a steam ambient at 900°C. Then the temperature was raised up to 1150°C during 25 minutes, and thermal densification of oxidized PS in wet oxygen-nitrogen atmosphere was performed to improve optical properties of the material obtained. Finally, the temperature was reduced up to 850°C during 30 min before unloading. These "soft" regimes together with proper choice of PS parameters provided low mechanical stress level of the waveguide structures, absence of cracks and dislocation slip lines. No special buffer layer was made in the experimental structure.

Optical waveguiding in the visible range was observed by observing the mirror-like cleaved end of a waveguide with a microscope. The light from a 0.633 μm He-Ne laser or tungsten lamp has been directed through a glass fiber on the cleaved end of the waveguide (Figure 5).

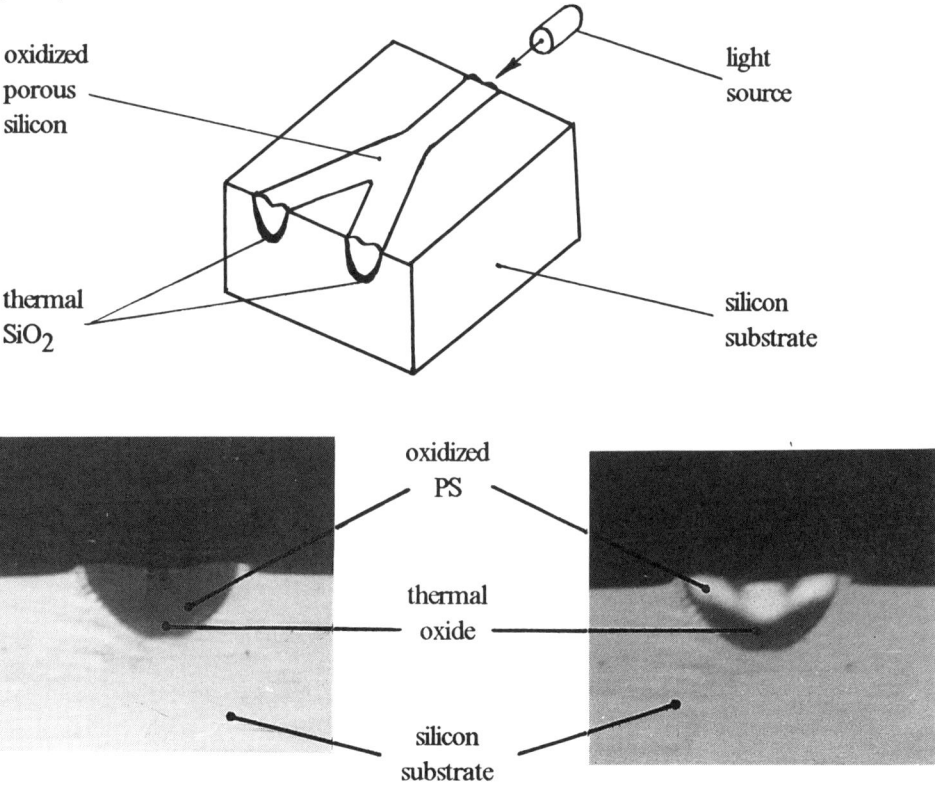

Figure 5. Schematic view (a) of optical waveguide (Y-splitter) based on oxidized porous silicon and microphotographs (b) of the waveguide cross section.

Waveguiding of both the white light of a tungsten lamp and the red light of an He-Ne laser were easily observed using the microscope. As seen from Figure 5, the light is guided in the center part of the cross section. In our opinion, thermal oxide layer is formed at the oxidized PS/Si interface and acts as a confining layer preventing light leakage from the waveguide into the Si substrate. Optical properties of oxidized porous silicon are very sensitive to its structure, composition, and density. In particular, its refractive index can be modulated by changing both thermal oxidation/densification regimes and PS porosity. The "core" part of the waveguide cross section has higher refractive index in comparison with the peripheral thermal oxide confining layer.

Besides, local mechanical stress field could influence light guiding through the waveguide. There exists a strong stress during silicon oxidation due to significant increase of oxide volume in comparison with the initial Si volume. In contrast, in the case of porous silicon oxidation, the growing oxide expends into pore volume thus decreasing the stress level. In our experiments measured optical loss less than 1 db/cm seems to be due to the absence of scattering centers on sidewalls, cracks and other defects reducing optical properties of the material obtained.

An optical waveguide with a cross section of 1-1000 μm can be easily fabricated using simple anodization technique and conventional thermal oxidation. There are a lot of freedom with regard to choice of waveguide cross section geometry and its arrangement in Si substrate. The waveguide can be arranged both planar to substrate and under electron circuit's components. In fact, any PS-based SOI technology (for instance, FIPOS or ISLANDS) can provide very attractive opportunities for lightguiding through oxidized porous silicon channels buried in Si substrate.

Optical properties of oxidized porous silicon were shown to be drastically changed by rare-earth elements doping and thermal treatment. Oxidized porous silicon doped with erbium from spin-on films exhibited sharp (FWHM of ~ 0.01 eV at 77 K) luminescence at 1.54 μm [15,16]. Such PS:Er waveguide structures are promising for integrated optical filters, amplifiers/modulators, light-emitting diodes and other optoelectronic devices.

Optical waveguides based on oxidized porous silicon have several advantages:
- any cross-section shape and topological design;
- buried in silicon, under microelectronic components;
- wide range of thickness (1-50 μm) and width (1-1000 μm);
- wide range of transmission (visible and IR) variable with rare-earth elements;
- change of transmission (rare-earth elements doping and thermal modification);
- compatibility with silicon-on-insulator structures.

4. Conclusion

The main peculiarity of porous silicon use in SOI technology is the unique possibility to form layers having very wide range of electrical and optical properties. Special technological treatments have to be used for these purposes. Not only silicon oxide but silicon carbide and silicon oxynitride can be formed by suitable treatment of porous silicon. Moreover, different metals and rare-earth elements can be introduced into porous silicon. Modification treatment of porous silicon allows to obtain local layers having semiconducting, conducting and dielectric properties. These layers can be incorporated into SOI structures.

From a futuristic point of view one can expect the development of the micro-optoelectronics based on PS SOI structures. This field seems to be very perspective and can be envisaged using the differences between the optical characteristics of the different layers (PS, modified PS, monocrystalline Si). We have demonstrated experimental samples of light emitting diodes of visible range and optical waveguides based on oxidized PS. The application of PS-based SOI may eventually allow fabrication of original integrated systems where the electron circuits and the optical elements can be fabricated by traditional silicon technology.

5. Acknowledgements

The authors have benefited from much useful collaboration and discussion with colleagues at Porous Silicon Laboratory of Belarussian State University of Informatics and Radioelectronics especially A.Bondarenko, V.Borisenko, L.Dolgyi, V. Filippov, N.Kazuchits, V. Labunov, S.Lazarouk, V. Levchenko, G.Troyanova, N. Vorozov, V.Yakovtseva.

6. References

1. Smith, R.L. and Collins, S.D. (1992) Porous silicon formation mechanisms, *J.Appl.Phys.* **71**, R1-R22.
2. Bondarenko, V.P., Bogatirev, Yu.V., Dolgyi, L.N., Dorofeev, A.M., Panfilenko, A.K., Shvedov, S.V., Troyanova, G.N., Vorozov, N.N., and Yakovtseva, V.A. (1994) 1.2 µm CMOS SOI on porous silicon (in this book). *NATO ARW,* Gurzuf, Ukraine, Nov.1-4.
3. Lehmann, V. (1992) Porous silicon preparation : alchemy or electrochemistry?, *Advanced Materials* **4**, 762-764.
4. Labunov, V., Bondarenko, V., Glinenko, L., Dorofeev, A., and Tabulina, L.(1986) Heat treatment effect on porous silicon, *Thin Solid Films* **137**, 123-134.
5. Labunov, V.A., Bondarenko, V.P., Borisenko, V.E., and Dorofeev, A.M. (1987), High-temperature treatment of porous silicon, *Phys.Stat.Sol.(a)* **104**, 193-198.
6. Herino, R., Perio, A., Barla, K., and Bomchil, G. (1984), Microstructure of porous silicon and its evolution with temperature, *Mater.Letters* **2**, 519-523.
7. Bondarenko, V.P., Borisenko, V.E., and Labunov, V.A. (1986) Arsenic diffusion through porous silicon under noncoherent light transient treatment. *Sov.Phys. Semicond.* **20**, 929-933.
8. Bondarenko, V.P., Borisenko, V.E., Zarovskii, D.I., and Raiko, V.A. (1989) Structure of cobalt

silicide films formed on porous silicon, *Rus.J.Surface* 1, 38-40.
9. Yon, J.J., Barla, K., Herino, R. and Bomchil, G. (1987) The kinetics and mechanism of oxide layer formation from porous silicon formed on p-Si substrates, *J.Appl.Phys.* 62, 1042-1048.
10. Bondarenko, V.P., Yakovtseva, V.A.,.Dolgyi, L.N., Vorozov, N.N., and Troyanova, G.N. (1994) SOI structures based on oxidized porous silicon, *Rus.J.Microelectronics* 23, 61-68.
11. Bondarenko, V.P., and Vorozov, N.N. (1988) AES investigation of thin films of nitridized and oxydized porous silicon, *Abstr. 2nd Int.Conf. Technol. EBT-88*, Varna, Bulgaria, Sect.6, 857-862.
12. Borisenko, V.E., Bondarenko, V.P., and Raiko, V.A. (1989) Electrochemical deposition of cobalt on porous silicon, *Doklady of Belarusian Academy of Sciences*, 33, 528-530.
13. Herino, R., Jan, P., Bomchil, G. (1985) Nickel Plating on Porous silicon, *J.Electrochem.Soc.*, 132, 2513-2514.
14. Borisenko, V.E., Bondarenko, V.P., Dorofeev, A.M., and Raiko, V.A. (1989) Structure of $CoSi_2$ films formed on porous silicon, *Proc.European Workshop on Refractory Metals and Silicides*, Leuven, Belgium, 129-130.
15. Kimura, T., Yokoi, A., Honguchi, H., Saito, R., Sato, A. (1994) Electrochemical Er doping of porous silicon and its room- temperature luminescence at 1,54 μm. *Appl.Phys.Lett.* 65, 983-985.
16. Dorofeev, A.M., Gaponenko, N.V., Bondarenko, V.P., Bachilo, E.E., Kazuchits, N.M., Leshok, A.A., Troyanova, G.N., Vorozov, N.N., Borisenko, V.E., Gnaser, H., Bock, W., Becker, P., Oechsner, H. (in press) Erbium luminescence in porous silicon doped from spin-on films, *J.Appl.Phys*.
17. Dorofeev, A.M. and Troyanova, G.N. (1994) Analysis of lattice misfit in "Diamond/Porous silicon" heteroepitaxial structure, *Proc. 2nd Int. Symp. Diamond Films (ISD-2)*, Minsk, 72-73.
18. Kovyazina, T., Kutas, A., Khitko, V., Gaiduk, P., Komarov, F., Solovev, V., Bondarenko, V.P., and Troyanova, G.N. (1993) Heteroepitaxy of GaAs on porous Si: The struture of the interface and defects of GaAs, *Abstr. 17th Int.Conf.Defects in Semiconductors*, Gmunden, Austria, 192.
19. Bondarenko, V.P., Vorozov, N.N., Dorofeev, A.M., Levchenko, V.I., Postnova L.I., and Troyanova, G.N. (1994) Porous silicon as a buffer layer for PbS epitaxy, *Tech.Phys.Lett.* 20, 51-55.
20. Raiko, V.A., Spitzl, R., Borisenko, V.E., and Bondarenko, V.P. (1994) MPCVD diamond deposition on porous silicon pretreated with the bias method, *Proc. 2nd Int.Symp.Diamond Films (ISD-2)*, 3-5 May 1994, Minsk, 101-102.
21. Canham, L.T. (1993) Progress toward crystalline-silicon-based light-emitting diodes, *MRS Bulletin*, 18, 22-28.
22. Lazarouk, S.,Bondarenko, V., Pershukevich, P.,La Monica, S., Maiello, G., and Ferrari, A. (1994) Visible luminescence from Al-porous silicon reverse bias diodes formed on the base of degenerate N-type silicon, as presented at *MRS'94 Fall Meeting*, Boston, Massachusetts, Nov.27-Dec.2.
23. Pickering, C.,Beale M.I.J., Robbins, D.J., Pearson, P.J., and Greef, R. (1984) Optical studies of the structure of porous silicon films formed in p-type degenerate and non-degenerate silicon, *J.Phys. C: Solid State Phys*. 17, 6535-6552.
24. Bondarenko V.P., Dorofeev A.M., Kazuchits N.M., Labunov V.A., Stelmakch V.F. (1993) Integrated optical waveguide fabricated with porous silicon, *Tech. Phys.Lett.* 19, 463-465.

DEFECT ENGINEERING IN SOI FILMS PREPARED BY ZONE-MELTING RECRYSTALLIZATION

E.I.GIVARGIZOV, V.A.LOUKIN, and A.B.LIMANOV
*Institute of Crystallography, Russian Academy
of Sciences, Moscow 117333, Russia*

ABSTRACT

Perfection of SOI Films prepared by different technologies play a crucial role in choosing of a given approach for an actual application. Possibilities to control the perfection of ZMR-SOI films are considered. Investigations in graphoepitaxy and in using a liquid sublayer are performed.

1. Introduction

By now, at least three SOI technologies have been extensively developed: SIMOX, BESOI (wafer bonding), and ZMR. Among these, SIMOX and BESOI wafers are produced commercially, while ZMR is still at the R&D stage. Each of these technologies will seemingly have its own niche for applications. Currently, the niches are still forming. In general, SIMOX, more suitable for preparation of ultrathin Si films, is used mainly for CMOS applications, while thicker BESOI films are used for bipolar, high-voltage, and sensor applications. ZMR is more universal in this respect.

Another factor, the perfection of the films prepared by different technologies, plays a very important role in the formation of the niches, too.

In this paper, possibilities to control perfection of ZMR-SOI films are considered. Experimental studies in graphoepitaxy and in using a liquid sublayer were performed and analyzed.

2. Experimental procedure

2.1. CRYSTALLIZATION PROCEDURE

ZMR was performed using an equipment where a linear molten zone was formed by a high-power ($300W$) CW YAG:Nd laser (wavelength $1.06\mu m$). A circular laser beam was transformed into a linear one by using $90°$-crossed cylindrical lenses. The linear spot exceeded $100\text{-}mm$ in length, and was focused at the surface of the sample.

The substrate was heated up to $1200 - 1300°C$ by a set of halogen lamps. The laser beam melted a $0.5 - 1mm$ wide zone of the polysilicon layer sitting on top of an oxidized silicon substrate. The film was encapsulated by a SiO_2 film in order to prevent agglomeration of the molten zone.

Crystallization was performed in a single pass of the molten zone across the poly-Si film. Scanning speed varied from 0.1 to $2mm/s$. Lateral temperature gradient was another changeable parameter. It was not measured directly, but could be controlled by adjusting the total laser power together with focusing/defocusing of the beam.

2.2. PREPARATION OF FILMS FOR ZMR

In usual multilayer structure for ZMR process (Si-substrate, SiO_2-insulator buried layer, poly-Si film, and SiO_2 anti-agglomeration cap layer) the following features were envisaged.

a) Straight rows of separate square-like seeding windows (approximately $10 \times 10 \mu m$, with center-to-center distances $20 \mu m$, with a period $5mm$) were made in the buried layer for lateral epitaxy [1]. Anti-reflection stripes intended for generation of periodic thermal relief [2] were made in the cap perpendicularly to the rows.

b) Inside the buried layer, a phosphosilicate glass (PSG) film, able to be soften at temperatures above $1000°C$, was created (fig.1). The poly-Si film intended for ZMR was separated from the PSG film by a Si_3N_4-SiO_2 layer. In the cap layer, regular anti-reflection stripes were made, too.

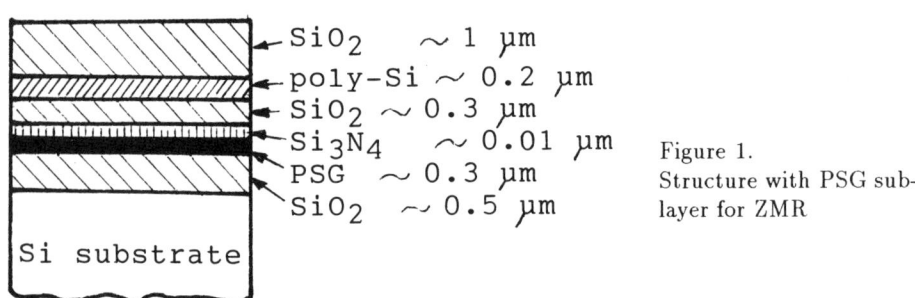

Figure 1. Structure with PSG sublayer for ZMR

2.3. INVESTIGATIONS IN ZMR-SOI FILMS

After recrystallization, the oxide cap was removed by treatment in an HF-contained solution. The ZMR-SOI film were then etched in selective Secco solution [3] during $1-2 min$ in order to reveal the microstructure of the films. The microstructure was studied by scanning electron microscope (SEM) with secondary-electron and back-scattering modes of operation.

3. Results and discussion

3.1. ZMR: STANDARD VERSION

As is known, branched subboundaries elongated in the direction of zone scanning are the most typical imperfections in ZMR-SOI films (fig. 2).

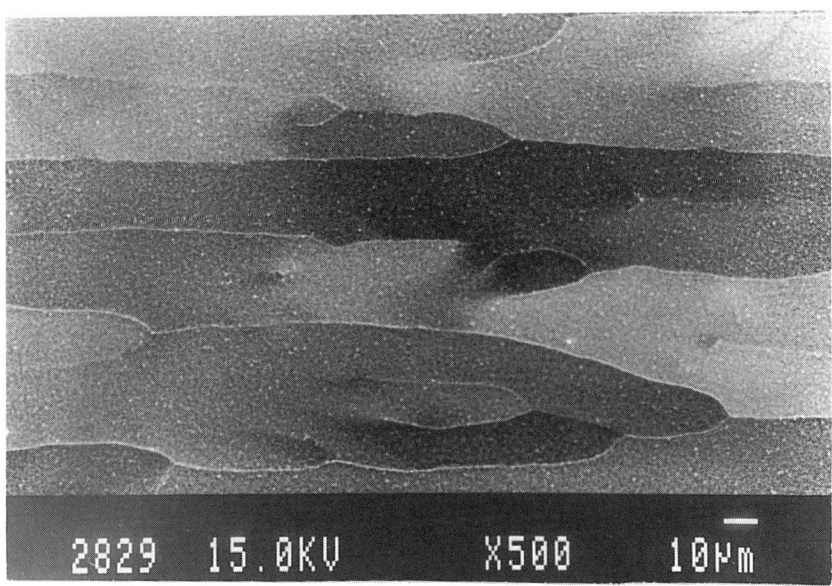

Figure 2. Typical microstructure of ZMR-SOI films. The contrast is caused by mutual misorientations between different grains and different ares of the grains. SEM micrograph in secondary-electron mode of operation.

It is known that the subboundaries separate slightly (for about $1 - 2°$) misoriented single-crystalline grains. This misorientation is measured with respect to the horizontal plane, it coincides with the scanning direction, and is perpendicular to the plane of the film. Accordingly, subboundaries consist of dislocations and/or bundles of dislocations that lie in both the plane of the film and perpendicularly to it (the latters are threading dislocations). The different contrast of various areas of the film reflects the level of the misorientation.

If a periodic thermal relief is superposed on the ZMR process, the subboundaries are localized in accordance with principles of graphoepitaxy [4], see fig.3. The period of the relief ($20\mu m$ in this case) was chosen as an average distance between the non-localized subboundaries in fig.2 that is inherent in given crystallization conditions (a temperature gradient, and a zone-scanning velocity). As is seen, the film is far more homogeneous in microstructure, and a misorientation remains only around the horizontal axis (indicated by the arrow).

3.2. ZMR WITH DEGENERATED SUBBOUNDARIES

Earlier, it was found that subboundaries can be degenerated: they are transformed into chains of dislocation bundles or even separate dislocations localized according to graphoepitaxy, provided that the period of the thermal relief is at least twice as small as that should be at given growth conditions [5].

Here, more detailed investigation in the phenomenon was undertaken.

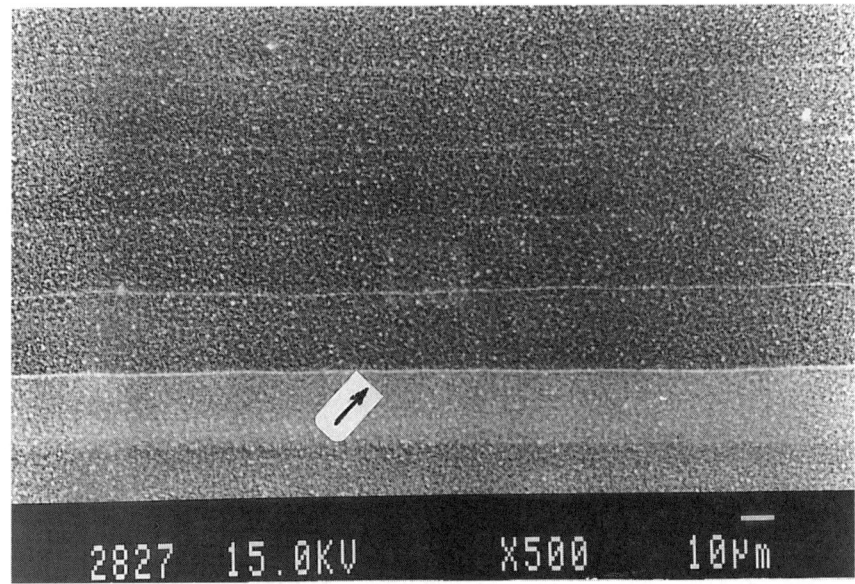

Figure 3. Localized subboundaries formed under action of periodic thermal relief.

In order to have reproducible orientation of the film, lateral-epitaxy approach was used that allowed one to grow SOI films with a given orientation (e.g., (100) in this case because Si-(100) served as a substrate).

The result is shown in fig.4. As it is known, in the lateral epitaxy process, smooth subboundary-free growth takes place at distances up to $30 - 50\mu m$ from seed windows as a result of strong temperature gradients established in growing film close to the windows [6]. This area is indicated by arrows. If, next, there is a regular thermal relief, the subboundaries are localized as is shown in fig.4.

Here, the period of the relief is $15\mu m$, i.e., slightly smaller then the nominal (for these crystallization conditions) value $20\mu m$. Accordingly, a part of the localized subboundaries are degenerated, while others are not.

In general, two types of constituents of subboundaries are obtained.

First, these are isolated dislocations as is shown in fig.5. Typically, they form chains with intervals between separate dislocations about $1\mu m$ or less. As a result of treatment of the film in Secco etch during 1 min, almost cylindrical holes with diameters $0.2 - 0.3\mu m$ were formed (see fig.5b).

In other cases, several near-parallel chains of such dislocations were found (fig.6). Some holes had an elliptical shape (indicated by single arrows); this means that they are inclined to the film plane. At sufficiently small distances between the holes, they are merged into quasi-continuous lines (indicated by double arrows).

Another type of the constituents is shown in figs.7 to 9. These are bundles of dislocations almost perpendicular to the film plane. At sufficiently large distances

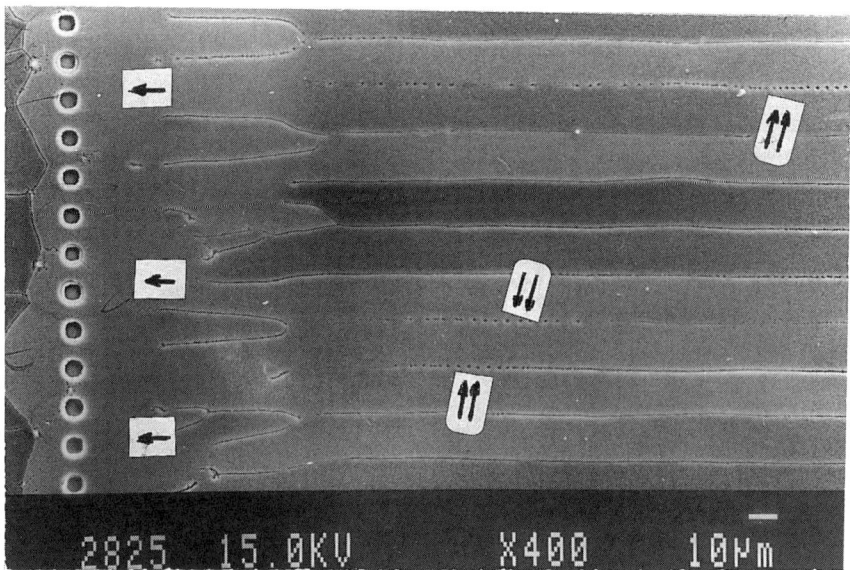

Figure 4. Morphology of SOI film formed in lateral epitaxy together with graphoepitaxy. Degenerated subboundaries are indicated by double arrows.

between them the bundles are not overlapped (fig.7). Comparative investigations of the same bundle in secondary-electron (fig.8a) and back-scattering (fig.8b) modes of operation show that most of holes in a given bundle are emanated from approximately the same point inside the film.

When the bundles have rather high density per unit length, they overlap each other (fig.9a) and even form a quasi-continuous line (fig.9b).

It is to note that all the pictures shown in figs.4 to 9 were obtained on the same sample subjected to the same etching procedure (duration about 1 min).

Earlier, defects similar to those shown in figs. 7 to 9 were observed, e.g., by optical microscopy (see fig. 3.7b in [7]) and were explained by the capture of droplets that consisted preferentially of Si, with subsequent solidification. It is well known that Si has an anomaly in solidification (it increases its volume and, so, causes constraints in the film). At high temperatures typical for near-growth-interface, the plasticity of Si leads to formation of a lot of defects (bundles of dislocations).

The capture of droplets is facilitated by formation of rather deep V-shaped "pockets" between protrusions of crystalline phase into melt ("cupolas") which are developed at the cellular growth interface (see fig. 2a in [5] and 2a in [8]). In the case of graphoepitaxy, the protrusions are formed along predetermined lines corresponding to the temperature minima, while the capture of the droplets takes place in between the temperature minima.

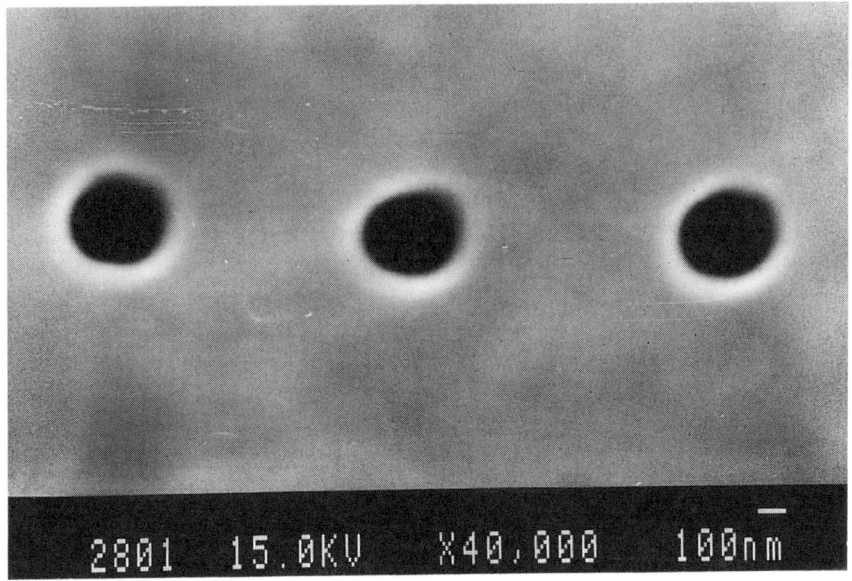

b

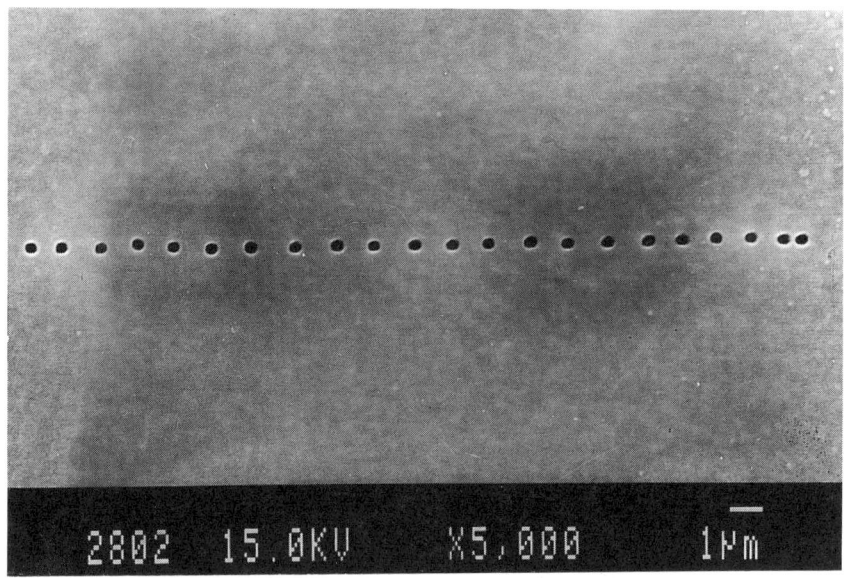

a

Figure 5. A chain of single dislocations (revealed by selective etching) localized by thermal relief.

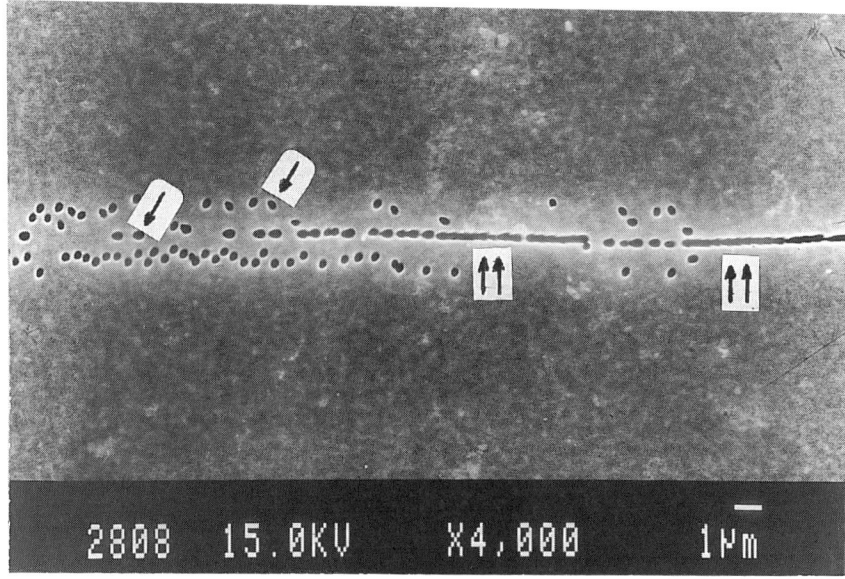

Figure 6. Chains of single dislocations. Some of them are merged into quasi-continuous lines.

Another factor that should be taken into account in the consideration of the defect formation is a rigid confinement of the growing SOI film by the substrate and the encapsulated layer. This factor can enhance the capture of the droplets.

Let's now return to the single dislocations shown in figs. 5 and 6. There are no dislocation bundles, i.e., most probably, no capture of droplets. The dislocations were formed at junction between neighbor grains, most probably as a result of slight misorientations between them.

3.3. ZMR OF FILMS WITH LIQUID SUBLAYER

In ZMR of multilayer structures with the PSG sublayer it was found that, at relatively low total heating, the perfection of SOI films improved drastically so that no dislocation bundles were detected, and only localized rows of single dislocations remained.

In fig.10, the morphology of such films prepared at relatively high zone-scanning velocities (about $0.5 mm/\sec$) is shown. The films were treated in Secco etch for a long period of time (about 2 min). As it is seen, a hardly-remarkable line (shown by an arrow), seemingly resulting from merging single dislocations, was formed at the subboundary.

At the lowest used velocity ($0.1 mm/\sec$), only traces rare dislocations along lines corresponding to subboundaries were observed (fig.11).

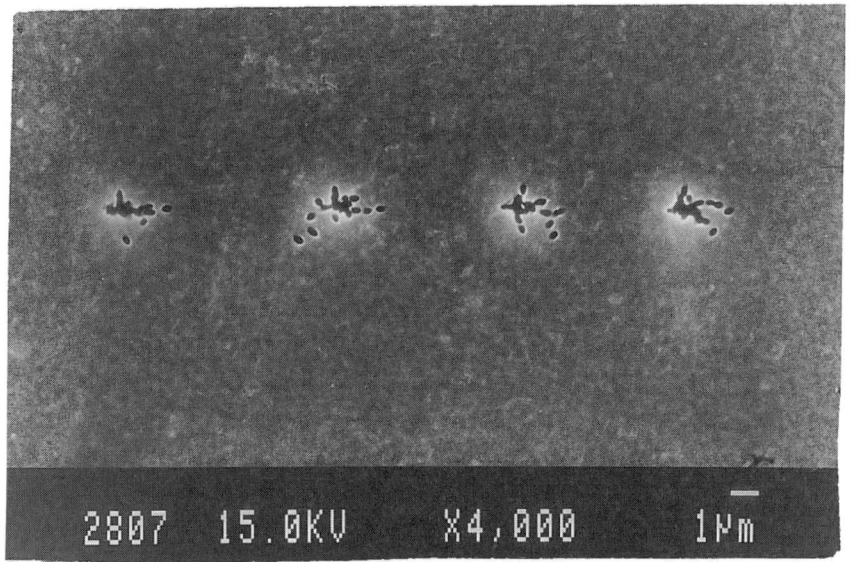

Figure 7. A row of dislocation bundles along of subboundary.

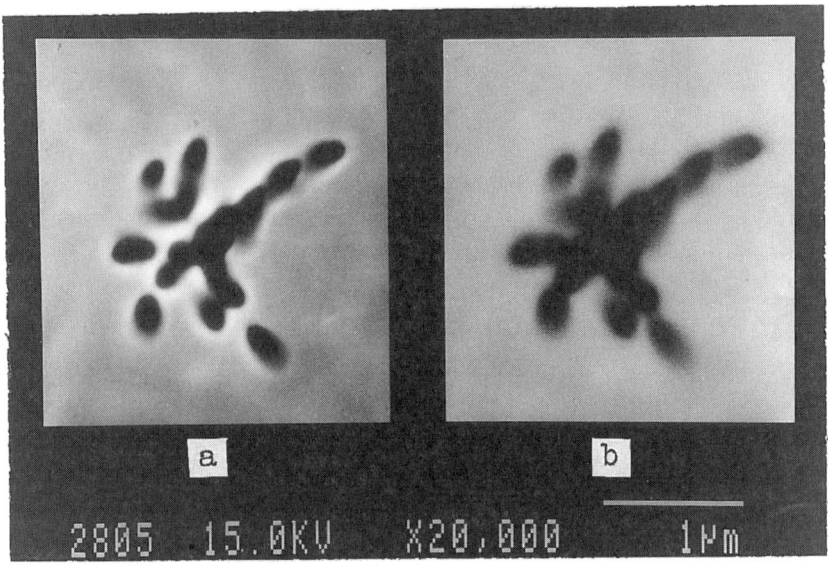

Figure 8. A bundle of dislocations at different SEM mode of operation.

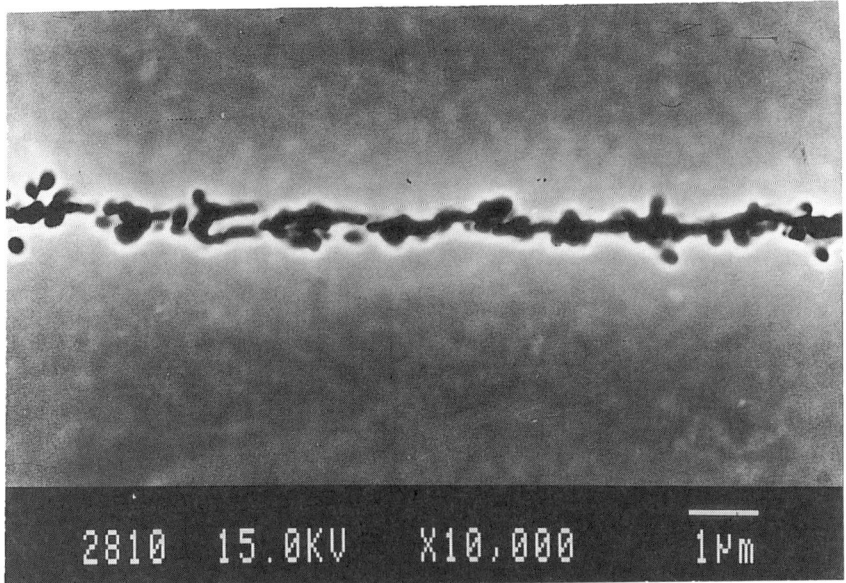
b

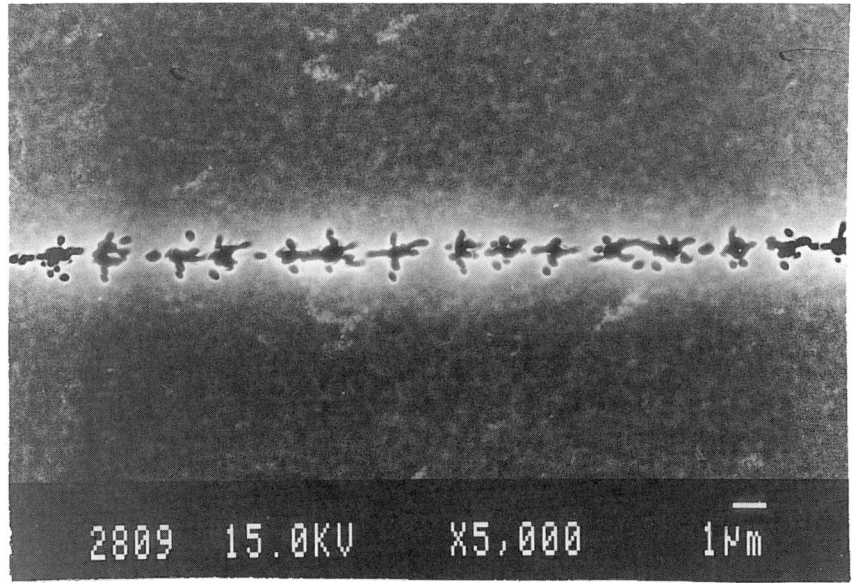
a

Figure 9. Chains of dislocation bundles merging into quasi-continuous lines.

Figure 10. Morphology of etched ZMR-SOI films prepared with liquid sublayer at relatively large zone velocity.

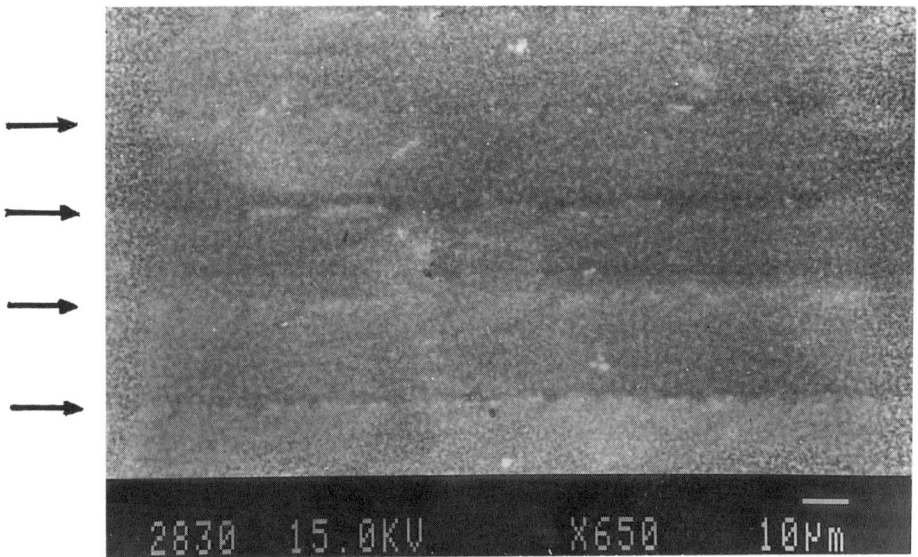

Figure 11. ZMR-SOI film prepared with liquid sublayer at a low velocity. The arrows indicate to traces of localized subboundaries.

Earlier, such an effect of drastic improvements of film perfection was observed in ZMR with single, narrow laser beams [9,10]. It was supposed that the softened sublayer relieved mechanical strains in the growing films.

In our case, we have in fact many parallel laser beams formed by the antireflection stripes.

The action of the liquid sublayer in our process can be explained in two ways. Firstly, it can, once again, relieve the mechanical strain in the growing film.

Secondly, thermal conductivity of the sublayer as a mobile substance can be larger then that of solid insulator; hence, the depth of the V-shaped pockets between the cupolas at the cellular growth interface should be smaller, and the capture of the droplets can be eliminated. As a result, no dislocation bundles, only single dislocations are formed in ZMR-SOI films prepared with the liquid sublayer.

4. Conclusions

The level of perfection of SOI films prepared by laser ZMR can be controlled using graphoepitaxial approach. Subboundaries in the SOI films are localized in accordance with periodic thermal relief organized by means of photolithographic mask. If the period of the relief is sufficiently small, the subboundaries are degenerated and transformed into chains of dislocations or dislocation bundles. The regularization of the topology of the imperfections improves homogeneity of growth conditions and, accordingly, of microstructural characteristics of the films.

ZMR of multilayer structures with low-temperature-softening glass sublayer gives films with drastically-improved perfection. In such a way, SOI films with no dislocation bundles are formed. This affect is attributed to mechanical strain relieve in the films and/or to decrease of in plane thermal gradients.

5. Acknowledgements

The authors thank Dr.B.A.Chubarenko for SEM studies of the films, and Mrs. O.B.Volskaya for assistance in preparation of the manuscript.

6. References

1. Fan, J.C.C., Geis, M.W., and Tsaur, B.-Y. (1981) Lateral epitaxy by seeded solidification for growth of single-crystal Si films on insulator, *Appl. Phys. Lett.* **38**, 365-367.
2. Colinge, J.-P., Demoulin, E., Bensahel, D., and Auvert, G. (1981) Use of selective annealing for growing very large grain silicon on insulator, *Appl. Phys. Lett.* **41**, 346-347.
3. Secco d'Aragona, F. (1972) Dislocation etch for (100) planes in silicon, *J. Electrochem. Soc.* **119**, 948-952.
4. Givargizov, E.I. (1991) *Oriented Crystallization on Amorphous Substrates*, Plenum Press, New York.
5. Givargizov, E.I., and Limanov, A.B. (1988) Artificial epitaxy (graphoepitaxy) as an approach to the formation of SOI, *Microelectronic Eng.* **8**, 273-291.
6. Lee, E.-H., and Rozgonyi, G.A. (1984) Models of growth stability breakdown in the seeded crystallization of microzone-melted silicon on insulator, *J. Crystal Growth* **70**, 223-229.
7. Im, J.S. (1989) Experimental and theoretical investigation of interface morphologies observed in directional solidification of thin Si films, PhD Theses, MIT.

8. Limanov, A.B., and Givargizov, E.I. (1983) Control of the structure in zone-melted silicon films on amorphous substrates, *Mater. Lett.* **2**, 93-96.
9. Pandya, R., and Martinez, A. (1988) Large-area defect-free silicon-on-insulator films by zone-melt recrystallization, *Appl. Phys. Lett.* **52**, 901-903.
10. Pandya, R., and Martinez, A. (1988) Elimination of defects in laser-crystallization SOI by stress relief, in J.C.Sturm, C.K.Chen, and L.Pfeiffer (eds), *Silicon-on-Insulator and Buried Metals in Semiconductors*, MRS Press, Pittsburgh, Proc. *MRS* **107**, 415-420.

ION BEAM PROCESSING FOR SILICON-ON-INSULATOR

WOLFGANG SKORUPA
Research Center Rossendorf Inc.
Institute of Ion Beam Physics and Materials Research
P.O.B. 51 01 19
D-01314 Dresden, Germany

0. Introduction

During the last few years ion beam processing penetrated very aggressively in many branches of advanced solid state technology. This holds also for the modern semiconductor technology in the area of VLSI (Very Large Scale Integration) and ULSI (Ultra Large Scale Integration). SOI (Silicon-on-Insulator) is one of the most discussed candidates of this branch offering the possibility to produce integrated circuits of high packing density, low power consumption and high speed.

Several techniques have been known for the production of SOI-structures using ion beam processing directly or as an efficient support, and these will be reviewed in this paper:

SILICON-ON-SAPHIRE (SPE-SOS) (combined with Solid Phase Epitaxy)

FULL ISOLATION BY POROUS OXIDIZED SILICON (HI-FIPOS)
(combined with hydrogen implantation)

LATERAL-ION BEAM INDUCED EPITAXIAL RECRYSTALLIZATION
(L-IBIEC)

SEED SELECTION THROUGH ION CHANNELING (SSIC)

SEPARATION BY IMPLANTED OXYGEN/NITROGEN
(SIMOX/SIMNI/SIMON)

MOLECULAR BEAM EPITAXY ON SOI (MBE-SOI)

ETCH STOP LAYERS FOR BONDED AND ETCHED SOI SRUCTURES
(ETCH-STOP-BESOI)

This is not a state-of-the-art-review of the described techniques. The first aim of this study is to take an advantage of the knowledge concerning the different approaches of using ion beams for the production of SOI-structures.

1. SPE-SOS

Apart from the HARRIS process of Dielectric Isolation, the well-known SOS-technology is the only one being implemented in production lines. An SOS-substrate consists of a single crystalline sapphire wafer on which a single crystalline silicon film is heteroepitaxially grown. The main disadvantages of this type of substrate are: (i) impurity outdiffusion from the sapphire substrate during the epitaxial growth taking place at high temperatures; (ii) formation of planar defects (stacking faults, twins) at the sapphire-silicon interface and in the silicon film due to lattice mismatch and

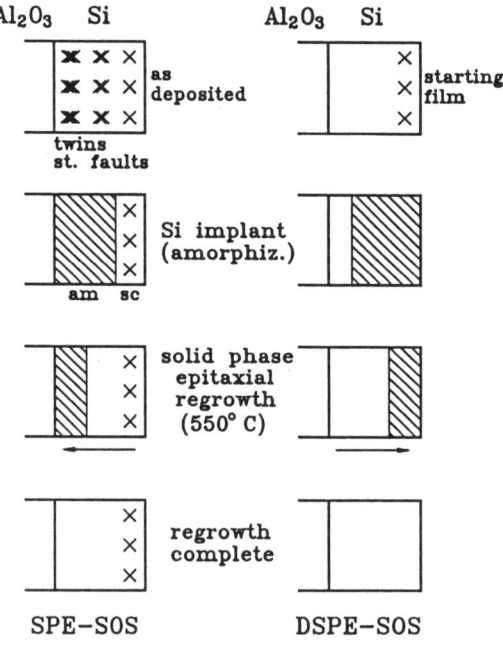

Fig.1: Schematic of the SOS-SPE process (after Roulet et al. [1]

different thermal expansion coefficients; and (iii) high substrate costs. Because of the latter fact electronic devices and circuits produced on such substrates are first of all used for space and military applications. The formation of planar defects during the heteroepitaxy is a problem which has been successfully solved by ion beam processing (see Fig. 1). A distinct reduction in the defect density of the silicon films can be achieved by amorphization of the interface sapphire-silicon region followed by SPE (Solid Phase Epitaxy) at 550-600°C [1]. Sometimes, an additional annealing step at 1000°C will be performed to further reduce the density of residual defects. It is clear, that during the amorphization step the surface region of the silicon film must remain single crystalline to serve as a seed during SPE. A reduction of the defects in the surface region is possible by amorphizing it during a second step (Double SPE-SOS) and repeating the SPE-regrowth using the lower part of the silicon film as a seed [2]. In this manner, a drastic lowering of the density of planary defects occurs leading to higher minority carrier lifetimes and, hence, to a lower junction leakage in MOS and bipolar devices. Both, ion implantation and furnace annealing as used for amorphization and SPE, respectively, are processes of high reproducibility and are widely used in the microelectronic technology. Consequently, the described method of defect reduction is state-of-the-art of modern SOS-technologies.

2. HI-FIPOS

The conventional FIPOS-process [3] suffers from several disadvantages [4]. In this process, n-type regions are formed at the surface of p-type silicon substrates by hydrogen implantation. During the anodization step necessary to form the porous silicon - only p-type material will be transformed - the current starts to flow in highly boron doped p^+-regions between the n-type regions and then spreads out in all directions within the bulk region. There, porous silicon with a higher density is formed because the density of porous silicon increases if the current density falls down. In this manner, material with mechanical stress is formed leading to dislocation densities up to $10^9 cm^{-2}$ within the silicon islands. On the other hand, due to the spreading out of the current, thick porous layers are needed to isolate even small islands and, moreover, stalks are formed at the bottom side of the islands so that the dielectric strength between the island and the silicon substrate is lowered.

Benjamin et al. [4] proposed the use of a buried n-type grid formed by ion beam processing to circumvent the above-mentioned problems. This configuration shown in Fig. 2 leads to a much more homogeneous current flow and thus to the formation of porous silicon with a homogeneous density. The stalks formed at the bottom side of the silicon islands are also avoided in this manner. The buried n-type grid is formed by masked hydrogen implantation. During the oxidation step necessary for the conversion of porous silicon to silicon dioxide and performed at temperatures higher than 800°C the implanted hydrogen diffuses out having no further influence on the SOI-structure. It was demonstrated that using this method silicon islands with a larger area and with much less mechanical strength can be formed [4].

FIPOS + Hydrogen implantation

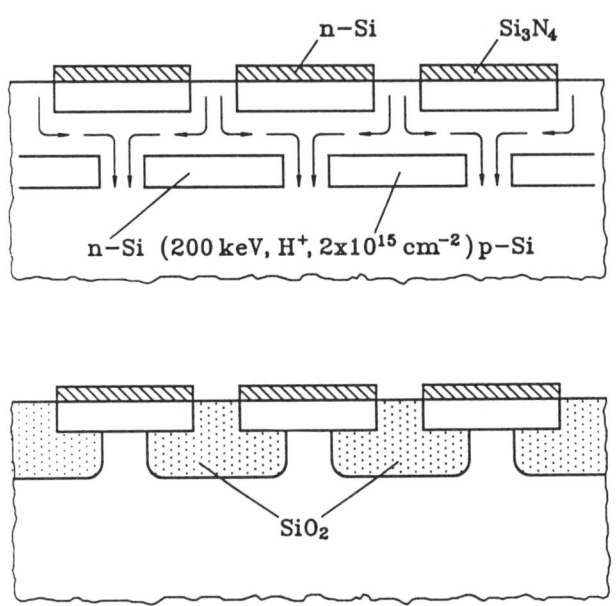

Benjamin et.al. RSRE 1986

Fig.2: Schematic of the HI-FIPOS process (after Benjamin et al. [4])

3. L-IBIEC

Recently, the basic IBIEC process has received a strongly growing interest and has been extensively reviewed and reported in the literature [5]. In brief, IBIEC is the well known process of thermally induced solid phase epitaxy (TH-SPE) being combined with ion beam processing for the purpose of depositing energy into nuclear processes. In this manner the activation energy of the TH-SPE process of about 2.7 eV can be lowered down to about 0.3 eV. TH-SPE normally needs a temperature of about 550°C or higher to take place in silicon. This limit can be lowered below 200°C by using IBIEC with appropriate parameters. By further lowering the temperature an inverse process of layer-by-layer interface amorphization can be induced [5]. The basic mechanism to explain these processes is not clear yet: if purely interface controlled [5] or related to point defect diffusion [6].

All the work concerning the basic mechanisms of IBIEC was performed on amorphous silicon layers produced by ion implantation directly on or within single crystalline silicon wafers. In 1988, the successful epitaxial recrystallization of an amorphous silicon layer deposited by chemical vapour deposition on a single crystalline silicon substrate was firstly demonstrated [7]. No special precautions against impurities and the natural silicon oxide layers were used in these experiments as necessary to recrystallize such layers by TH-SPE. This type of processing can be principally used to produce layer systems with steep doping gradients. On the other hand, such a vertically driven recrystallization of a deposited layer from a single crystalline seed is an important prerequisite to drive an epitaxial growth front laterally on an insulator. In this way an SOI-structure can be formed which was already demonstrated for the case of TH-SPE [8]. Recently, our group showed for the first time a laterally oriented growth of a silicon layer by IBIEC to be possible [9]. The SOI-structure consisted of

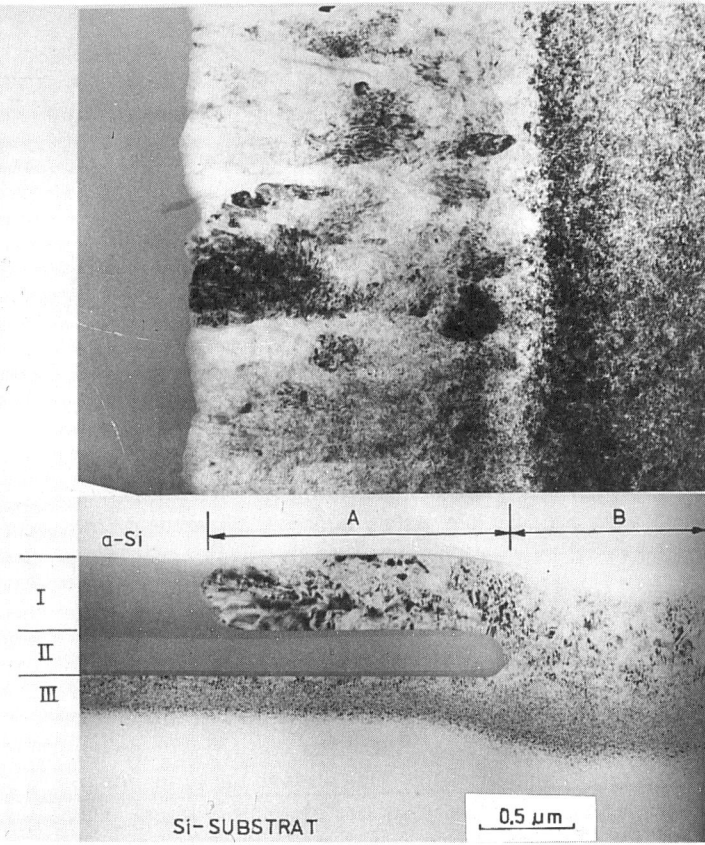

Fig.3: TEM micrographs (plan view in the upper part and cross section in the lower part with a correlated depth scale) of an SOI-structure after L-IBIEC [9]. For preparation details, see text.

of a thermally grown silicon dioxide layer (200 nm) with seeds defined by photolithography and wet chemical etching and a silicon layer (400 nm) deposited upon the silicon dioxide layer by chemical vapour deposition. The ion irradiation was performed using 330 keV silicon ions with doses in the range between 10^{17}-10^{18}cm^{-2}. The temperature during IBIEC was 400°C simply realized by the ion beam itself. It was demonstrated by Transmission Electron Microscopy in both plan view and cross sectional samples that a texture-like lateral growth of the silicon layer occurs as shown in Fig.3. In this case the IBIEC-dose was 5×10^{17}cm^{-2}. The letters A and B mark the overgrown area, and the seed region, respectively. It is worth noting that despite the high IBIEC-dose there is no indication of random nucleation in the amorphous silicon layer (left of A). Of course, the quality of this material is not yet satisfactory. Further efforts are awaited from a better understanding of the phenomena taking place at the interface between the substrate and the deposited layer as well as from an appropriate design of the mask edge between the seed and the insulating layer.

Nevertheless, this technique offers the possibility of producing SOI-structures in the low temperature regime of solid phase epitaxy combined with the advantages of ion beam irradiation which allows the use of lower process temperatures as compared to TH-SPE, as well as to perform the epitaxial growth only in selected regions of the wafer-both laterally and/or vertically. In this manner, L-IBIEC is a potential candidate for three-dimensional integration.

4. SSIC

The topic of SOI includes not only methods of producing single crystalline but also polycrystalline films on any type of insulating material. Such substrates are mainly used for the production of flat panel displays including thin film transistors and related devices. Especially, it is necessary to prepare material with a large grain size to minimize the influence of grain boundaries on the electronic transport properties. An interesting idea to produce not only grains with larger size but also with nearly homogeneous crystal orientation related to the surface was proposed by Reif and coworkers [10,11]. It was demonstrated that by skilfully utilizing the channeling phenomenon of ions in single crystalline material silicon films with the above-mentioned properties can be formed.

This process (see Fig.4) begins with low pressure-chemical vapour deposition of a polycrystalline silicon layer on a silicon wafer covered with a thermally grown silicon dioxide layer serving as an insulator. This leads to the formation of a silicon film with randomly oriented grains with a grain size of about 0.08 μm. Then, ion implantation of silicon ions follows by performing the irradiation under at a predetermined angle. In this manner, grains oriented with their main crystal axis parallel to the ion beam will be scarcely damaged due to the ion channeling effect whereas all other grains that are not aligned with the ion beam will be amorphized. The ion beam irradiation is performed with silicon ions to exclude any impurity or doping effects. Subsequent annealing at 600°C for several hours leads to recrystallization of the irradiated silicon film by seeding at the remaining grains with uniform orientation. Using this method, polycrystalline silicon layers with (100)-oriented grains and a size of 0.8 μm were

SEED SELECTION THROUGH ION CHANNELING

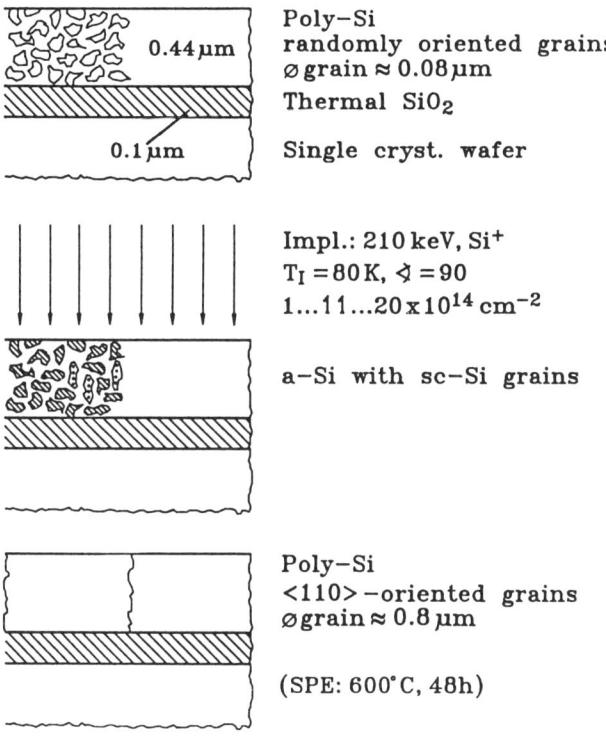

Fig.4: Schematic of the SSIC process (after Reif et al.[10-12])

produced [11]. The maximum angle deviation from the (100)-axis was 4°. It was demonstrated by using of X-ray diffraction that an optimum irradiation dose exists where most of the grains are of the same orientation after the recrystallization process. This is caused by the energy deposition into nuclear processes occuring during implantation of the silicon ions. At too low doses the energy deposition is not sufficient to render the misoriented grains amorphous. If the implanted dose is too high the nuclear energy deposition reaches the critical level for the amorphization of single crystalline silicon.

Thin film transistors produced in all three types of material did show the lowest threshold voltages and the highest channel mobilities for the silicon layers implanted with the optimum dose [11]. Moreover, it was observed that this effect is first of all related to the better surface properties of the (110)-oriented grains as compared to that of randomly oriented grains [12]. On the other hand, the grain boundary effect was demonstrated to be of lesser importance.

5. SIMOX / SIMNI / SIMON

This technique involving high dose implantation of oxygen and/or nitrogen has reached a high level of maturity in both research and production environments [13-15]. A few years ago, the design of high current implanters able to deliver oxygen currents in the 100 mA range was undertaken. At present, several laboratories and companies have installed such machines for the pilot production of SOI-substrates and, of course, integrated circuits upon them. Furthermore, the possiblity of producing SOI-substrates with high quality silicon layers of a thickness of about 100 nm has made the SIMOX technology a powerful candidate for the fabrication of ULSI devices with high performance [15]. Beyond the traditional speed, power consumption and rad-hard advantages of SOI-devices, a MOS transistor in a thin film SOI-substrate shows a higher short-channel threshold voltage stability, an improved subthreshold slope, an increased saturation current and reduced hot electron degradation. Additionally, the well known disadvantage of an SOI-transistor - the kink-effect - can be avoided [16]. All these advantages are related to the operation of the transistors as fully depleted devices [15]. So, this technique can be really named the most promising future SOI-technology for producing devices with submicron dimensions. After several years of hesitant waiting the industry is now at the point of introducing SIMOX (in competition with the wafer bonding technique) in the driving technology of DRAM production. The limits of optical lithography have brought SIMOX to this level of interest. The reason is that for a higher packing density of the devices the insulating area between them has to be lowered. This needs the SOI concept as a more advanced method to lower parasitic capacitances. A more detailed review concerning the evolution of SIMOX-technology is presented by Peter Hemment in these Proceedings.

Further in this chapter I would like to remind the SOI community that looking at the big success of SIMOX one should not forget that the topic of ion beam synthesis for SOI concerns also the use of nitrogen alone or in combination with oxygen. As a matter of fact, the properties of the pure SIMNI system are not so ideal for an SOI-structure as that of a SIMOX system:

- silicon film with low defect density
- sharp and smooth semiconductor-insulator interfaces
- highly resistive buried insulator.

On the other hand the most revolutionary step in the history of SIMOX (from my point of view) was firstly observed in SIMNI structures. This was the finding that high temperature annealing leads to a drastic decrease of the defect density in the silicon top layer as well as to the formation of sharper interfaces between the buried compound layer and the surrounding silicon. This was firstly found and documented both at the University of Surrey [17] and in Rossendorf [18]. From the viewpoint of the basics of ion beam synthesis the use of two elements to form a buried layer is much more interesting than the use of only one element like oxygen or nitrogen. Several pioneering studies were done during the eighties to form SIMON (Separation by IMplanted

Oxygen and Nitrogen) by NO-implantation [19] or by sequential implantation of oxygen before or after nitrogen [20]. A detailed investigation of the microstructure after sequential oxygen/nitrogen implantation into silicon was published by DeVeirman et al. [21]. The most encouraging result was that upon annealing up to about 1400°C at 52 min no crystallization of the buried silicon oxynitride layer occurred. On the other hand, the microstructure of the silicon top layer and the interfaces between the buried layer and the neighbouring silicon were not yet in a state known for advanced SIMOX systems

A new approach to use the best properties of silicon dioxide (formation of sharp and planar interfaces with silicon, low leakage current) and silicon nitride (diffusion inhibition against and gettering of unwanted impurities) was the formation of a buried stacked insulator [22]. This new type of a buried compound layer consists of a layer rich in oxygen (silicon dioxide, silicon oxynitride) above a layer rich in nitrogen (silicon nitride, silicon oxynitride). In this manner, the upper part of the buried layer should form an interface of good quality with the silicon top layer where electronic devices will subsequently be formed. This layer determines also the insulation properties of the buried layer. The lower part of the stack should perform diffusion inhibition against and gettering of unwanted impurities originating from contamination of the substrate and/or processing steps. In Fig.5 nitrogen and oxygen profiles of a stacked insulator structure are shown taken by Auger Electron Spectroscopy in combination with sputter profiling [23]. The implantation was performed with $^{14}N^+$, 310 keV, $10^{18}cm^{-2}$ before $^{16}O^+$, 250 keV, $10^{18}cm^{-2}$, followed by annealing at 1390°C for 30 min.

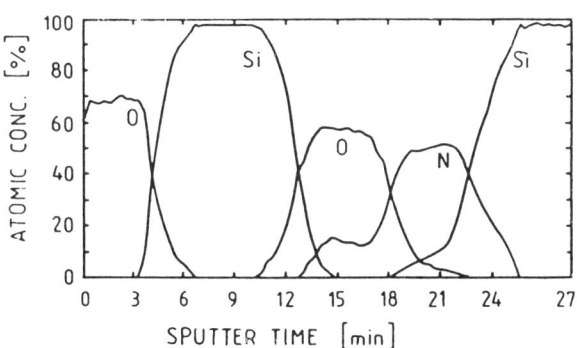

Fig.5: AES sputtering depth profile of a stacked layer system [23]

An important feature of modern semiconductor technologies not yet taken into account for ion beam synthesised SOI-structures in the appropriate manner as being necessary is the topic of gettering of unwanted impurities. This is first of all related to metallic contamination, mainly from the group of 3d transition elements (Fe, Cu, Ni,etc.). The first experiment was performed by our group in 1984 [24]. It was shown that in epitaxial layers above a buried SIMNI layer the generation lifetime of the minority carriers was higher by 1-2 orders of magnitude as compared to bulk silicon on the same substrate. Accordingly we found that this effect is related to the diffusion inhibition ability of silicon nitride [25] as well as to an agglomeration of the metallic contaminant in the silicon nitride-silicon interfaces [26].

The first investigations of SIMOX materal concerning the behaviour of metallic contaminants were reported by Delfino et al. [27] and Kamins and Chiang [28]. They demonstrated that intentionally introduced copper is gettered beneath the buried oxide layer after diffusing through it. Recently, Jablonski et al. [29] showed a distinct gettering of copper and nickel at the bulk side of the buried oxide independent of the source of incorporation: surface or back of the wafer (Fig.6). On the other hand, the

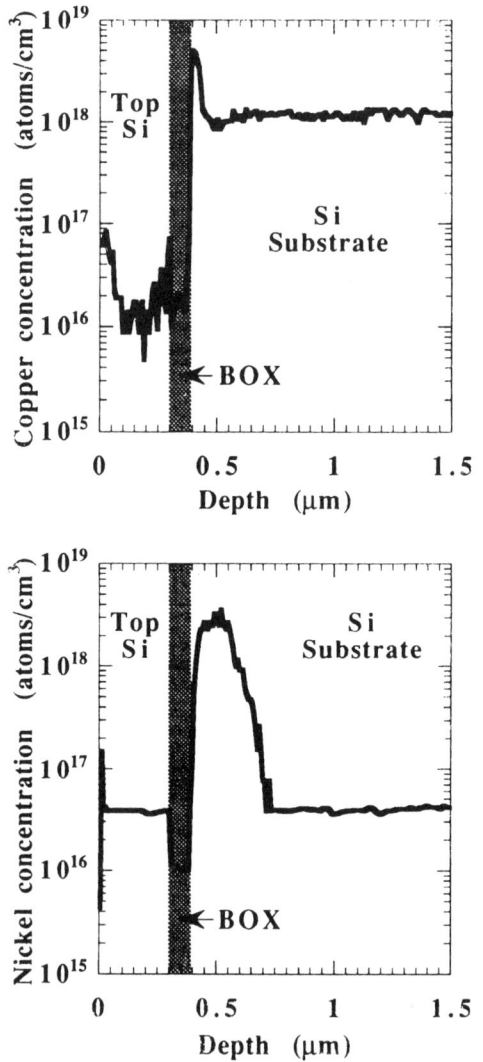

Fig.6: SIMS profiles of copper (top) and nickel (bottom) in SIMOX structures [29]

concept of proximity gettering, i.e. gettering at sites located very close to the electronic devices, is becoming increasingly important in combination with ion implantation at MeV energies. An SOI structure is a special case of proximity gettering because both interfaces of the buried layer or the layer itself can act as gettering sites. Taking into account that the two main sources of metal contamination in modern semiconductor fab-lines are plasma etching, reactive ion etching and high dose ion implantation [30], it turns out that most of the contamination is introduced from the device side of the wafer. This demonstrates the importance of proximity gettering. The main task for an SOI structure is in this respect to getter the metallic contaminants very efficiently at an appropriate distance from the active device region. In the ideal case, this active region should also include the buried insulating layer. This means for SIMOX-structures to locate gettering sites at or below the lower interface of the buried layer where they should be stabilised against further transformation during subsequent annealing treatments. This goal could also be achieved by a buried stacked insulator where gettering at the transition region between the upper silicon oxide and the lower silicon nitride should be favourable. A further solution to be thought about is MeV implantation of carbon below such a buried insulating layer. It was shown that such a carbon-rich layer formed with doses in the range of 10^{15}-10^{16}cm^{-2} is a very efficient gettering site against gold and iron [31] (see Fig.7). Recently, the formation of a buried layer of helium filled bubbles by ion implantation transformed to cavities by subsequent annealing was also shown to be a an efficient site for proximity gettering [32].

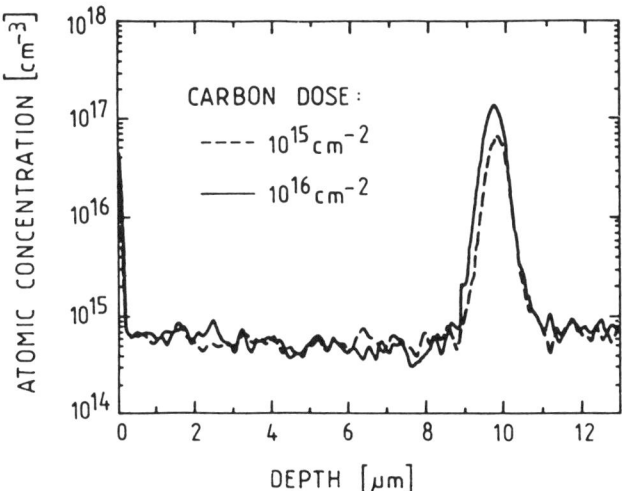

Fig.7: SIMS profile of iron in silicon after 10 MeV C-implantation and intentional iron contamination (10^{13}cm^{-2}) from the wafer backside [31b]

6. MBE-SOI

It may be somewhat bold to extend the topic of ion beam processing up to the molecular beam epitaxy technique were most of the particles are not charged but are transported in form of a directed vapour beam. This aspect but also taking into account the Ion Cluster Beam technique [33] as an alternative for MBE should satisfy the discussion of two interesting MBE-applications for SOI in this paper.

The first one concerns immediately the methods to reduce the dislocation density of SIMOX structures recently reported [13,14]. Adapting an experimental finding of compound semiconductor superlattices, Kamins et al. [34] tested the influence of SiGe-superlattice structures on the dislocation density of epitaxial silicon layers deposited above SIMOX wafers. To describe shortly the mechanism, a strain near heteroepitaxial interfaces produces a shear force which is able to bend dislocations parallel to the superlattices rather than allowing them to propagate to the surface. In this manner, close dislocations then also interact and will be annihilated or they may terminate at the edge of the sample. Of course, it is necessary that the superlattice should be properly designed.

One must however state that the results were not very encouraging. Although the bending of some dislocations could be observed by XTEM in such SIMOX-SiGe-structures most of the dislocations with a density of about 10^9 cm^{-2} propagated to the surface. However, it must be claimed that the SOI-structures used were not optimised ones from the viewpoint of damage annealing. A 1250°C, 6h-anneal step was used so that apart from dislocations also oxide precipitates still remain.

Whereas much more effective and also economic techniques are known to reduce the dislocation density of SIMOX-silicon films, e.g. the multiple implant-anneal-cycle [13,14], the formation of a single crystalline compound semiconductor film on a SOI-structure could combine all the advantages of both heteroepitaxy and SOI and, maybe, some additional ones. This should be of importance for the production of electronic and optoelectronic devices on the same wafer. First successful experiments to use MBE for the heteroepitaxial growth of GaAs on SQI-structures were reported by Das et al. [35] for the case of SIMOX structures. The growth of the GaAs layer was dominated by hillock formation at microscopic mechanically damaged regions of the silicon surface. The cause for this mechanical damage was not reported. However, between the hillocks an uniform layer growth was already observed. Microtwins, threading dislocations, twinning dislocations and antiphase domain boundaries were the predominant defects in these layers. Generally, this is a problem of heteroepitaxy, but on the other hand it must be stated again that the quality of the SIMOX-structures was not the best one due to the anneal step used: 1200°C, 3 h. In this manner, a high density of defects still remains within the silicon top layer, especially silicon dioxide precipitates. Undoubtly, optimised SOI-substrates from the viewpoint of defect engineering, e.g. SIMOX substrates annealed at 1300°C for 6 hours, should lead to much more encouraging results.

7. Etch Stop Layers for BESOI

In this SOI technology (see Fig.8), a thin layer of silicon dioxide present on one or both two silicon wafers is sandwiched between these wafers at room temperature followed by annealing at elevated temperatures. Then one of the wafers has to be thinned (the so called seed or device wafer) to produce a single crystalline silicon film of a desired thickness. This film is isolated from the substrate (or handle wafer) by the

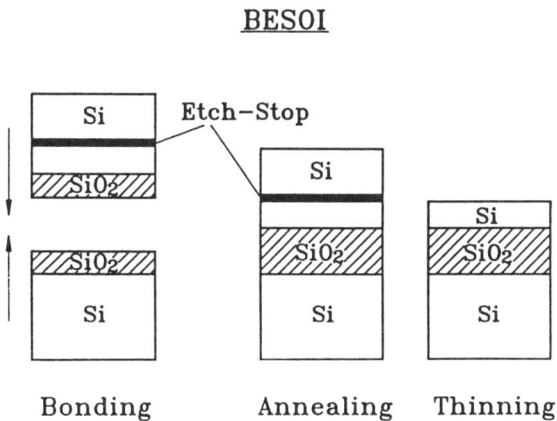

Fig.8: Schematic of the main steps of BESOI

silicon dioxide layer grown at the beginning. One of the basic process steps in this technique is the controlled thinning of the seed wafer.

The most frequently used selective method, the Bonded and Etch back -SOI (BESOI), involves chemical removal of a certain part of the device wafer by etching. This method is capable of fabricating high quality silicon films with thickness variations of about 7% at thicknesses as low as 100 nm [36]. In this case, a highly selective etch stop layer is formed in the seed wafer prior to bonding, usually by ion implantation. One method is the implantation of boron up to volume concentrations of $10^{20} cm^{-3}$. The etch-back procedure results in the removal of almost all of the seed wafer except for the thin layer ahead of the peak of the implant profile. For medium implantation energies (<200 keV) this layer contains boron of unacceptably high concentrations, which necessitates the growth of an undoped epitaxial silicon layer on the implanted surface. This layer becomes then the device region.

One method to overcome this disadvantage of additional and expensive processing is the use of high energy implantation with MeV-energies [37]. Also, buried oxide and nitride layers in silicon formed by ion beam synthesis have been successfully employed for etch-stop purposes [38]. Recently, it has been demonstrated that carbon implantation into silicon at medium energies (<200 keV) and doses of the order of $10^{16} cm^{-2}$ can produce efficient etch stop layers suitable for both, BESOI and micromechanical applications [39].. The advantages of this approach arise from the facts that carbon in silicon is electrically inactive (although carbon implantation may generate defect-related energy levels which in turn may cause changes in the carrier lifetime and resistivity), and a carbon-depleted zone to support the device action can be formed under standard implantation conditions [40]. This eliminates the necessity of growing an epitaxial layer of undoped silicon.

8. Conclusions

1. Nearly all types of ion-solid-interaction causing mass transport and energy deposition into nuclear and electronic processes leading to

- doping (HI-FIPOS, BESOI),
- compound formation (SIMOX/SIMNI/SIMON, BESOI),
- materials deposition (MBE-SOI),
- damage/amorphization (SPE-SOS, SSIC, L-IBIEC), and
- dynamical annealing (SIMOX/SIMNI, L-IBIEC)

were successively used to produce SOI-structures itself or to support efficiently other SOI-technologies. Moreover, the specialty of particle channeling in single crystalline matter was also successfully applied. Sputtering is the only process that still remains to be used in a constructive manner for SOI.

2. Several techniques reported above are still in the period of early beginning (L-IBIEC, MBE-SOI) whereas SPE-SOS is a well established technology. SIMOX in competition with BESOI is now a powerful motor to bring SOI to ULSI.

3. Gettering in SOI structures which are coming now on a mainstream for application in technology -like SIMOX and BESOI- should now be paid more attention than during the last fifteen years.

9. ACKNOWLEDGEMENTS

I would like to thank the many collegues who shared with me their enthusiasm in the work on materials research for SOI structures: H.Bartsch, A.Danilin, A.DeVeirman, R.Grötzschel, N.Hatzopoulos, P.L.F.Hemment, E.Hensel, U. Kreißig, J.Vanhellemont, M.Voelskow, R.Weber, K.Wollschläger, and R.A.Yankov. The critical reading of the manuscript by R.A.Yankov is gratefully acknowledged.

10. REFERENCES

1. Roulet, M.E., Schwob, P., and Golecki, I. (1979), Electron. Lett.**15**, 527.
2. Yoshii, T., Taguchi, S., Inoue, T., and Tango, H. (1982), Jpn. J. Appl.Phys.**21**, 175.
3. Imai, K., and Unno, H. (1984), IEEE Trans. Electron. Dev. **ED-31**, 297.
4. Benjamin, J. D., Keen, J.M., Cullis, A.G., Innes, B., and Chew, N.G. (1986), Appl. Phys. Lett. **49**, 716.
5. Priolo, F., and Rimini, E. (1990), Materials Science Reports **5**, 319.
6. Heera, V. (1989), phys. stat. sol. (a) **114**, 599.
7a. La Ferla, A., Rimini, E., and Ferla, G. (1988), Appl. Phys. Lett. **52**, 712.
7b. Skorupa, W., Voelskow, M., Matthäi, J., and Knothe, P. (1988), Electron. Lett. **24**, 876.
8. Ishiwara, H., Asano, T., Tsutsui, K., Lee, H.C., and Furukawa, S. (1988), Mat. Res. Soc. Symp. Proc. **107**, 241.
9. Voelskow, M., Skorupa, W., Wollschläger, K., Matthäi, J., Knothe, P., Heinig, K.-H., and Bartsch, H. (1989), Appl. Surf. Sc. **43**, 196.
10. Reif, R., and Knott, J.E. (1981), Electron. Lett. **17**, 587.
11. Kung, K.T.-Y., and Reif, R. (1987), J. Appl. Phys. **62**, 1503 and references therein.
12. Kung, K.T.-Y., and Reif, R. (1988), J. Appl. Phys. **63**, 2131.
13. For state-of-the-art see: (1988), Mat. Res. Soc. Symp. Proc. **107**; Biennal SOI-Symposia of the Electrochemical Soc.1984-1994; SOS-SOI workshops of the IEEE (Institution of Electrical and Electronic Engineers).
14. Skorupa, W., Grötzschel, R., and Bartsch, H. (1989), phys. stat. sol. (a) **1 1 2**, 661.
15. Colinge, J.P. (1991), Silicon-on-Insulator: Materials to VLSI, Kluwer Academic Publishers, Dordrecht.
16. Colinge, J.P. (1988), IEEE Electron. Dev. Lett. **9**, 97.
17. Hemment, P.L.F., Peart, R.F., Yao, M.F., Stephens, K.G., Arrowsmith, R.P., Chater, R.J., and Kilner, J.F. (1985), Nucl. Instr. Meth. **B6**, 292.
18. Skorupa, W., Kreißig, U., Wollschläger, K., Hensel, E., and Bartsch, H. (1985), ZfK-Annual Report 1984 **559**, 96.
19. Reeson, K.J., Hemment, P.L.F., Meekison, C.D., Marsh, C.D., Booker, G.R., Chater, R.J., Kilner, J.A., and Davis, J. (1988), Nucl. Instr. Meth. **B32**, 427.
20. Skorupa, W., Wollschläger, K., Grötzschel, R., Schöneich, J., Hentschel, E ., Kotte, R., Stary, F., Bartsch, H., and Goetz, G. (1988), Nucl. Instr. Meth. **B32**, 440.
21. DeVeirman, A., Van Laduyt, J., and Skorupa, W. (1991), Philosopical Magazine **A64**, 513.
22. Skorupa, W., Schöneich, A., Grötzschel, R., Wollschläger, K., and Vöhse, H. (1992), Mat. Sc. Eng. **B12**, 63.
23. Skorupa, W., Grötzschel, R., Wollschläger, K., Albrecht, J., and Vöhse, H (1992), Mat. Res. Soc. Symp. Proc. **235**, 127.

24. Skorupa, W., Kreissig, U., Hensel, E., and Bartsch, H. (1984), Electron. Lett. **20**, 426.
25. Skorupa, W., Knothe, P., and Grötzschel, R. (1988), Nucl. Instr. Meth. **B 3 4**, 523.
26. Skorupa, W., Knothe, P., and Grötzschel, R. (1988), Electron. Lett. **24**, 464.
27. Delfino, M., Jaczynski, M., Morgan, A.E., Vorst, C., Lunnon, M.E., and Maillot, P. (1987), J. Electrochem. Soc. **134**, 2027.
28. Kamins, T.I.and Chiang, S.Y. (1989), J. Appl. Phys. **58**, 2559.
29. Jablonski, J., Miyamura, Y., Imai, M., and Tsuya, H. (1994), Silicon-on-Insulator Technology and Devices, S. Cristoloveanu (ed.), ECS-Proc. **94-11**, The Electrochem. Soc., Inc. Pennington.
30. Borland, J.Ogawa (1994), personal communication.
31a. Skorupa, W., Kögler, R., Schmalz, K., and Bartsch, H. (1991), Nucl. Instr. Meth. **B55**, 224.
31b. Skorupa, W., Kögler, R., Schmalz, K., Gaworzewski, P., Morgenstern, G . , and Syhre, H. (1993), Nucl. Instr. Meth. **B74**, 70.
32. Myers, S.M., Follstaedt, D.M., and Bishop, D.M. (1994), Mat. Res. Soc. Symp. Proc. **316**, 33.
33. Yamada, I. (1990), Energy Pulse and Particle Beam Modification of Materials, K. Hohmuth and E. Richter (eds.), Physical Research **13**, Akademie Verlag, Berlin.
34. Kamins, T.I., Wang, K.L., Park, J., and Davis, G.E. (1989), J. Appl. Phys. **65**, 1505.
35. Das, K., Humphreys, T.P., Posthill, J.B., Parikh, N., Tarn, J., ElMasry, N . , Bedair, S.M., Chu, W.K., and Wortman, J.J. (1988), Mat. Res. Soc. Symp. Proc. **107**, 513.
36. Iyer, S.S. (1994), Th Electrochem. Soc. Spring Mtg. **94-1**, Ext. Abstr.No.439.
37. Maszara, W.P.,Pronko, P.P.,McCormick, A.W. (1991), Appl.Phys.Lett.**68**,2779.
38. Stoev, I.G., Yankov, R.A., and Jeynes, C. (1989), Sensors and Actuators **19**, 183.
39 Feijoo, D., Lehmann, V., Mitani, K., and Gösele, U.M. (1992), J.Electrochem. Soc. **139**, 2309.
40. Tong, Q.-Y., You, H.-M., Cha, G., and Gösele, U.M. (1993), Appl. Phys. Lett. **62**, 970.

SEMI-INSULATING OXYGEN-DOPED SILICON BY LOW TEMPERATURE CHEMICAL VAPOR DEPOSITION FOR SOI APPLICATIONS

J.C. STURM
Department of Electrical Engineering
Princeton University
Princeton, NJ 08544 USA

P.V. SCHWARTZ
Department of Electrical and Computer Engineering
University of Iowa
Iowa City, IA 52242 USA

Z. LILIENTAL-WEBER
Center for Advanced Materials
Materials Science Division
Lawrence Berkeley Laboratory 62/203
Berkeley, CA 94720 USA

Abstract

In this paper the growth of crystalline silicon layers with large amounts of oxygen ($\sim 10^{20}$ cm^{-3}) by low temperature chemical vapor deposition is described. These layers have semi-insulating electrical properties with resistivities as large as 10^6 Ω-cm at room temperature and exhibit the classic characteristics of a space-charge limited current in an insulator with many deep traps. Single crystal SOI layers without oxygen were grown on top of the semi-insulating layers. P-channel MOSFET's fabricated in these films had characteristics above threshold similar to bulk control samples, although their subthreshold characteristics were degraded.

1. Introduction

The advantages of SOI, such as reduced substrate capacitances, reduced process complexity, and improved scaling and subthreshold slopes (Colinge effect) are well known. However, all of the conventional processes for the formation of

SOI layers, such as SIMOX, ZMR, bond-and-etch, etc, suffer from one or more of several technological or economic problems. The primary economic problem is the time for processing (such as high oxygen doses or long laser scanning times), leading to high wafer costs. Besides defects, other key technical problems associated with the buried oxide and active silicon layer are the control, uniformity, and flexibility of choice of the film thicknesses. These problems will probably only become more significant as the wafer size increases.

One technical approach towards an SOI formation process which addresses the issues of large-area capability, low-cost, and good layer thickness control would be an all epitaxial process. First an epitaxial insulator on the silicon substrate would be grown, followed by the SOI layer. This approach has been limited by the availability of an epitaxial insulator lattice-matched to silicon. Considerable work in the past has been done using CaF_2 and related materials as insulators [1]. While considerable progress was made, this effort suffered from differing thermal expansion coefficients, facetting of the CaF_2 surface, and incompatibility of CaF_2 with conventional Si processing.

In this paper, we describe the growth, electrical properties, and SOI application of a new material, oxygen-doped crystalline silicon (Si:O) [2], as an epitaxial semi-insulator on (100) silicon substrates. The motivation for the work was the fact that O is known under certain conditions to introduce levels 0.4 eV and 0.6 eV above the valence band edge in silicon [3]. The incorporation of large amounts of oxygen into silicon might then lead to a pinning of the fermi level near midgap, leading to semi-insulating characteristics.

Previous attempts to incorporate large amounts of oxygen into silicon included Semi-Insulating POlycrystalline Silicon (SIPOS), which was formed by CVD and could contain up to several tens of percent of oxygen [4]. The application of these polycrystalline films was for surface passivation [5] and for a wide bandgap emitter in silicon-based heterojunction bipolar transistors [6]. OXygen-doped Silicon Epitaxial Films (OXSEF) have also been grown as a crystalline substitute for wide bandgap emitters [7]. These crystalline layers were grown by MBE between 300 and 700 °C, and surprisingly could contain on the order of 10% oxygen [8]. When heavily doped with arsenic, the layers were conducting.

2. Growth of Oxygen Doped Silicon (Si:O) by Chemical Vapor Deposition (CVD)

During conventional silicon epitaxy (e.g. > 900 °C), the presence of small amounts of oxygen or water vapor (e.g. < 10 ppm) in the growth chamber has little effect on the quality of the layers. The oxygen concentration in the epitaxial layers in this case is typically below that of the maximum solid solubility (~2 x 10^{18} cm^{-3}). For higher amounts of contaminant in the source gases, first heavily defected layers (eg. with many stacking faults and a very hazy surface) are found and then polycrystalline layers result. Therefore we have focussed our work on low-temperature CVD (700-750 °C) with the hope that excessive SiO_2 precipitation and the defect formation which leads to polycrystalline growth could be suppressed. The layers were grown in a lamp-heated CVD chamber equipped with a load-lock and gas purifiers, so that the background oxygen and water concentrations were <100 ppb. The source gases were either silane or dichlorosilane in a hydrogen carrier at 6 torr. Oxygen was introduced by bleeding small controlled amounts of a dilute oxygen in argon mixture into the growth chamber. Oxygen was chosen instead of water vapor because water vapor was found to stick to the quartz chamber walls, and the water partial pressure could take days to decrease after the water vapor source was shut off. No such problem was observed using an oxygen source.

The material was initially characterized by infrared absorption (FTIR), X-ray diffraction (XRD), and SIMS. It was found that up to 10^{20} cm^{-3} of oxygen could be incorporated into the films while they still remained crystalline and without any evidence of haze on the sample surface. The amount of oxygen incorporated into the films was two orders of magnitude lower than that expected from sticking coeffiiceints determined in UHV experiments. This was attributed to the passivating effect of a hydrogen layer on the sample surface present during CVD, and also due to the presence of a boundary layer during CVD growth [9]. Fig. 1 shows a typical X-ray diffraction curve of a sample containing 5 x 10^{19} cm^{-3} of oxygen and a thickness of 5 microns. No polysilicon peaks are present and only strong single crystal peaks are seen. A typical FTIR spectrum of a similar sample with a slightly higher oxygen level (~10^{20} cm^{-3}) is shown in Fig. 2. A broad peak around 1010 cm^{-1} is observed, which is at an energy considerably lower than those normally observed for interstitial O in Si (1107 cm^{-1}) and for various forms of SiO_x precipitates (1100-1230 cm^{-1}). This spectrum looks very similar to that previously observed in OXSEF samples [8]. The exact microstructural configuration of the oxygen in our Si:O layers is not known at present.

3. Electrical Characterization and SOI Applications

To avoid complications with parallel conduction from the underlying substrate, the layers were evaluated mostly using vertical transport measurements. In some cases the layer directly on top of the Si:O was a metal contact, and in other

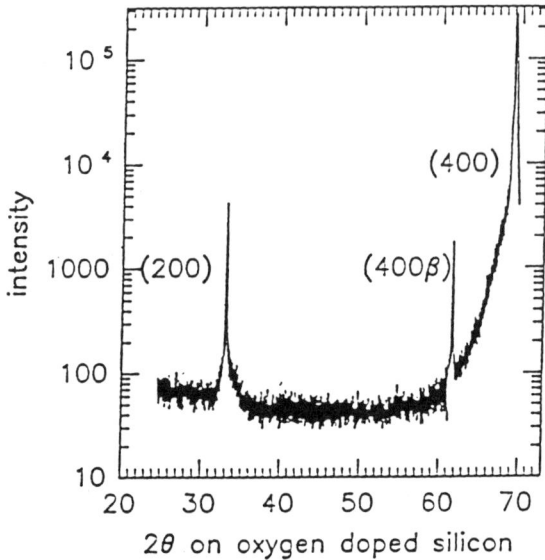

Fig. 1. X-ray diffraction (using Cu-K$_\alpha$ and Cu-K$_\beta$ radiation) of a 5-μm thick Si:O film containing 5×10^{19} cm^{-3} of oxygen, showing no evidence of any polycrystalline silicon.

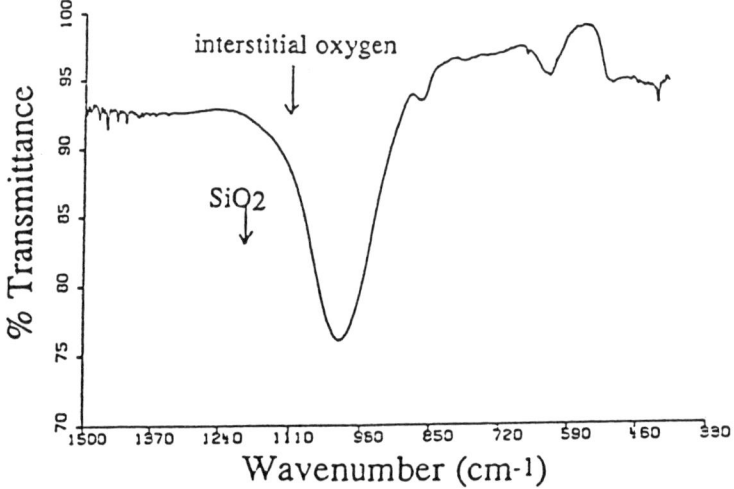

Fig. 2. FTIR transmission spectrum of a 5 μm thick Si:O layer containing ~10^{20} cm^{-3} of oxygen.

cases it was a crystalline silicon layer without oxygen grown on top of the Si:O. This was accomplished simply by turning off the oxygen flow, although in most cases the growth temperature was also then raised to 1000 °C to give a faster growth rate. These top Si layers were either doped p-type or n-type by introducing dopants during the epitaxial growth.

A typical I-V curve at room temperature from an aluminum Schottky barrier on undoped Si:O (thickness 10 microns) on an n-type substrate is shown in Fig. 3. The current in forward bias was limited by the resistance of the Si:O layer, from which the resistivity of the SiO layer was then extracted. These measurements were made as a function of temperature for layers containing total oxygen concentrations on the order of both 5×10^{19} cm^{-3} and 2×10^{20} cm^{-3} (Fig. 4). For both samples a resistivity of $\sim 10^6$ Ω-cm was found near room temperature, with the value for the sample with more O somewhat higher. Assuming mobilities near that of bulk Si, such a resistivity implies a fermi level pinned near midgap. Further evidence that the fermi level in the material was pinned near midgap is the activation energy of the resistivity which in both cases was 0.6 eV, approximately half of the bandgap of silicon. As the nature of the microstructure of the oxygen is not known, the exact mechanism which contributes to the fermi level pinning (surface states of oxide precipitates, levels due to isolated O atoms, etc) is also not known.

More complex are the I-V characteristics observed in p-Si/undoped Si:O/p-Si structures at higher electric fields. In materials with very low intrinsic carrier densities, such as insulators or semi-insulators, there are not sufficient carriers present in the material to conduct large amounts of current. At higher current levels, excess carriers which flow in from the contacts must be present, giving the material a net charge. This charge in the material causes the I-V characteristics to deviate from a simple ohmic relationship. In the simplest case (Fig. 5(a.)), the log I - log V characteristics have a slope of 1 at low currents where the background carrier density is sufficient to carry current, but a slope of 2 at higher currents where there is a large injected carrier density. If traps are present in the material, the situation can become more complex [10]. In this case, a jump in the log I - log V characteristics can be observed as the quasi-fermi level passes through the trap level, both because of changes in scattering and because all new injected carriers will then be available for conduction (Fig. 5 (b.)). Such characteristics are indeed observed from experimental structures (Fig. 6), confirming that deep levels are indeed present in the Si:O material. From a quantitative analysis of Fig. 6, a trap density on the order of 10^{19} cm^{-3} can be inferred [2]. This is consistent with high donor and acceptor doping experiments of Si:O in our lab, as at very high doping levels (e.g. concentrations of phosphorus or boron $\gg 10^{19}$ cm^{-3}), the Si:O films are no longer semi-insulating and conduct well. Hall measurements of doped layers yield a hole mobility which is somewhat lower

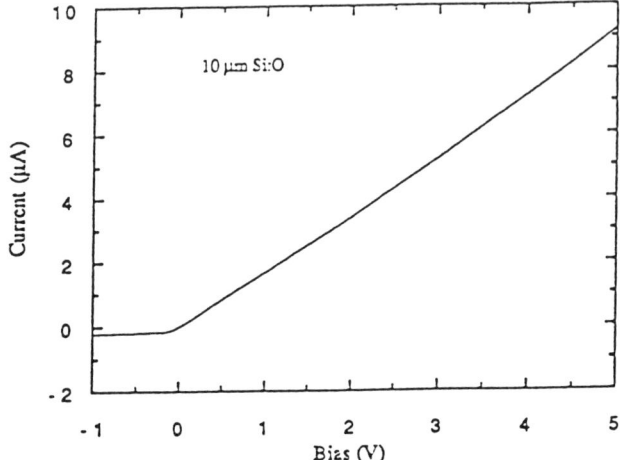

Fig. 3. Current-voltage characteristic of an Al/10-μm undoped Si:O/n-type Si Schottky barrier at room temperature.

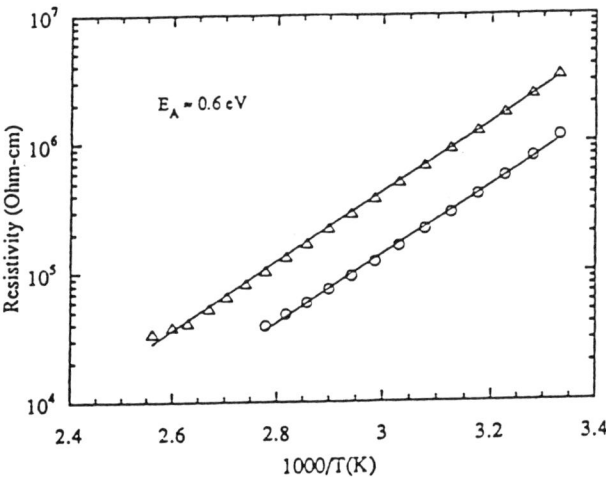

Fig. 4. Resistivity vs temperature of Si:O layers for fields less than 5×10^3 V/cm. Triangles indicate oxygen concentrations of $\sim 2 \times 10^{20}$ cm^{-3} and circles indicate oxygen concentrations of $\sim 5 \times 10^{19}$ cm^{-3}.

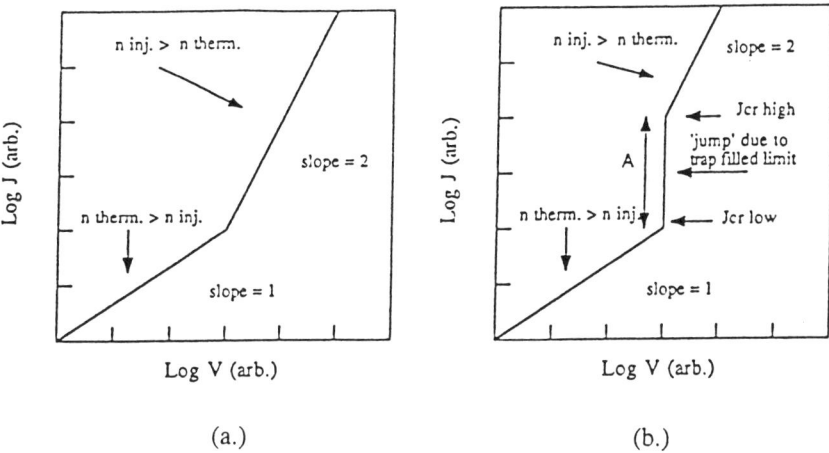

Fig. 5. Schematic log I - log V curves of space charge limited current in an insulator (a.) without traps and (b.) with traps.

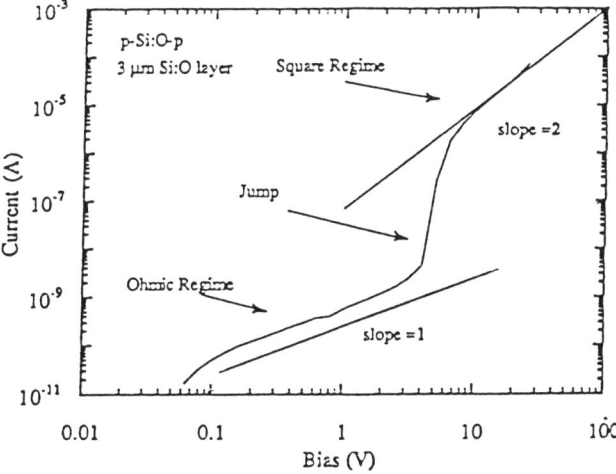

Fig. 6. Log I- log V characteristics of a p-Si/Si:O/p-Si structure. The Si:O thickness is 3 μm.

than bulk material. p-Si/undoped Si:O/n-Si structures in forward bias show a current-controlled hsyterisis (Fig. 7). This is also characteristic of a semi-insulating material with a high trap density [10].

The lifetime of the material was probed using an all optical pump and probe technique in a thick (10 μm) Si:O layer with approximately 10^{20} cm^{-3} oxygen [11]. First a pump pulse excites carriers in the Si:O layer, and their recombination is monitored through their effect on the reflectance of a second probe pulse, whose delay from the pump pulse can be adjusted. From the exponential decay of the reflected probe signal vs. delay, a lifetime of ~ 50 ps may be extracted (Fig. 8). Such a low lifetime is to be expected given the large number of traps in the material.

4. SOI FET's

SOI FET's were fabricated by growing ~0.6 um of crystalline Si n-type (~10^{16} cm^{-3}) without oxygen doping at 1000 °C on top of 1.5 um of Si:O. P-channel MOSFET's were then fabricated by a standard self-aligned process using simple mesa isolation for the SOI islands. The gate oxide was thermally grown to a thickness of ~40 nm. Typical SOI FET characteristics and those of simultaneously fabricated bulk control FET's are shown in Fig. 9. Well-behaved characteristics were obtained, with a slightly higher threshold voltage in the SOI case, possibly due to a lower doping in the control wafer. Long-channel devices showed similar mobilities for both the SOI and bulk devices. N-channel devices in the SOI films were also well behaved above threshold, but a processing failure in the control devices prevented any comparison with bulk devices.

The subthreshold characteristics of the PMOS devices are shown in Fig. 10. An unknown process problem caused fairly poor subthreshold slopes (~250 mV/decade) and high leakage current in the bulk devices. The subthreshold performance of the SOI devices was still worse, however (~1000 mV/decade). It is not known if this poor performance is due to defects in the SOI films (or at the top SiO$_2$ interface) itself, or is due to full depletion resulting from poor control of the intended doping of the SOI film during epitaxial growth. If full depletion occured, the many deep levels in the underlying Si:O layer would be expected to lead to a poor subthreshold slope.

5. Future Directions

Recent TEM results [12] have shown that the Si:O layers with oxygen con-

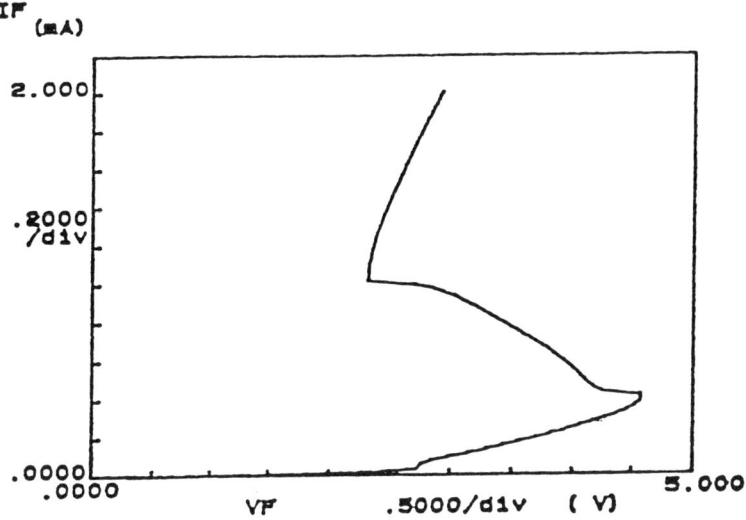

Fig. 7. Current-Voltage characteristics of a double-injector (p-Si/Si:O/n-Si) structure in forward bias.

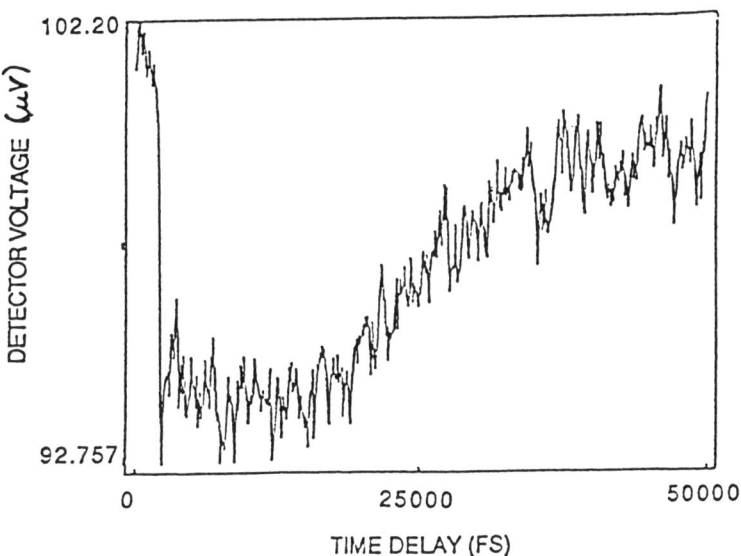

Fig. 8. Reflectance vs time of a 10 mm Si:O layer after excitation by a 20 ns laser pulse. A recombination lifetime of ~50 ps can be extracted.

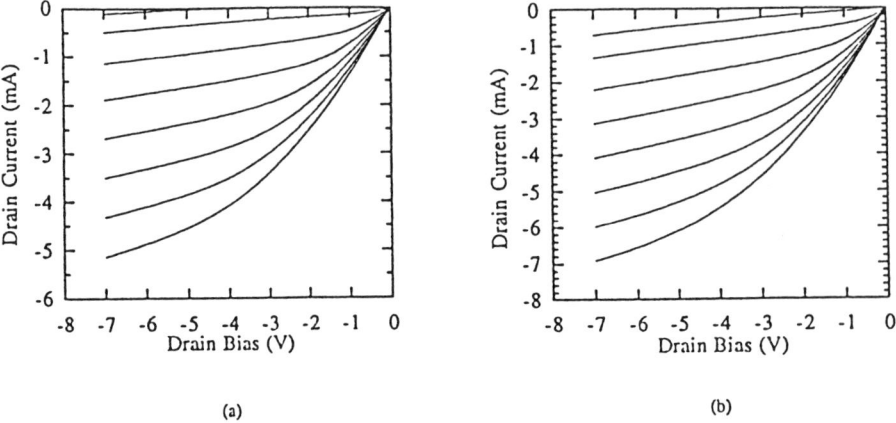

Fig. 9. I-V characteristics of (a.) SOI and (b.) control (bulk Si) FET's with W = 50 μm and L = 2 μm. The gate voltage is stepped from 0 V to -7 V in 1 V increments.

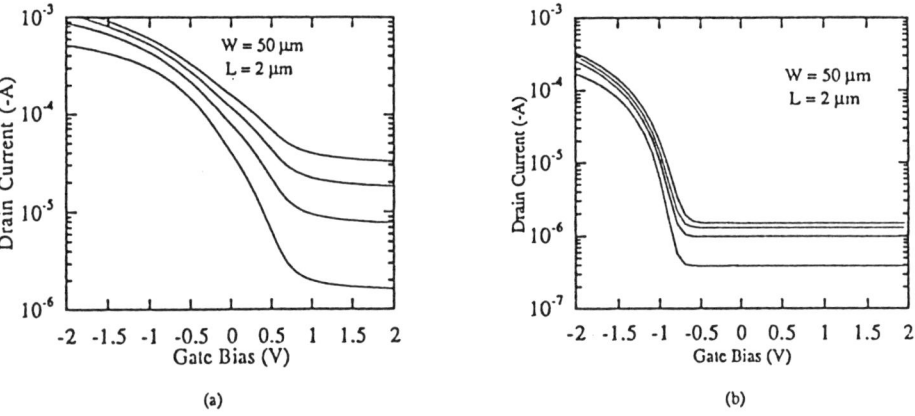

Fig. 10. Subthreshold characteristics of (a.) SOI and (b.) control MOSFET's. The drain voltages were -0.5, -1.0, -1.5, and -2.0 V.

centations on the order of 1 - 5 x 10^{19} cm^{-3} contain hairpin dislocations originating at the lower Si/Si:O interface, but with a density not exceeeding 10^7 cm^{-2}. SiO$_x$ precipitates were also found at the lower Si/SSi:O interface (about 2-3 nm diameter), as were smaller precipitates uniformly throughout the film. For oxygen levels an order of magnitude higher, a high density of stacking faults (10^{11} cm^{-2}) and microtwins was seen in the Si:O as well as much larger precipitates (~ 15 nm) in the entire layer.. For the two cases of high and low oxygen concenration, defect densites (stacking faults and dislocations) in overlying SOI films without O were ~10^7 and 10^5 cm^{-2}, respectively. This indicates that still lower growth temperatures are required to surpress oxygen precipitation and related defects during the CVD process. It should be noted that due to unintentional contamination, SiGe layers grown by CVD on Si at ~640 °C with ~ 10^{20} cm^{-3} of oxygen have been reported [13]. Despite this high oxygen density, no defects or precipitates could be observed in these layers by TEM, or in overlying Si layers grown at higher temperature without O contamination . Although the films were conducting due to their high dopant concentration (~ 10^{19} cm^{-3} to ~ 10^{20} cm^{-3}), they demonstrate the advantages in crystal quality which can be obtained at lower growth temperatures when large amounts of oxygen are present.

6. Summary

Semi-insulating crystalline silicon layers with ~ 10^{20} cm^{-3} of oxygen can be grown by low temperature CVD. The layers have a resistivity of ~ 10^6 Ω-cm at electric fields below 10^4 V/cm, and exhibit the classic characteristics of space-charge limited current in an insulator with traps at higher current levels. Crystalline Si layers without oxygen can be grown on top of these semi-insulating layers. FET's in the SOI layers have good mobilities, but poor subthreshold slopes and high leakage levels. The key to improved material quality appears to be lowering the growth temperature of the Si:O layers.

7. References

1. T. Asano and H. Ishiwara, J. Appl. Phys. **55**, 3566 (1984).

2. P.V. Schwartz, C.W. Liu, and J.C. Sturm, Appl. Phys. Lett. **62**, 1102 (1993).

3. S.M. Sze, *Physics of Semiconductor Devices*, 2nd. ed. (Wiley, New York, 1981), p. 27.

4. H. Mochizuki, T. Aoki. H. Yamaoto, M. Okayama, and T. Ando, Suppl. Jpn. J. Appl. Phys. **15**, 41 (1976).

5. T. Matsushita, T. Aoki, T. Ohtsu, H. Yamoto, H. Yayashi, M..Okayama, and Y. Kawana, IEEE Trans. Electron Device **TED-23**, 826 (1976).

6. T. Matsushita, N. Oh-uchi, H. Hayashi, and H. Yamoto, Appl. Phys. Lett. **35**, 549 (1979).

7. M. Takahishi, M. Tabe, and Y. Sakakibara, IEEE Electron Device Lett., **EDL-8**, 475 (1987)

8. M. Tabe, M. Takahashi, and Y. Sakakibara, Jpn. J. Appl. Phys. **26**, 1830 (1987).

9. P.V. Schwartz and J.C. Sturm, J. Electrochem. Soc. **141**, 1284 (1994).

10. M.A. Lampert and P. Mark, *Current Injection in Solids* (Academic, New York, 1970).

11. P.V. Schwartz, C.W. Liu, J.C. S turm, T. Gong, and P.M. Fauchet, presented at the Elec. Mat. Conf., Cambridge, MA, USA (June, 1993).

12. Z. Liliental-Weber, P.V Schwartz, C.C. Wu, and J.C. Sturm, J. Vac. Sci. Tech. **B12**, 2511 (1994).

13. C.A. King, J.L. Hoyt., and J.F. Gibbons, IEEE Trans. Elec. Dev. **TED-36**, 2093 (1989).

DIRECT FORMATION OF THIN FILM NITRIDE STRUCTURES BY HIGH INTENSITY ION IMPLANTATION OF NITROGEN INTO SILICON

ROSSEN YANKOV*
Research Center Rossendorf Inc.
Institute of Ion Beam Physics and Materials Research
P.O.B. 51 01 19
D-01314 Dresden, Germany
and
FADEY KOMAROV
Institute of Applied Physical Problems
7 Kurchatova Street, 220064 Minsk, Belarus

*On leave from the Institute of Electronics,
Bulgarian Academy of Sciences, 1784 Sofia,
Bulgaria

1. Introduction

The technique of ion beam synthesis of buried oxide and nitride layers in silicon has been successfully used to fabricate SOI material suitable for VLSI device applications. To date the bulk of the research conducted along this line has been in the field of SIMOX [1]. In comparison to SIMOX, the technique of SIMNI has received less attention mainly because of the fact that nitrogen atoms do not redistribute during conventional high dose implantation involving low beam current densities, typically less than 20 μA cm^{-2}, and substrate temperatures in the range of 400 to 600°C. Subsequent annealing at 1200°C for 2 h results in the formation of complex structured nitride layers which for doses less than the critical dose D_c include silicon islands, and for doses above D_c inevitably comprise a porous region of nitrogen bubbles [2]. This zone in particular is a serious problem as it proves to be a source of mechanical weakness in SIMNI substrates. Moreover, high leakage currents have been measured even through nitride layers formed with doses well above D_c [3], so the use of standard implantation and annealing conditions has been strongly questioned.

During recent years however, apparent progress has been made in improving the quality of SIMNI substrates. Current developments include the fabrication of ultrathin buried nitride layers with high quality interfaces [4], and the use of high intensity ion implantation (HIII) of nitrogen which enables one to produce directly stoichiometric Si$_3$N$_4$ layers of good structural integrity [5]. HIII of nitrogen appears to have similarities to SIMOX process in that a flat topped profile can be formed while maintaining good crystalline quality in the upper silicon layer. This is a technologically important issue which may warrant a reconsideration of the standard SIMNI technique with a view to obtaining device worthy SOI substrates.

The aims of this research are to assess the aplicability of HIII of nitrogen into silicon with respect to the direct formation of stoichiometric layers of Si$_3$N$_4$, and to help establish the relative importance of the dose rate, implant temperature and dose influences on the resulting structures, thus outlining a basis for further optimisation of the technique.

2. Experimental

In our experiments molecular nitrogen ions were implanted into (111) silicon samples using a stationary beam from a Van de Graaff accelerator. The particle energy was 1 MeV mol^{-1} (500 keV at^{-1}), the dose range was 7×10^{17} - 2.1×10^{18} cm^{-2}, the beam current density was in the range 15 - 150 μA cm^{-2} and the incident ion beam was normal to the wafer flat. The samples were clamped on a metallic disk holder and were heat sunk by means of a thermally conducting silicon grease. Consequently, conduction cooling was the dominant heat loss mechanism. As the implants were performed at high beam power densities, the temperatute of the sample holder (as well as that of the samples) showed a discernible rise above room temperatute (RT) during irradiation. The actual sample temperature T_i was not directly measured. Instead, a model taking into account the main effects of ion beam heating, radiation cooling, conduction cooling and re-radiation was developed to estimate the surface temperature of the sample holder for the particular mounting configuration employed. Partial temperature measurements were done using a chromel/alumel thermocouple mounted in contact with the sample holder, and it was accordingly found that the calculated and actual temperatures agreed to $\pm 10^{\circ}$C over the temperature range from RT to 300°C.

Two series of experiments were carried out. The first series of implants was designed to trace the influence of the initial T_i on the crystallinity of the top Si layer. For this purpose three samples were mounted on the sample holder and were consecutively implanted to a dose of 7×10^{17} cm^{-2} each, using a beam current density of 50 μA cm^{-2}. The implantation started at RT and after 40 min an equilibrium substrate tenperature of about 300° C was reached, so that the first two samples were irradiated under non-equilibrium regimes of temperature rise, while the third one was implanted at a nearly constant temperature of 300° C.

The second implantation series was performed to study the effect of dose rate on the nitrogen profile. The beam current densities used were 15, 70, 100 and 150 μA cm^{-2} and the doses ranged from 1×10^{18} to 2.1×10^{18} cm^{-2}.

The as-implanted substrates were cleaved into small specimens which were analysed by Rutherford backscattering and channelling (RBS/C) of 1.5 MeV He$^+$ ions using a surface barrier detector with an energy resolution of better than 13 keV and a scattering angle of 170°.

3. Results and discussion

Figure 1 shows the calculated heating curve of the sample holder for a beam current density of 50 μA cm^{-2}. Three different temperature regimes of implantation are indicated, corresponding to the three implants carried out. In regions I and II the implants were conducted under non-stationary conditions of sample heating, whereas in region III an almost steady state T_i of 300° C was reached at which the third implant was carried out.

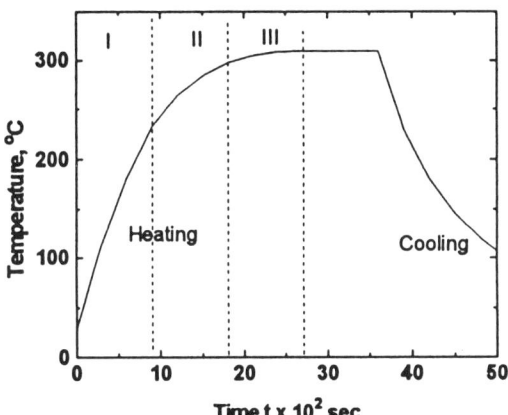

Fig. 1: Calculated surface temperature of the sample holder as a function of time, for a beam current density of 50 μA cm^{-2}. Regions I, II and III correspond to implants conducted under different conditions of sample heating.

Figure 2 shows the aligned RBS spectra for as-implanted samples from the above set (curves 1, 2 and 3 pertain to implants in regions I, II and III, respectively). As can be seen, increasing the T_i from RT to 300° C improves the near-surface crystallinity and reduces the thickness of the amorphised region. As distinct from the case of standard nitrogen implants which unavoidably result in the upper Si layer being highly defective ($x_{min} > 50\%$), the use of HIII at sufficiently high energies makes possible the dynamic restoration of the crystallinity of the top Si layer in situ even at relatively low T_i not exceeding 300° C. Still, the presence of comparatively large surface peaks in the spectra designated by (a), (b) and (c) implies that the T_i ought to be optimised in order to further reduce the damage level in the near-surface region.

Most important however is the fact that HIII enables good material quality to be maintained in the Si overlayer at T_i down to 300° C as confirmed by the low value of x_{min} (8%) measured just behind the surface peak (Ch 255) for the highest T_i of 300° C. This is quite the opposite to the case of conventional nitrogen implantation where the resulting as-implanted top Si layers are either highly deffective for T_i in the range of 400 to 800° C or amorphised for T_i lower than 300° C.

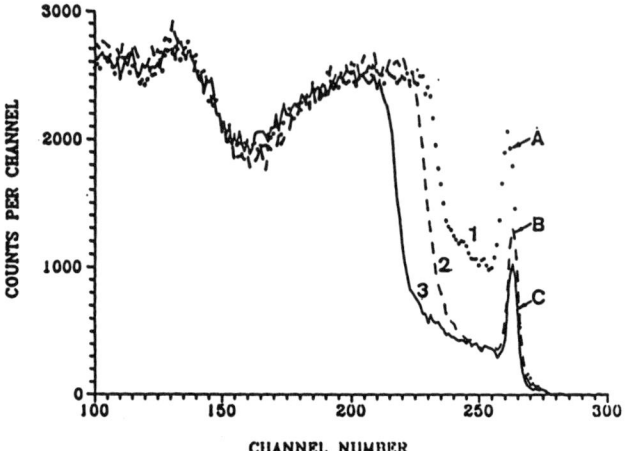

Fig. 2: Channelled RBS spectra from as-implanted samples which display the improvement in the crystalline quality of the Si overlayer with increasing substrate temperature. Curves 1, 2 and 3 correspond to implants in regions I, II and III, respectively, as indicated in fig. 1. For each implant the beam current density was 50 μA cm^{-2} and the dose was 7×10^{17} cm$^{-2} < D_c$.

Figure 3 shows the as-implanted nitrogen profiles from implants at different doses and beam current densities of (a) 15 and (b) 150 μA cm^{-2}. It is evident that at sufficiently high dose rates and for doses above D_c a redistribution of the excess nitrogen occurs and flat topped profiles are formed, which is an implication of the absence of a nitrogen bubbled layer and is consistent with our previous results [5].

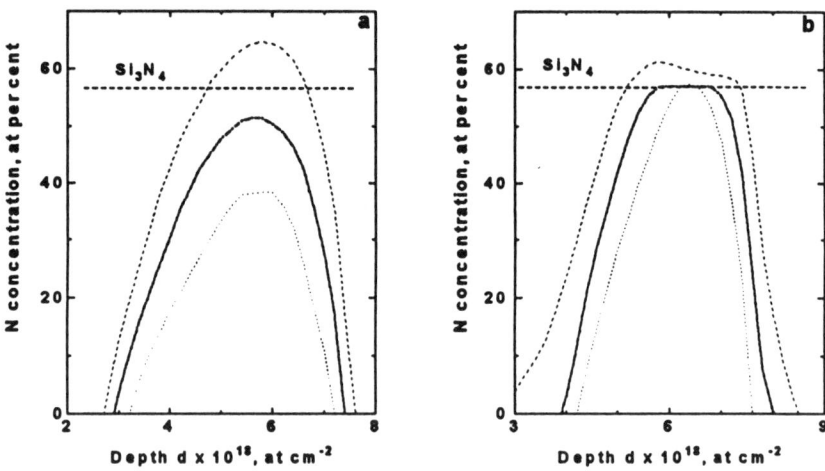

Fig.3 : Nitrogen depth profiles in samples implanted with a beam current density of (a) 15 μA cm^{-2} to doses of 1.0×10^{18} (dotted line), 1.6×10^{18} (dot-dashed line) and 2.1×10^{18} cm^{-2} (dashed line); and (b) 150 μA cm^{-2} to doses of 1.1×10^{18} (dotted line), 1.4×10^{18} (dot-dashed line) and 1.7×10^{18} cm^{-2} (dashed line).

Figure 4 shows the influence of the beam current density on the overall nitrogen distribution for implants at a dose of 1×10^{18} cm^{-2} < D_c. Two features can be noticed. Firstly, the profiles increase in height with increasing dose rate and for the highest one (100 μA cm^{-2}) the composition at the peak of the distribution approaches closely the stoichiometric ratio for Si_3N_4.

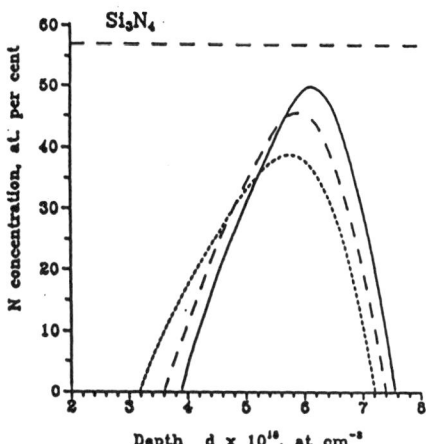

Fig. 4: Variation of nitrogen depth profile with beam current density for implants to a dose of 1.0×10^{18} cm^{-2} < D_c; 15 (dotted line), 70 (dashed line) and 100 μA cm^{-2} (solid line).

This may be a consequence of more effective nucleation of nitride precipitates favoured by the beam enhanced mobility of the nitrogen. Secondly, the peak of the unsaturated profile moves to greater depths for higher current densities, which may be attributed to more pronounced swelling.

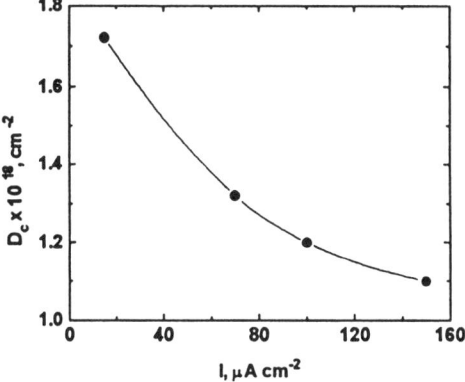

Fig. 5: Variation of the critical dose D_c with beam current density.

Prehaps the most intriguing result in this study is that the threshold dose D_c for direct formation of a stoichiometric buried layer of Si_3N_4 appears to be dose rate dependent, as shown in Figure 5. One can see that the values of D_c decrease with increasing beam current density. This reduction is similarly thought to be associated with the accelerated process of segregation which is favoured by the nitrogen diffusion enhancement at higher dose rates.

All these results once again reinforce the inference that the overall nitrogen distributions depend critically on the combined effect of dose rate and implant dose.

4. Conclusions

(i) The quality of the near surface silicon layer is sensitive to substrate temperature and improves with increasing T_i. Implant temperatures not less than 300°C are evidently required to achieve good crystallinity in the silicon overlayer.

(ii) The direct synthesis of a buried nitride layer depends crucially on the dose rate and implant dose.

(iii) The critical dose for direct formation of a buried Si_3N_4 layer is dose rate dependent.

(iv) The width of the amorphised region shrinks with increasing buried layer thickness.

(v) An apparent benefit from the HIII of nitrogen into silicon is that it produces as implanted material which is superior to that achieved by standard low flux nitrogen implants.

References

1. Proceedings of the 6th International Symposium on SOI Technology and Devices (1994), S. Cristoloveanu (ed.), Proc. Vol. 94-11, The Electrochem Soc.
2. C.D. Meekison, G.R. Booker, K.J. Reeson, P.L.F. Hemment, R.F. Peart, R.J. Chater, J.A. Kilner and J.R. Davis (1991) A transmission electron microscope investigation of the dose dependence of the microstructure of silicon-on-insulator structures formed by nitrogen implantation of silicon, J. Appl. Phys. **69**, 3503 - 3511.
3. R. Kwor, R.J. Matson, M. M. Al-Jassim, S. Polchlopek, P.L.F. Hemment and K.J. Reeson (1989) EBIC study of silicon-on-insulator structures formed by high dose nitrogen implantation, J. Electrochem .Soc .**136**, 876 -878.
4. R. Schork and H. Ryssel (1992) Formation of ultra thin SOI-layer by implantation of nitrogen, IEEE Int. SOI Proc., 17-18.
5. R. Yankov, F. Komarov and S. Petrov (1993) RBS, RHEED and THEED studies of SIMOX and SIMNI structures formed by ion beam synthesis, Vacuum **44**, 1077-1084.

STIMULATED TECHNOLOGY FOR IMPLANTED SOI FORMATION

V.G. Litovchenko, B.N. Romanyuk, A.A. Efremov, and V.P. Mel'nik

Institute of Semiconductor Physics
Academy of Sciences of Ukraine
45 Prospect Nauki
252028 Kiev, Ukraine

1. Introduction

For the fabrication technology of SOI structures the following problems are of importance: (i) the decrease of annealing temperatures; (ii) the improvement of sharpness and quality of the interfaces (IF); (iii) the decrease of the thickness of the uniform and single-phase insulating layer (IL) and, hence, the decrease of the implanted dose; (iv) the improvement of the quality and stability of the IL and of the perfection of the upper semiconductor layer.[1] In this work we will try to demonstrate that the least part of these important questions can be solved using different types of stimulating factors.

2. Theoretical approaches

1. First of all, it is possible to use an additional (using O^+ implantation) generation of point defects with their subsequent separation, due to the absorption of highly mobile interstitials beyond the active region, and the coupling of vacancies into V-clusters (up to the creation of micropores) in those places where oxygen precipitation is required. So, such V-clusters can play the role of new centers of precipitation of the insulating phase.

2. Another possibility for the creation of additional special centers of precipitation consists int the introduction (for example, using ion implantation [2]), of elements such as C, N, B, Al, Cl, F, H and others, which are chemically active with respect to oxygen and have a rather small atomic radius. The same result may also be achieved by δ-doping under deposition of the upper semiconductor layer. This factor can substantially facilitate the creation of an homogeneous, stoichiometric and thin IL.

3. It is known that at the melting point T_m elastic constants $E(T_m) \to 0$ and $v(T_m) \to 0.5$ [3], where E is young's modulus and v is the Poisson ratio of the material. So, during annealing in the framework of conventional SIMOX technology the role of mechanical stresses are not essential. However, during the "hot" implantation step (T~300-600°C and under an annealing at moderate temperatures (~1100°C) the following relations can be applied:

$$E(T) = E_o - a_1 T - b_1 T^2 \quad (1)$$

and
$$\nu(T) = \nu_o + a_2 T + b_2 T^2 \qquad (2)$$

where E_o and ν_o are the "room temperature" values of the Young modulus and the Poisson ratio. a_1, b_1, a_2 and b_2 are interpolation coefficients. Under these conditions, E remains large enough. This is proved by the influence of mechanical stresses on Si oxidation rate in SiO$_2$ film structures, as reported in [4,5]. In [6] it has been show that, during an oxidation, stresses in Si are equal to about $1-5 \times 10^8$ dyne/cm^2. So, in this case mechanical stresses can also be used as the driving force and facilitate the rate of the oxidation reaction to the successful control of the IL thickness and the IF sharpness. This factor becomes substantial when the absolute value of the drift flux of atoms ($J_{Dr} = (D/kT) \cdot \Delta v \cdot \nabla P \cdot N$) generated by some hydrostatic pressure $P = -1/3 \cdot \Sigma \sigma_{mm}$, (corresponding to a tensor field σ) is comparable to a diffusion flux ($J_{Df} = -D \cdot \nabla N$) value. Here Δv is the atomic volume mismatch of a foreign atom (or defect), $\Delta v \cdot \nabla P$ is a mechanical force acting upon the particle. So, only uniform internal stress is of importance in this case. it appears, for example, from a non-uniform distribution of atoms or defects N(z) during ion implantation [7]. For a thick Si wafer we can ignore the numbers describing the mechanical reaction of a substrate as a whole and obtain:

$$\sigma_{xx} = \sigma_{yy} \cong -\beta \cdot E \cdot N(z)/1-\nu) \text{ and } \sigma_{zz}=0 \qquad (3)$$

Here $\beta = \Delta v/v$ is a coefficient the lattice expansion induced by a foreign atom. As for the external mechanical field, it may be effective only in rather small vicinities of the so-called concentrators of mechanical stress, where values of P may be theoretically infinite [8]. Dislocations, clusters of defects or silicon dioxide inclusions and thin regions near interfaces may play the role of such concentrators. In the last two cases the mechanical stress influences the precipitate growth through the change of equilibrium critical size of precipitates and modification of atomic transport in their vicinity.

Experimental results presented below concern (i) the stress-induced atomic concentration distribution; (ii) the precipitate growth mechanisms, and (iii) the role of both the internal and external mechanical stresses in the superficial silicon layer perfection.

To clarify these effects we perform some computer simulations (CS) and calculations [9] based on theoretical approaches and equations developed in [7] for such cases as:

a) stimulation of the 1D growth (one-dimensional) of IL to obtain a non-rough flat layer, by means of 3D growth of dispersed SiO$_2$ inclusion suppression,

b) enhancement and compensation of the mechanical stress fields of different cases of SOI stimulated technology, including self-consistent atomic drift in stress fields, created by the impurities themselves,

c) the influence of the external stresses on the precipitation.

3. Experiment

The samples were prepared by implanting 150 keV O+ ions into Si(100) wafers (Dose=$1-10 \times 10^{17}$ cm^{-2}, at T=350°C, annealing - 3h at 1100°C). As for the combined implantations we used nitrogen N$^+$, carbon C$^+$ and carbon monoxide CO$^+$ ions at an energy of 150 keV and large enough doses (10^{17}-10^{18} cm^{-2}). The samples were studied by TEM, AES and SIMS. Defects and stresses in the surface Si layer have been determined by electroreflectance (ER).

4. Results and discussion

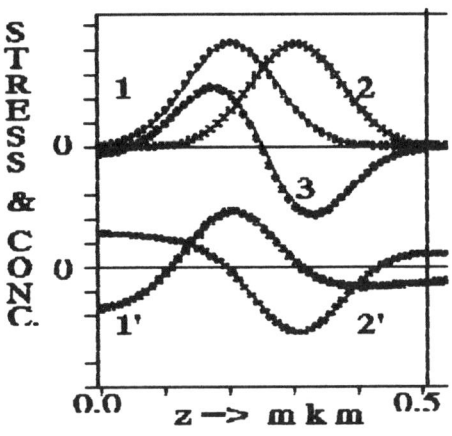

FIGURE 1: The calculated stress distribution over the sample depth in the case of a combined implantation. 1,2: concentration of C and O; 1',2': respective stress distribution for each impurity; 3: resulting stress distribution.

1. After combined O+ and C+ implantation and annealing (850°C and 1150°C) the effect of oxygen accumulation is revealed in the regions where C+ was implanted preliminary (AES).
2. The redistribution of implanted O^+ in the presence of implanted N^+ and C+ accumulation are observed near the interface (SIMS).
3. The effect of external mechanical stresses on the properties of the superficial layer was also studied on SOI structures prepared by high-dose implantation of O^+ into Si wafers deformed by the growth of a backside SiO_2 layer followed by an annealing step (550°C, Ar).

In the latter case the structural perfection of the superficial silicon layer is substantially higher for an SOI structure formed in an elastically deformed wafer than for a structure formed in an unstressed wafer. it should be noted that the interference oscillations of ER spectra are observed only for SOI structures created in deformed wafers. The interference is caused by the existence of a sharp atomic interface between IL and the silicon overlayer. Let us discuss these results.

Internal mechanical stress (IMS) induced by volume misfit between the matrix and impurity atoms (Eqn 4) influences the atomic transport at the initial stage of implantation. As a result achievement of both the growth of precipitates and high supersaturation becomes difficult. In order to avoid this situation, combined implantation is used when impurities with small radii, such as N or C, are implanted in addition to O^+. In this case the built-in field appears according to CS (Figure 1) and redistribution of the implanted impurities takes place in the first and second experiments.

The third experiment shows the important role of the stress-controlled precipitate formation. In is shown that the external stresses may, in some special cases, be used to filter the precipitate size: small precipitates of SiO2 are decomposed and, after this, 1D growth of the layer with a sharp upper IF is obtained. Indeed, according to [10], in an isotropic case, we have the following expressions for the variation of Gibbs' free energy ΔG during the formation of inclusions in the solid matrix and for the critical number of particles n_c in the nuclei:

$$\Delta G = n \cdot (-\Delta g_v + \Delta g_\varepsilon) + \eta \cdot \sigma_s \cdot n^{2/3} \qquad (4)$$

and

$$n_c = (2\eta/3) \cdot \sigma_s / (\Delta g_v - \Delta g_\varepsilon)^3 \qquad (5)$$

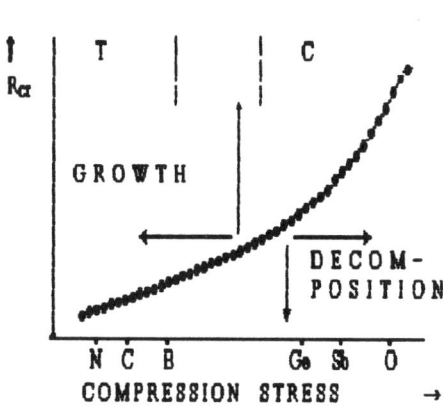

where n is the current number of particles (it is proportional to the volume of an inclusion), η is a shape factor (=area of nucleus surface/$n^{2/3}$), σ_s is a surface energy per unit area, Δg_v is an energy gain (per particle) due to a new phase formation (the driving force of the reaction), Δg_ε is an elastic energy. If $\Delta g_\varepsilon \geq \Delta g_v$, the nucleation of new isolated precipitates becomes impossible, and old ones tend to dissolve. So, on the initial stages of the IL formation it is necessary to reduce the value of Δg_ε to give the inclusions a chance to form. But after the formation of a cluster of inclusions having an infinite size ($n_c \to \infty$), the role of Δg_ε should vanish.

FIGURE 2: Qualitative dependence of the critical radius Rcr of the precipitates on stress created by the silicon matrix (external and internal). T and C are tensile and compressive stresses, respectively.

The elastic energy (for a rigid inclusion with an ellipsoid shape) may be presented as:

$$\Delta g_\varepsilon = (2\mu_m \varepsilon^2 / 3 v_m) \cdot E(y/R) = \text{Peff} \cdot \varepsilon \qquad (6)$$

where μ_m is the shear modulus of the matrix (silicon), $\varepsilon = v - v_m$ is the volume mismatch, v_m and v are the volumes of the old and the new phase, respectively, E(y/R) is the inclusion anisotropy function [10], {R,R,y} are the ellipsoid half-axes. If y=R (sphere), then E(y/R)=1.

For smaller values of y (disc-like inclusions), E(y/R)≅y/R. Peff may be treated as some effective pressure (compressive in the case of SiO_2 precipitates in Si) which depends both on the matrix reaction and on the shape and dimension of the inclusions. It is possible to reduce this pressure by means of (i) an increase of the annealing temperature ($\mu_m \to 0$) - this is the classical way to do it; (ii) the introduction of small radius impurities or vacancies (Fig. 2) - this results in the creation of an opposite sign stress field which is equivalent to a decrease of ε; (iii) a change in the shape of the precipitates because a disk-like shape is more favourable (Fig. 3), (iv) if $R \to \infty$ but remains finite (by means of δ-doping, or through the creation of a highly defective implanted layer [2]) the layer-by-layer 1D growth may be realized from the very beginning, and in this case E(y/R)→0 as well (Fig. 3).

On the other hand, the application of compression stress after the inner single-phase sublayer creates favourable conditions for the complete decay of small isolated spherical inclusions (Fig. 2) and leads to a pure layered subsequent 1D growth (Fig. 3). According to computer simulations, the right side of the IL is formed under oxygen surplus conditions during the annealing, but vacancies predominate at the left side.

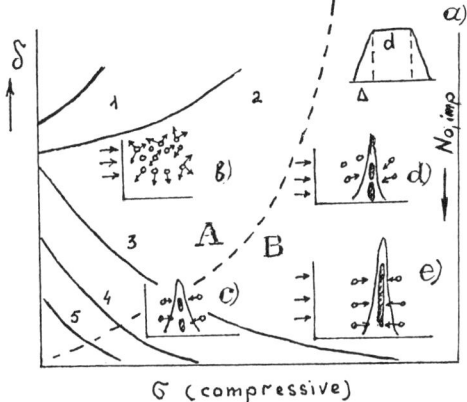

As a consequence, an asymmetrical IL is observed: its left side contains small inclusions, but the right side consists of large but stressed precipitates because of the relaxation of stress is difficult (Fig. 4). SIMS data support this conclusion (Fig. 5). According to the simulations, some changes of the stress distribution over the sample is predicted in the central part of the IL when the continuous layer begins to form (Fig. 4).

FIGURE 3: The dependence of (layer sharpness)$^{-1}$ =δ/d (solid lines) on the compression stress σ, at different concentrations N_{impl} of implanted O^+ (isoconcentration curves) according to computer simulation. The N_{impl} value increases with the curve number (1-5). a) IL configuration: d is the thickness of an uniform part of the layer, δ is a width of a region where isolated inclusions remain; b) {σ - N_{impl}} region of the diagram, where the precipitation is impossible; c) and d) {σ - N_{impl}} regions where δ/d>0 (smooth IF); e){σ - N_{impl}} region of predominating 1D growth. The dashed line separates the regions of predominating layered {B} and spherical {A} growth.

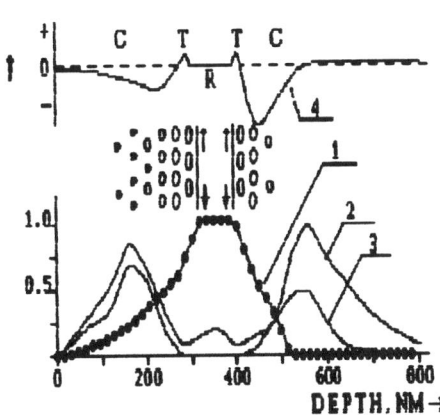

The elastic free energy of a system decreases and so further deformation of the sharp dielectric layer becomes energetically unfavourable. Beside this, confinement effects take place in the IL [11]. In order to avoid this, both buried nitride and back oxide layers may be used, as it was demonstrated in the third experiment. It should be noted, however, that the average stress produced by wafer bending is nor sufficient for respective transformations. For this reason the true mechanism should be connected to the stress concentration near the isolated precipitates right after their formation. So, only a flat layer formed from coagulated plate-like inclusions begins to grow in a silicon matrix under these conditions.

FIGURE 4: The IL configuration and stress didtribution according to simulation. C, T and R- are compressed, tensed and relaxed regions. 1: R(z) distribution of the precipitate radius; 2: weakly bonded oxygen; 3: vacancy distribution; 4: stress distribution. Right-side precipitates are larger and more compressed than left ones. Here, the oxygen concentration is higher than the vacancy concentration.

5. Conclusion

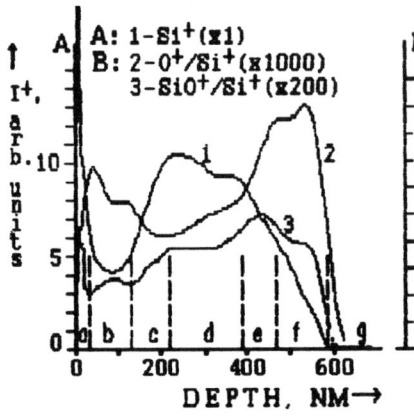

FIGURE 5: SIMS data for the insulating layer configuration. At the right side increased values for the SiO$^+$/Si$^+$ ratio are observed (indicator of a disordered SiO$_2$ matrix). O$^+$/Si is correlated with the unbonded oxygen distribution.

Using computer simulation, some perspective approaches based on stimulated technology were demonstrated. The proposed approaches allow one to control such parameters of the buried layer as its localization and thickness, the sharpness and the quality of the interface, and so on. Results predicted by computer simulation are confirmed by SIMS and Auger depth profiling experiments. The proposed methods for the control of the insulating buried layer are promising for the fabrication of high-speed IC's and for different applications in the field of vacuum microelectronics (field-emission cathodes, ...).

6. Acknowledgements

This work was done under support of the Ukr. Nat. Committee of Sci. and Technol. and the J. Soros International Science Foundation (Project Nr. X273.1, Proposal Nr. 30849). Also our thanks to Prof. A.G. Revesz.

7. References

[1] V.G. Litovchenko et al., (1994) "The effect of mechanical stress ...", Extended abstracts, **94.1**, (Spring Meeting San Francisco, CA, May 22-27), Electrochem. Society., 868

[2] L.F. Gilles et al (1994), "Nucleation of oxidation induced ...", Extended abstracts, **94.1**, (Spring Meeting San Francisco, CA, May 22-27), Electrochem. Society, 874

[3] N.N. Nikitenko (1983), "The theory of heat and mass transfer", Nauk. Dumka-Pub, Kiev

[4] Sh. Yamazaki (1971), "A study of the Si-SiO$_2$ system", Jap. J. Appl. Phys. **10**, pp. 1555

[5] S.A. Litvinenko et al. (1985), "The influence of mechanical ..." Opt. & Sem. Tech, **8**, 40

[6] B.J. Mrstik, A.G. Revesz et al. (1985) "Structural and strain-related effects during growth of SiO$_2$ films on silicon", J. Electrochem. Soc., **134**, 2020

[7] B. Ya. Lubov (1981), "Diffusion processes in inhomogeneous solid media", Nauka Pub, Moscow

[8] S.M. Hu (1978), "Film-edge induced stress in silicon structures", Appl. Phys. Lett. **32-1**, 5.

[9] V.G. Litovchenko, B.N. Romanyuk and A.A. Efremov (1993), "Simulated formation and interface Si properties of the buried ...", in B. Lengeler *et al.* (Eds), proc. of Conference ICFSI-4, World Sci. Pub., London, 389

[10] J.W. Christian (1975), "The theory of transformations in metals and alloys", Pergamon Press, Oxford, New York-Toronto.

[11] A.G. Revesz (1995), "The defect structure of buried oxide layers in SIMOX and BESOI structures", this volume.

BEHAVIOUR OF OXYGEN AND NITROGEN ATOMS SEQUENTIALLY IMPLANTED INTO SILICON

A.B. DANILIN

Centre for Analysis of Substances, Moscow Physical Society
9, Elektrodnaya St., 111524 Moscow, Russia

This paper is a brief review of recent experiments dealing with elaboration of a new SOI technology based on sequential implantation of oxygen and nitrogen ions into silicon. We have shown that annealing causes mutual redistribution and coaggregation of the atoms. Implantation with similar energies of ions ensures a high efficiency of ion beam synthesis in a thin layer, produces a buried dielectric layer at lower implantation doses and requires shorter annealing time and lower temperature than for SIMOX technology. The process of heterogeneous ion beam synthesis proves to be quite sensitive to the parameters and conditions of implantation and annealing. Methods of optimising ion beam synthesis have been proposed. The phase composition of the dielectric layers has been characterised.

1. Introduction

The high performance, small size, and stability against temperature and radiation effects which is inherent to IC based on thin SOI structures show good promise for microelectronics [1].

At present, the basic technique of SOI fabrication is SIMOX. However, the necessity of superhigh implantation dose and long-term, high-temperature annealing (1300-1405 °C and more than 5 h [2]) drastically increases the cost of SOI structures and, hence, precludes their general use in IC fabrication.

Therefore, it is a pressing problem to find an alternative technique for producing buried dielectric layers. Recently, a number of studies have dealt with low-dose synthesis of SOI structures by SIMOX [3-6]. This technique imposes a number of limitations; in particular, it requires a long-term high-temperature annealing. The desired structural perfection of superficial silicon layers requires oxygen implantation temperature be not lower than 500 °C [7]. In turn, the high implantation temperature produces silicon oxides directly during implantation [8]. The coalescence of the microprecipitates into a continuous SiO_2 layer requires long-term annealing at temperatures close to the silicon melting point [9]. The other limitation is caused by the fact that during high-temperature annealing which follows the low-dose implantation, the coalescence of microprecipitates and single atoms in the buried SiO_2 layer involves the formation of large silicon precipitates [10], which increase the leakage currents and reduce the breakdown voltage of SOI structures.

This work considers SOI fabrication technique with low implantation doses by sequential implantation of oxygen and nitrogen atoms. The technique is referred to as SIMON (Separation by IMplanted Oxygen and Nitrogen).

2. Results and Discussion

We consider ion beam synthesis of a new phase buried layer. Let us assume that a continuous layer of the phase is to be produced the thickness of which is considerably smaller than that of the impurity concentration profile. For conventional ion beam synthesis using ion implantation, the concentration of reactive impurity in a large part of the layer is sufficient for the formation of a continuous layer during subsequent annealing. This method can be spoken about as a stoichiometric synthesis. However, substoichiometric ion beam synthesis is the best tool for lowering the implantation doses. During this process, implantation provides for only the formation of stable new phase precipitates, and the continuity is achieved during annealing due to the mass transport from the concentration profile "wings" to the center (Fig. 1). P.L.F. Hemment

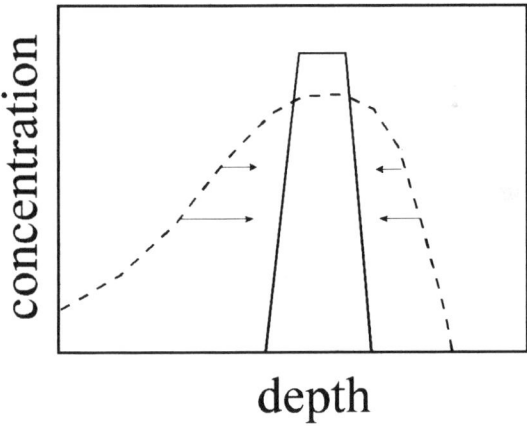

Figure 1. Redistribution of implanted impurity during annealing.

et al. [10] were among the first to propose this synthesis technique. If one implantation-annealing cycle fails to produce a continuous layer, the cycle should be repeated. Since the surplus oxygen atoms are removed from the SiO_2 phase during annealing and are then involved into the phase formation at the $Si-SiO_2$ interface [11], the process can easily be performed. However, the first attempts at accomplishing ion-beam synthesis in one implantation-annealing cycle failed because, due to the different specific volumes of SiO_2 and Si, the formation of the dielectric phase occurred in the layer where vacancial-type defects were generated, this layer being at a smaller depth than the oxygen concentration profile maximum. Thus, there were two SiO_2 precipitate layers, and it was quite difficult to produce a single continuous layer [10]. Naturally, one could lower the implantation dose in order to reduce the supersaturation at the main concentration maximum and to enhance the effect of the radiation defects. In that case the new phase formed in the layer where the radiation defects were generated [12], but the required number of implantation-annealing cycles became very large and hindered the commercial use of the method. It seems to be promising to use substoichiometric ion beam synthesis at low implantation energies in order for the defect and impurity concentration maxima to be closer to each other.

It appeared that substoichiometric ion beam synthesis, which is mainly based on mass transport of a reactive impurity to the phase formation region, can easily be performed by sequentially implanting oxygen and nitrogen ions.

The nature of this phenomenon is that these atoms coaggregate in silicon during annealing [13]. However, the effect also occurs if oxygen and nitrogen are implanted into silicon with doses sufficient for phase formation [14]. Figure 2 presents concentration profiles of oxygen and nitrogen implanted into silicon at energies of 150 keV for oxygen and 100 keV for nitrogen and doses of $6 \cdot 10^{16}$ and $4 \cdot 10^{16}$ cm^{-2}, respectively, and annealed at 1200 °C for 5 min in a nitrogen atmosphere. It can be

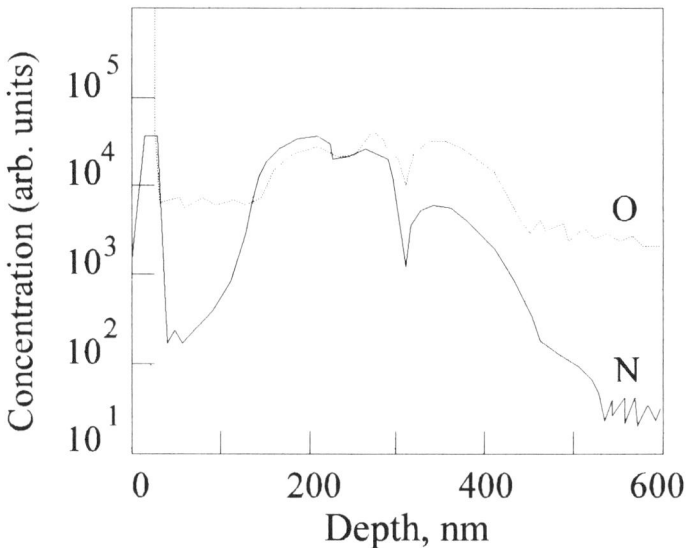

Figure 2. SIMS depth profiles of oxygen and nitrogen.

seen (Fig. 2) that annealing leads to an intense mutual redistribution of nitrogen and oxygen. The nature of this effect has not yet been understood, and we therefore present data from the literature based on which some models can be proposed:

(1) Oxygen precipitation is enhanced by nitrogen [15];

(2) Nitrogen and oxygen dissolved in Cz-Si mutually redistribute and coaggregate [13];

(3) SiO_2 phase formation produces surplus interstitial silicon [16];

(4) Nitrogen which segregates onto interstitial-type defects may stabilise them [17];

(5) Oxygen atoms may leave the solid solution and segregate on interstitial-type defects [18].

These facts suggest the following model. The behavior of oxygen and nitrogen ions sequentially implanted into silicon with substoichiometric doses largely depends on radiation defect reactions. Assuming that nitrogen may neutralise the effect of interstitial-type defects on oxygen, one reasonably concludes that oxygen precipitation and SiO_2 phase formation are the most probable in the nitrogen-rich layer. The precipitates are sinks for unbonded oxygen atoms which migrate toward these centres through large distances. Moreover, if the nonstoichiometric silicon nitride precipitates which form during implantation are sources of elastic strain, then vacancial-type defects, including A-centres, will move toward the precipitates. These are the major effects causing oxygen to migrate toward nitrogen-saturated regions. In turn, nitrogen atoms segregate at the $Si-SiO_x$ interface [19], and, hence, a large amount of nitrogen moves toward the oxygen concentration maximum. One may also assume that nitrogen atoms may be captured by interstitial-type defects forming near the SiO_x precipitates due to the generation of interstitial atoms during oxidation. These effects may cause nitrogen migration toward the oxygen concentration maximum. Since the silicon nitride precipitates which form during implantation [20] are diffusion barriers for unbonded impurity atoms [21], it is reasonable that the impurity mass transport toward the nitrogen concentration maximum is less intense.

The mass transport of oxygen and nitrogen toward their concentration profile maxima during annealing is the most intense if the implantation energies of the ions are similar (Fig. 3) [22]. One may assume that the redistribution is due to both the well-known coaggregation, whose role is doubtlessly great at an early stage of synthesis, and the fact that the new phase also acts as an inner self-supporting getter. There are indications [23] that oxygen atoms hinder the Si_3N_4 crystallisation. On the other hand, silicon-nitrogen compounds are a good diffusion barrier [20] and preclude the redistribution of unbonded silicon and oxygen. The system is, therefore, fine-grained, and, due to the high total area of the interfaces, it is a good impurity getter

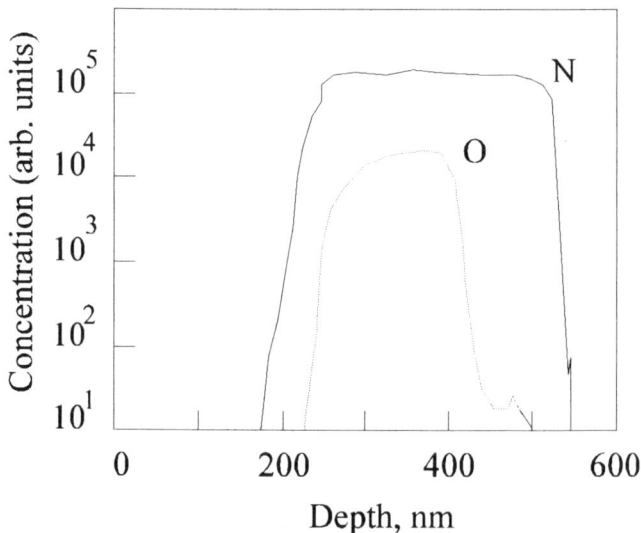

Figure 3. SIMS depth profiles of oxygen and nitrogen implanted with 150 keV and doses of $1\cdot10^{17}$ and $5\cdot10^{17}$ cm^{-2}, respectively (annealing 1200 °C for 2 h in a nitrogen atmosphere)

and provides for an intense mass transport of reactive species to the phase formation region. In turn, oxygen and nitrogen mass transport allows the fine structure to be retained. Apart from the high efficiency of this synthesis mode, its additional advantage is the fact that the blocking of the unbonded silicon atoms by the forming compound avoids the formation of large precipitates that would deleteriously affect the insulating properties of the buried layer.

The model has been experimentally tested. Buried dielectric layers of complex composition were produced by sequential implantation of oxygen and nitrogen ions into silicon at an energy of 150 keV and a total dose of $(4-6) \cdot 10^{17}$ cm^{-2}. The structures were annealed at 1200 °C for 2 h. Electron microscopic data showed the dielectric layer width to be 0.18 to 0.21 μm. The interfaces were sharp, and the dislocation density in the superficial silicon layers was max. 10^7 cm^{-2}. XPS study revealed the chemical composition of the buried layers. These were disordered molecular networks of silicon tetrahedra, in which oxygen and nitrogen were randomly distributed in accordance with their concentrations [24]. Measurements made at 10 V over the 20-100 °C range showed the dielectric strength of the buried layer is about $5 \cdot 10^6$ V/cm and the leakage current density under 10^{-9} A/cm^2.

The main obstacle to the practical use of oxynitride layers has been the high donor concentration in the superficial silicon layer (about 10^{17} cm^{-3}) leading to a high free electron concentration [25]. The problem was solved by optimizing the O/N dose ratio and implantation temperatures for both ions. In recent SOI structures, the free electron concentration does not exceed $2 \cdot 10^{16}$ cm^{-3}, but for the best samples it was $7.5 \cdot 10^{15}$ cm^{-3} [26].

MIS structures were made with 200×50 and 200×30 μm gates formed from highly doped polycrystalline n-type silicon. A contact to the silicon layer was made with a highly doped n^+ silicon region surrounding the gate electrode and aluminum was melted into this contact layer. This form of contact corrects for effects due to spreading resistance of the superficial silicon layer and the capacity of the surface space charge region.

Measurements showed that SOI MISFET devices with a buried silicon oxynitride layer exhibit normal characteristics. The threshold voltage variation is less than 10% at temperatures up to 160 °C and with gamma radiation doses of 10^5 rad. The corresponding values for conventional MISFET devices are only 80°C and $5 \cdot 10^3$ rad [26].

This ion beam synthesis of buried dielectric layers proved to be very sensitive to parameters and conditions of implantation and annealing, their choice being therefore important for commercially suitable SOI structures. The major requirements are as follows.

(1) Oxygen should be implanted the first. Otherwise, its ions will be included into a layer containing substoichiometric silicon nitride, which is a good diffusion barrier. The buried layer will therefore be porous, its outer interface will be rough, and the superficial silicon layer will be much damaged (Fig 4, a). Cycle ion beam synthesis, which is attractive at a first glance, is inefficient for the same reason (Fig. 4, b) [27].

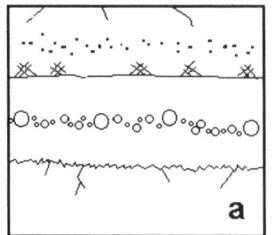 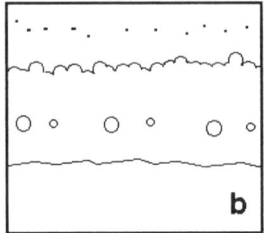

Figure 4. Schematic of typical XTEM images of specimens obtained using (a) sequential implantation of nitrogen and oxygen (nitrogen implanted the first) and (b) cycle synthesis.

(2) The optimum ratio of nitrogen and oxygen implantation doses depends on implantation energy, and at an energy of 150 keV it usually ranges from 3 to 5.

(3) Implantation temperature should avoid SiO_2 precipitation both during implantation and annealing since the precipitation makes the layer inhomogeneous and incontinuous [28]. Thus, implantation temperature should not be higher than 400 °C. On the other hand, to avoid saturation of the buried layer with vacancial-type defects and SiO_2 precipitation during annealing, implantation temperature should be higher than 300 °C.

(4) Annealing temperature and time should be sufficient for complete mass transport of impurities to the phase formation region and should provide a sufficient structural perfection of SOI structures. On the other hand, these parameters should preclude crystallisation of the buried layer, which would eliminate its insulating properties. The best annealing mode is 1200 °C for 2 h.

(5) The best annealing atmosphere is an inert gas with max. 10% oxygen. During annealing, oxygen atoms of the medium diffuse to the outer interface of the buried layer and, being involved into the ion beam synthesis, increase the structural perfection of this interface [29].

3. Conclusion

The behaviour of oxygen and nitrogen atoms implanted into silicon (coaggregation) and the phase formation process, during which the buried dielectric layer acts as an efficient getter, facilitate SOI formation.

Acknowledgements

The present data are the result of investigations carried out by the large group of researchers in the period of time from 1987 to 1993. I express my sincere gratitude to A.F. Petrov for carrying out implantation, O.I. Vyletalina and K.A. Drakin for developing the ion beam synthesis technique, A.A. Malinin for electron microscopic investigation, A.F. Borun for Auger and XPS analyses, V.V. Saraikin and V.K. Smirnov for SIMS measurements, V.N. Mordkovich for fruitful comments and discussions, and to A.W. Nemirovski for help in preparation of this manuscript. I would also like to thank everybody who has helped us in the performance of this work.

REFERENCES

1. Chen, D.C.E. (1988) SIMOX Devices and Circuits, *SIMOX Workshop 1:* University of Surrey, Nov. 1988, 1-8.
2. Hemment, P.L.F. and Reeson, K.J. (1989) Hidden Depth to Devices, *Inst. of Phys.: Physics World*, **2**, No. 6, 39-42.
3. Robinson, A.K., Marsh, C.D., Bussman, U., Kilner, J.A., Vanhellemont, Y.Li., Reeson, K.J., Hemment, P.L.F., and Booker, G.R. (1991) Formation of Thin Silicon Films Using Low Energy Ion Implantation, *Nucl. Instr. and Meth. in Phys. Res.*, **B55**, 555-560.
4. Namavar, F., Cortesi, E., Kalhoran, N.M., Manke, J.M., and Buchanan, B.L. (1990) Characterisation of Low-Energy SIMOX (LES) Structures, *Proc. 1990 IEEE SOS/SOI Technology Conf.*, Key West, Florida, p.49.
5. Nakashima, S. and Izumi, K. (1990) *Electron. Lett.*, **26**, 1647.
6. Cefolini, G.F., Bertoni, S., Meda, L., and Spraggiari, C. (1994) Formation and Stability of Continuous Buried SiO_2 Layers in SIMOX, *Nucl. Instr. and Meth. in Phys. Res.*, **B84**, 234-237.
7. Holland, O.W., Screen, T.P., Faty, D., and Narayan, J. (1984) Influence of Substrate Temperature on the Formation of Buried Oxide and Surface Crystallinity During High-Dose Oxygen Implantation, *Appl. Phys. Lett.*, **45**, No. 10. 1081-1083.
8. Reeson, K.J., Marsh, C.D., Chater, R.J., Kilner, J.A., Robinson, A.K., Christensen, K.N., Hemment, P.L.F., Harbeke, G., Steigmeier, E.F., Booker, G.R., and Celler, G.K. (1988) The Role of Implantation Temperature and Dose in the Control of the Microstructure of SIMOX Structures, *Microelectronics Enineering*, **8(3/4)**, 163-174.
9. Celler, J.K., Hemment, P.L.F., West, K.W., and Gibson, J.M. (1986) High-Quality Si-on-SiO_2 Films by Large-Dose Oxygen Implantation and Lamp Annealing, Appl. Phys. Lett., **48**, No. 8., 532-534.
10. Hemment, P.L.F., Reeson, K.J., Kilner, J.A., Chater, R.J., Marsh, C., Booker, G.R., Celler, G.K., and Stoemenos, J. (1986) Ion Beam Synthesis of Thin Buried Layers of SiO_2 in Silicon, *Vacuum*, **36**, No. 11-12, 877-881.
11. Mayama, S. and Kajiama, K. (1982) Surface Silicon Crystallinity and Anomalous Composition Profiles of Buried SiO_2 and Si_3N_4 Layers Fabricated by Oxygen and Nitrogen Implantation in Silicon, *Jap. J. Appl. Phys.*, **21**, No. 5, 744-751.
12. Stoemenos, J., Reeson, K.J., Robinson, A.K., and Hemment, P.L.F. (1991) *J.Appl.Phys.*, **69**, 793.
13. Hockett, R.S., Fraundorf, P.B., Reed, D.A., and Wayne, D.H. (1986) Oxygen, Carbon, Hydrogen and Nitrogen in Crystalline Silicon, in Mikkesen, J. C., Jr., Pearton, S.J., Corbett, J.W. and Pennycook S.J., (eds.), *Mat. Res. Soc. Symp.*, **59**, 433.
14. Borun, A.F., Danilin, A.B., Mordkovich, V.N., and Temper, E.M. (1988) Behaviour of Oxygen and Nitrogen upon Simultaneous Substoichuiometric Implantation into Silicon, *Rad. Eff.*, **107**, 9-13.
15. *VLSI Science and Technology* (1984) Bean, K.E., and Rozgonyi, G.A., (eds.) Electrochem. Soc., Pennington, N.J., 1984, p.59.
16. *Defects in Semiconductors* (1981) Narayan, J., and Tan, T.Y., (eds.) North-Holland, Amsterdam. 333p.
17. *Semiconductor Silicon* (1981) Huff, H.R., Kriegler, R.J., and Takeishi, Y. (eds.) Electrochem. Soc., Pennigton, NJ, p.126.
18. Sumino, K. (1987) *2nd Int. Autumn Meeting Proc. Gettering and Defect Engeenering in Semiconductor Technology (GADEST 87)*, Garzau, Germany, 1987. Akad. Weisenschaften DDR, Frankfurt (Oder), p.218.
19. Josquin, W.J.M.J., (1983) The Application of Nitrogen Ion Implantation in Silicon Technology, *Nucl. Instr. and Meth. in Phys. Res.*, **209/210**, 581-590.
20. Pavlov, P.V., Kruze, T.A., Tetelboum, D.I., Zorin, E.I., Shitov, E.V., and Gudkova, N.V. (1976) Electron Microscopic Studies of Silicon Layers Implanted with High Doses of Nitrogen Ions, *Phys. Stat. Sol.(a)*, **36**, No. 1, 81-88.
21. Samsonov, G.V. (1969) *Non-Metallic Nitrides*, Metallurgiya, Moscow (In Russian) 258p.

22. Danilin, A.B., Drakin, K.A., Kukin, V.V., Malinin, A.A., Mordkovich, *Nucl. Instr. and Meth. in Phys. Res.*, **B58**, 191-193.
23. Edelman, F.L. (1980) *Structure of the LSIC components,* Nauka, Novosibirsk (In Russian), 256p.
24. Borun, A.F., Danilin, A.B., Galayev, A.A., Goriounova, I.I., Parkhomenko, Yu.N., Toropova, O.V., and Vyletalina, O.I., (1995) Study of Low-Dose Ion Synthesis with Separate and Sequential High and Low-Temperature of Implantation of Oxygen and Nitrogen into Silicon, *Mat. Sci. and Eng.* (to be published).
25. Davies, D.E. and Adamski, J.A. (1988) Nitrogen-Related Doping with Implant Si_3N_4 Formation in Si, *Appl. Phys. Lett.*, **48**, No. 3., 347-349.
26. Semiconductor International, March 1994, SOI MISFETs Produced with Buried Silicon Oxynitride Layers, p. 20.
27. Danilin, A.B., Drakin, K.A., Malinin, A.A., Mordkovich, V.N., Petrov, A.F., and Vyletalina, O.I. (1991) Behaviour of Implanted Nitrogen in Si with the Buried Layer of SiO_2 Precipitates, *Solid State Phenomena*, **19&20**, 405-410.
28. Vyletalina, O.I., Danilin, A.B., Drakin, K.A., Malinin, A.A., Mordkovich, V.N., Petrov, A.F., and Saraikin, V.V., Structure of Buried Dielectric Layers Produced in Silicon Using Cycle Ion Beam Synthesis (1991) *Poverkhnost'*, No.4, 90-94 (in Russian).
29. Schork, R. and Ryssel, H. (1994) Ion Beam Synthesis of Buried Nitride Layers With High-Quality Interfaces, *Xth International Conference on Ion Implantation Technology, Abstracts*, Catania, Italy, P-3.61.

SOI FABRICATION BY SILICON WAFER BONDING WITH THE HELP OF GLASS-LAYER FUSION

N.I.KOSHELEV, A.I.ERMOLAEVA, V.Z.PETROVA
Moscow State Institute of Electronic Engineering,
103498 Moscow, Russia

Silicon-on-insulator (SOI) technology by silicon wafer bonding with the help of the specially synthesized glass dielectric with the thermal expansion coefficient matching that of the single-crystal silicon has been considered. Selective electrochemical etching was used for the device wafer thinning.

1. Introduction

Wafer bonding is one of the leading techniques for the formation of SOI structures, because of the high quality device silicon layer and the flexibility in controlling the thickness of the insulating layer.

The wafer bonding method is based on bonding of the support silicon substrate with the device wafer through the dielectric layer and subsequent thinning of the device wafer up to the required thickness by chemomechanical polishing and selective electrochemical or chemical etching (bond-and-etch-back silicon-on-insulator- BESOI) [1-3].

Silicon wafer bonding can be realized with the help of the activated by special processing thin SiO_2 film [1-3]. However, especially high quality of bonding silicon wafer surfaces is required to receive high integrity bond by this method. Besides this method does not permit dielectric layers thicker than 2 μm because of the considerable mismatching of the thermal expansion coefficient (TEC) of Si and SiO_2. For high-voltage devices it is desirable to have a thick insulating layer between the support substrate and the device silicon layer.

Another perspective method of bonding silicon wafers is based on the application of multicomponent glass dielectrics with TEC matching that of single-crystal silicon and lower flow temperature in comparison with SiO_2 [1,2,4].

The latter method was used in present work. Selective electrochemical etching was used for device wafer thinning.

2. Experimental

A specially developed glass dielectric in system $BaO-Al_2O_3-SiO_2$ - glass BAS-35- was used for the fabrication of SOI structures. In order to receive high purity of glass materials their synthesis was conducted in the inductive furnace in the platinum-rhodiume crucible and in RF inductive plasmatron. The main properties of glass BAS-35 are represented in Table 1.

TABLE 1. Main properties of glass dielectric BAS-35

Thermal linear expansion coefficient	$35 \cdot 10^{-7}$ K^{-1}
Flow temperature	1100 °C
Resistivity at 20 °C	$>10^{15}$ $\Omega \cdot cm$
Breakdown voltage of thin films at 20 °C	$8 \cdot 10^6$ V/cm
Dielectric constant at 20 °C and 1 MHz	8
Dielectric loss tangent at 20 °C and 1 MHz	$20 \cdot 10^{-4}$

Developed glass dielectric BAS-35 have the thermal expansion coefficient matching that of the single-crystal silicon that permits to obtain stressless dielectric-silicon structures with the camber nearly zero. Differential thermal analysis and X-ray diffraction investigation showed that dielectric BAS-35 is noncrystallizable glass.

Highly Sb-doped (111) silicon wafers ($\rho=0.01$ $\Omega \cdot cm$) with the low-doped epilayer (P-doped, $\rho=1.0$ $\Omega \cdot cm$) with 3-10 μm thickness were used for the fabrication of SOI structures. SiO_2 film (0.5-1.0 μm thickness) was formed by chemical vapor deposition on the initial epiwafer. Next, glass BAS-35 films (0.5- 4.0 μm thickness) were deposited by RF magnetron sputtering on the epiwafer and the support silicon substrate. Targets used were 120 mm diameter disks of glass BAS-35. Substrate temperature during the deposition was 200° C. The total atmosphere gas pressure ($Ar+20\%O_2$) was 0.1 Pa. A typical deposition rate was 15 nm/min. The electron probe X-ray microanalysis was used to determine the glass layer composition. It was found that the composition of the glass layers was almost equal to that of the target within 5% accuracy. X-ray diffraction analysis indicated that the as-deposited and annealed (800-1200° C) glass layers were amorphous.

After that the epiwafer and the support silicon substrate were bonded in the diffusion furnace at 1100-1200° C and pressure of 10 Pa using special fused quartz fixture.

In order to investigate the quality of the bonds SEM investigations of a cracked wafer were performed. No microscopic voids could be observed. It was impossible to observe any interfaces between the insulating layer and the silicon layer.

Preliminary device wafer thinning was realized by grinding and chemomechanical polishing. The remaining n^+-layer (10-30 μm thickness) was removed by selective electrochemical etching based on different rates of anodic dissolving of n^+- and n-type silicon in HF-based electrolytes [5].

Electrochemical etching was realized in the aqueous-ethylene glycol solution of HF by electropolishing of the low-resistivity substrate. The introduction of ethylene glycol into the electrolyte substantially displaces the electropolishing region of the n-type epilayer to higher potentials and permits to combine in one electrochemical process etching of the n^+-type silicon substrate and polishing of the low-resistivity n-type epilayer. Figure 1 represents voltage-current characteristics of highly Sb-doped (ρ= 0.01 $\Omega\cdot$cm) and low-P-doped (ρ= 1.0 $\Omega\cdot$cm) silicon in the 5% HF aqueous-ethylene glycol solution.

Selective electrochemical etching was realized using potentiostatic condition and the anode-cathode voltage of 7V. As can be seen from Figure 1 in these conditions anode dissolving current of the low-doped silicon is two orders of magnitude lower than that of the highly doped silicon. In order to provide full etching of the n^+-layer on all surface of the device wafer electrochemical etching was realized with gradual dipping of the structure into the electrolyte (1-5 mm/min rate depending on n^+-layer thickness) by a special mechanism. After electrochemical etching transient layer was removed by chemical etching in the solution of HF, CH_3COOH and $KMnO_4$, reducing the silicon film to the desired thickness.

3. Experimental results

SOI structures of 76 and 100 mm in diameter were obtained by silicon wafer bonding and selective electrochemical polishing for device wafer thinning. The silicon device layer was within the range of 1-10 μm thickness with the total thickness variation not exceeding 10% across the whole wafer. The thickness of insulating BAS-35 layer was 1-8 μm, while the thickness of masking SiO_2 layer- 0.5-1.0 μm. The presence of masking SiO_2 layer in silicon structures obtained prevents the glass component diffusion into the device silicon layer. The warpage value of SOI structures obtained was no more than 20-30 μm. It must be noted that in order to obtain the silicon film with high uniformity of thickness and quality of the surface it is necessary to use initial epitaxial structures with donors concentration in the wafer not lower than $3\cdot 10^{18}$ cm^{-3} and in the epilayer not higher than $2\cdot 10^{16}$ cm^{-3}. Another necessary condition is high structural perfection of using the epiwafer.

Structural quality of thin isolated silicon layers in SOI structures obtained was estimated by the analysis of RHEED patterns and rocking curves. RHEED patterns from the surface of the initial epitaxial structures and silicon layers of SOI structures are similar. Figure 2 represents rocking curves with respect to (333) plane (CuK_α - radiation) from the n-layer surface of initial epitaxial structure and silicon films (10 μm thickness) of SOI structures. The full width at half-maximum (FWHM) of rocking curves of the initial epitaxial structure n-layer and SOI structure is 15" and 30", respectively. The heat treatment of the SOI structures at 1200° C (30 min) reduces rocking curve FWHM up to 25". Small values of FWHM of rocking curves are evidence of the low defect density in SOI device layer. Etch pit densities in the silicon layers produced was $5\cdot 10^3$ cm^{-2}

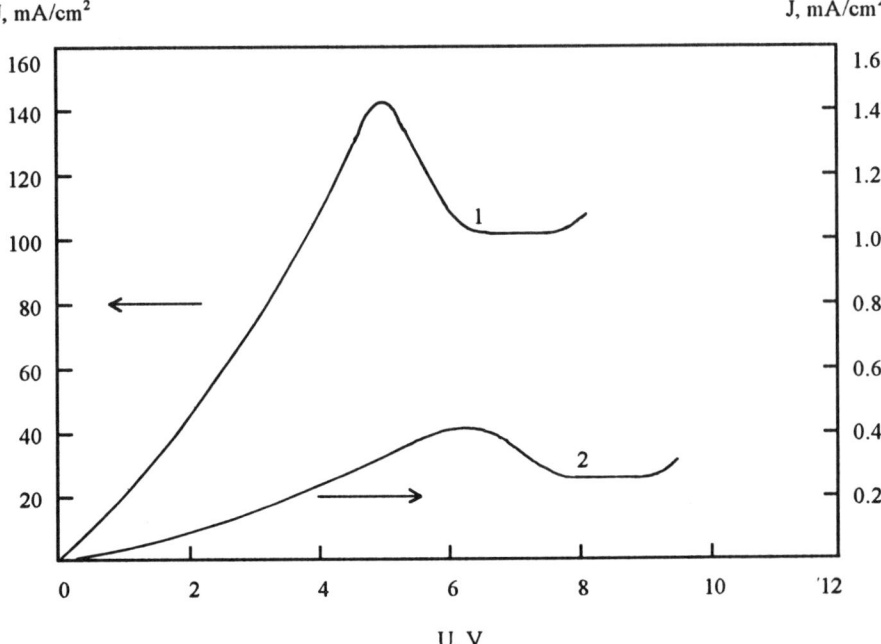

Figure 1. Voltage-current characteristics of silicon:
1) Sb-doped, $\rho = 0.01\ \Omega\cdot\text{cm}$,
2) P-doped, $\rho = 1.0\ \Omega\cdot\text{cm}$.

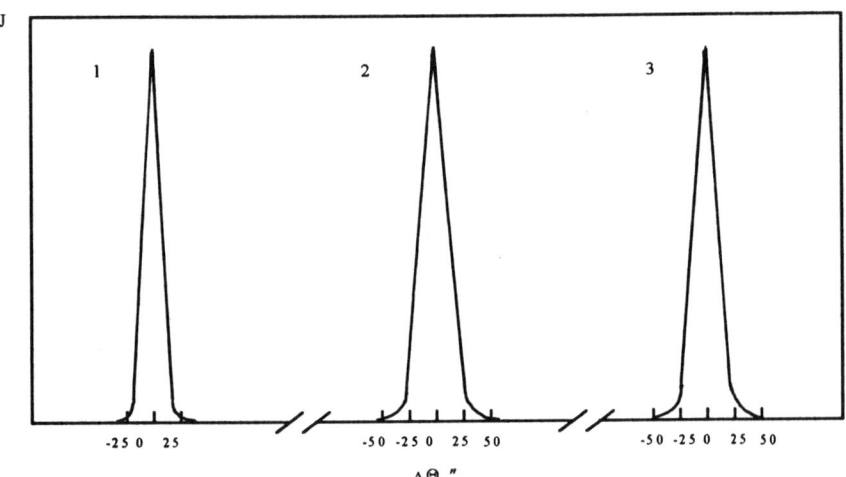

Figure 2. Rocking curves of SOI device layers:
1) initial epitaxial structure,
2) as-fabricated SOI structure,
3) heat treated SOI structure (1200° C, 30 min).

The electrical characteristics (resistivity, majority carrier mobility) of SOI device layers were in good agreement with the initial epitaxial layers parameters.

4. Conclusion

Technological processes for high quality SOI structure formation by silicon wafer bonding with the help of glass-layer fusion and subsequent device wafer thinning by selective electrochemical etching have been developed. The device single-crystal silicon layer thickness in SOI structures developed is in the range from 1 to 10 microns.

Obtained SOI structures can be used for the fabrication of high-voltage, high-power, high-temperature and radiation-hardened integrated circuits.

The developed technological processes of bonding silicon wafers and selective electrochemical etching of n/n^+-epitaxial structures can be used for the fabrication of thin silicon membranes and cavities or channels in silicon for sensors and other micromechanical devices.

5. References

1. Bengtsson, S. (1992) Semiconductor wafer bonding: a review of interfacial properties and applications, *J. Electronic Materials* **21**, 841-862.
2. Maszara, W.P. (1991) Silicon-on-insulator by wafer bonding: a review, *J. Electrochemical Society* **138**, 341-347.
3. Maszara, W.P., Goetz, G., Caviglia, A. and McKitterick, J.B. (1988) Bonding of silicon wafers for silicon-on-insulator, *J. Applied Physics* **64**, 4943-4950.
4. Sawada, R., Watanabe, J., Nakada, H. and Koyabu, K. (1991) Soot bonding process and its application to Si dielectric isolation, *J. Electrochemical Society* **138**, 184-189.
5. Theunissen, M.J.J., Appels, J.A., and Verkuylen, W.H.C.G. (1970) Application of preferential electrochemical etching of silicon to semiconductor device technology, *J. Electrochemical Society* **117**, 959-965.

CRYSTALLIZATION OF a-SI FILMS ON GLASSES BY MULTIPULSE-EXCIMER-LASER TECHNIQUE

A.B.LIMANOV
Institute of Crystallography, Russian Academy of Sciences, Moscow 117333, Russia

1. Introduction

Excimer-laser recrystallization is one of the best suitable processes for large-area-electronics applications [1-3]. A short pulse of a highly absorbed excimer laser heats a-Si film and causes its melting and further transformation to poly-Si material without noticeable heating of substrate. Such a process is of importance when one deals with ultrathin a-Si films on inexpensive glass such as Corning 7059.

In this paper, we present experimental results on multipulse-scanning crystallization of a-Si films by long and narrow beam of excimer laser. Possible mechanisms of excimer crystallization of a-Si films are discussed. To clarify the mechanisms, the results of numerical simulation of the process are presented.

2. Crystallization Technique

Corning Glass 7059 substrates were used in these experiments. To decrease the temperature in the substrates and to avoid damage, a buffer layer of SiO_2 or Si_3N_4 of thickness $200 nm$ was deposited on them. A-Si films of thickness about $40 nm$ were deposited on the substrates by PECVD process. We used the a-Si films with high (a-Si:H) and low (a-Si) hydrogen content (the latter were prepared from the former ones by annealing at $450°C$ in vacuum [4]).

XeCl excimer laser having pulse duration $30 ns$, pulse energy about $200 mJ$, and repetition rate up to $40 Hz$ was used. To obtain a large area film with uniform microstructure without regions of line-to-line overlapping, a laser beam incident onto the sample was transformed in line 1 to $2mm$ width and several centimeters in length [4]. Along the line a laser radiation was uniform, its distribution in cross-section was close to Gaussian one. The pulse energetic was attenuated by neutral density filters. The sample was scanned under the beam in Gaussian direction with velocities about $0.5 mm/s$ in air ambient at room temperature. In such a way, each point of Si film was treated by a pulse sequence having increasing energy, then decreasing one. A shot density was about 100 pulses/point. Microstructure of grown films was studied by TEM. Film uniformity was tested by Raman-spectroscopy.

3. Experimental Results

After a single passage of the laser beam across the substrate, a broad line was recrystallized. The line of several cm in width had relatively uniform polycrystalline microstructure, as it was confirmed by Raman measurement. Grain size depended on energy density (E) of the laser pulses. The higher E, the larger grains were formed in the films. However, over some threshold of energy density (E_m), grown films contained amorphous material with separate large grains. In this case, a roughness developed on film surface, and voids were formed in the film.

a) b) 500nm c)

Figure 1. Microstructure of grown films.

The results obtained at values E below E_m are present in Fig.1 for a-Si:H (a), and a-Si (b) films. Grain sizes are $50 - 80nm$ for (a) case, and $200 - 500nm$ for (b) one. It should by noted, that E_m for a-Si films was higher then E_m for a-Si:H films. The microstructure of grown film obtained at E over E_m is shown at Fig.1c for a-Si:H layer (This result is similar to that obtained for a-Si films at slightly higher energy). Large grains of width up to $500nm$, separated by amorphous material, are seen here as striped. Dark round stains are protrusions formed due to an expansion of the material frozen at a last moment. We believe that its reflect density of solid phase nucleation in the melt.

4. Crystallization Mechanisms

There are three most important concerns in excimer crystallization of a-Si films: high hydrogen content, control of seeding, and high undercooling of the melt.

4.1. HYDROGEN IN AMORPHOUS Si FILMS

Rapid evolution of hydrogen into a-Si:H films provides voids and microcraters in grown films [1]. In general, hydrogen, as well as oxygen and nitrogen, may deteriorate wetting conditions of the melt, and decrease the value of E_m. For example, the best result in excimer crystallization of a-Si films (channel mobility up to $329 cm^2/Vc$) was obtained by especially decreasing of the gases content [2].

4.2. CONTROL OF THE SEEDING

The problem of seeding is connected with threshold energy density E_m. It is considered [5] that a thin bottom a-Si layer remains unmelted when the energy densities are below E_m. Unmelted material provides seeds for crystal growth: it is shown [6], that liquid-amorphous interface is a preferable place for heterogeneous nucleation of crystalline phase under the process conditions. In this case, grain sizes are determined by competitive grain growth up to a film surface, and are limited by a thickness of the film. Unmelted bottom part of the film is too transformed into crystalline state by explosive crystallization [7].

Separated islands of unmelted materials at the film substrate interface are considered as best condition for seeding [5]. (It is also important that the islands cement the film to substrate and prevent a transport of liquid material, that is possible in completely melted films.) The grain sizes are determined here by a distance L between the islands. The bigger L, the larger are the grain sizes. A value of L may by controlled by appropriate choice of energy density of laser treatment. On the other hand, if L is too large, the melt between the islands have a time to cool down strongly during lateral film growth, and amorphous (or fine-grained) phase is formed between large grains [5], as it is seen in fig.1c. (In this case, L should be smaller than the distance between the dark stains.) The formation of amorphous (or fine-grain) phase between the islands may be reduced or even prevented by substrate preheating [2,5].

4.2.1. Influence of Wetting Conditions on Survival of the Seeds.

Survival of seeds-islands during film treatment, as well as value of L should be strongly determined by wetting conditions of the melt. The wetting controls a contact angle at a meeting point of melt, substrate, and edge of the island, and, hence, control a curvature of island surface. According to the Gibbs-Thompson effect, melting point of convex surface of silicon is lowered for a value ΔT as it is shown by eq.(1):

$$\Delta T = T_m \Gamma / R, \qquad (1)$$

where T_m is the equilibrium melting point of Si, Γ is the capillary constant ($\Gamma = \gamma/L = 10^{-10} m$ for crystalline Si; γ is the surface tension and L is the latent heat), R is the radius of the island.

Due to strong influence of the effect in nanometer scale, melting temperature of the island at poor wetting may by decreased for tens of degrees. Therefore, in the case of a-Si:H films, just-formed islands should be melted due to a high curvature of their surface. As it is seen in fig.1c, only few island seeds are survived under conditions when L is relatively large.

The better the wetting, the lower is the island curvature, and the higher will be the melting point of the islands. At a good wetting, inherent in a-Si films, stable survival of seeds is obtained at distance L as large as $500 nm$ (see fig.1b).

4.2.2. Solidification of Completely Melted Films

If a-Si film is melted completely (at $E > E_m$), there are no places for heterogeneous nucleation in the film. A homogeneous nucleation in the melt is needed for crystalline growth. Some time is passed until the nucleation begins, and the melt is cooled down to $1100 - 1200K$ [8]. The higher the cooling rate, the lower is the nucleation temperature [8,9]. Under such conditions, embryo density is very high, and grown film consists of fine-grained material. At very high cooling rates common for film thicknesses $< 50nm$, an amorphous phase is formed from the melt [10] because atomic transport is reduced due to increase of viscosity [11].

4.3. UNDERCOOLING OF THE MELT

The third problem is attributed to an energy stored in amorphous Si. The energy lowers melting point of a-Si, T_{ma} up to $1440K$. Therefore, after a-Si melting the melt become undercooling on $245K$ in respect to melting point of crystalline Si, $T_{mc} = 1685K$. Under such conditions, an interface-propagation velocity is controlled by crystallization kinetics [11] rather then by heat dissipation into substrate (e.g., the velocity is determined only by interface temperature and may reach $15m/s$). As a result, crystallization here proceeds under extremely nonequilibrium conditions, when any substantial enlargement of grains is impossible.

4.4. ADVANTAGES OF MULTIPULSE TECHNIQUE

The problems listed are of special importance for single-pulse excimer crystallization of a-Si films. The problems can be also solved by multipulse crystallization technique when each point of the film is treated by a sequence of pulses with increasing energy density E. This approach is realized, for example [1,3,4], by scanning of sample along the beam having Gaussian (or stepped one) energy density distribution as it is used in this paper. It should be noted that each pulse of the pulse sequence is energetically independent: the film has sufficient time between the pulses to cool down to an initial temperature. In the same time, differences of hydrogen content and of film microstructure caused by previous pulses are principal factors in the technique. For example, hydrogen content of a-Si:H film in a given point is decreased after first pulses of the treatment [1,3], and further hydrogen release does not influence strongly on grown film quality.

In order to provide seeds for film regrowth, the energy density of the most energetic pulse should be chosen just slightly lower then E_m. Such a pulse melts the film on maximum depth and, hence, determines, in general, microstructure of the film. It is important that previous pulses have already transformed a-Si film into crystalline state, either by cycles of melting-solidification or by explosive crystallization [7]. Therefore, the pulse melts Si film at a temperature close to T_{mc}, and film regrowth begins at this temperature. As a result, crystallization velocity is controlled by heat dissipation into the substrate, and film growth proceeds here under most equilibrium conditions as far as it is possible during excimer processing.

5. Simulation technique

We will consider one-dimensional heat flow inside the layered structure of $50nm$ a-Si film on glass substrate. The a-Si film is divided for 10 sublayers of equal thickness m_j. Upper part of the substrate is divided for five unequal sublayers: a thickness m_j of the first sublayer is chosen to be $20nm$, a thickness of each next sublayer in five times larger then previous one. M is the number of last sublayer.

Temperature field T_{ij} of the column m_j is determined by eqs. (2) written here in finite-differences formulation:

$$\Delta T_{i,j-1,j}\Delta k_{j,j-1}/\Delta m_{j,j-1} + \Delta T_{i,j+1,j}\Delta k_{j+1,j}/\Delta m_{j+1,j}+$$
$$+p_{i,j} = \Delta T_{i,i+1,j}m_j C_{i,j}/\Delta t + V_g L_p; \qquad (2)$$
$$V_g = F(T_{ij} - T_m);$$
$$\Delta T_{i,0,-1}\Delta k_{0,-1}/\Delta m_{0,-1} = 0; T_{i,M+1} = T_0;$$
$$T_{0,j} = T_0;$$

where k is the thermal conductivity, C is the specific heat, L_p is the latent heat of current phase. T_0 is the environment temperature (300K). The materials data are depended on temperature, and on phase state as well as on state fraction in partially melted cells. These data are chosen from paper [12].

Velocity of growing interface V_g is determined by the interface response function, which has been analytically obtained by Stiffler et al. [11]. According the function, as T deviates from T_m, V_g increases; then as temperature decreases V_g increases up to maximum about $15 m/s$ at $1500 K$, then V_g decreases to zero at 1000 or at $1100 K$ depending on parameters chosen by authors of [11].

Laser-pulse power is considered as linear dependence of time in first duration w ($w = 30 nc$), then as Gaussian one. w is HWHM of power distribution. Power heat source $p_{i,j}$ is determined by film surface reflectivity and by absorption in the sublayers.

6. Simulation results

Simulation results of separate acts of melting-solidification processes are presented in fig.2. The temperature fields are shown as isotherms. Power profile of the laser pulses are presented in the same temporal scale in upper part of drawings. The isotherms cross 10 sublayers of Si film as well as a half of the first substrate sublayer. The isotherms are different on $50°$. Some values of them are presented in drawings. Stepped lines show liquid-solid interfaces. Dotted ones are isotherms of T_{ma} for fig.1a,b, and of T_{mc} for fig.1c. Let's consider processes discussed above.

Identical features of the processes may be seen in figs.2. A rapid heating of solid Si films is significantly retarded during the films melting due to a latent heat release. This effect stabilizes the films temperature during crystallization. A temperature of propagated interfaces is not constant, and stepped lines of the interfaces are shifted from T_{ma} (T_{mc}) isotherms due to kinetic effect. The shift is seen most markedly in a case of crystalline film (fig.2c) because of the thermal conductivity of c-Si (about 0.25) is higher than one of amorphous material ($0.025 W/cm^2 K$).

In fig.2a, one can see melting of a-Si film down to bottom part of last Si sublayer followed by regrowth of crystalline phase from unmelted part of the Si sublayer. As the temperature in this case is close to T_{ma}, the film regrowth begins at very high velocity. Due to intensive evolution of latent heat during growth of three

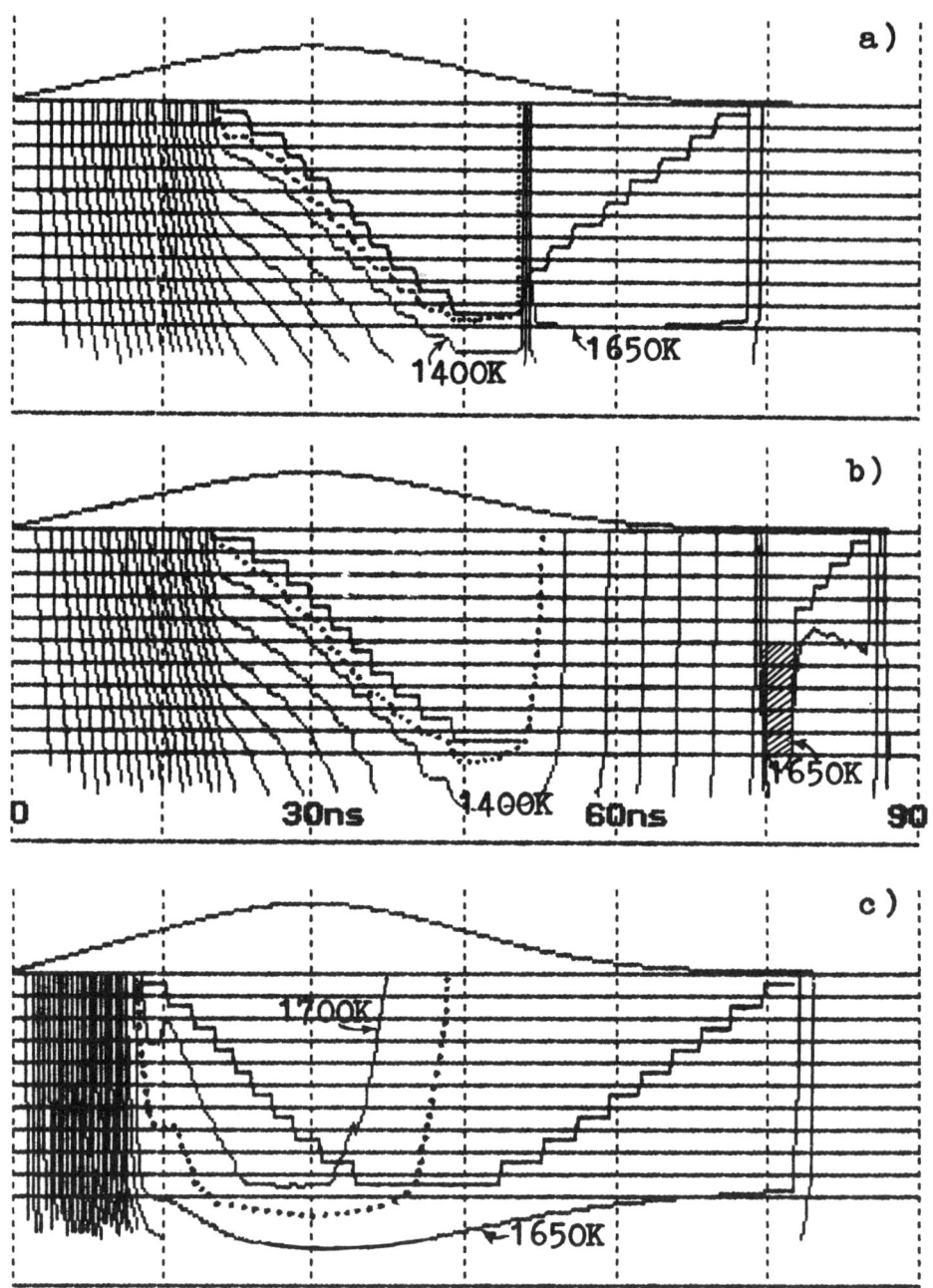

Figure 2. Simulations results of treatment of amorphous (a, b), and of crystalline (c) Si films. Energy densities of the pulses are 220 (a), 224 (b), and $280 mJ/cm^2$ in (c) case.

sublayers at film bottom, the temperature rapidly increases close to T_{mc}. (Several isotherms are not shown here because of their high density.) As a result, growth velocity decreases to about $2m/s$.

In fig.2b, a-Si film melts completely, then cools down to nucleation temperature T_n about $1130°C$. The T_n is determined here by temperature drop velocity, according to the paper [9]. A nucleation takes place in six bottom sublayers, in the sequence beginning from bottom sublayer. Then, a rapid growth of spherical embryos occurred in a dashed rectangular. It should be noted, that latent heat evolution caused by growth of embryos in first bottom layer doesn't prevent at first moment temperature drop of the film, because of: (1) small dimensions of the embryos; (2) relatively low growth velocity at temperature T_n. Then, as embryos size increase, the latent heat release increases, too, and film temperature increases up to T_{mc}. It prevents nucleation in four surface sublayers. Their growth is performed by crystallization front after complete coalescence of the crystallites in six bottom sublayers. It should be noted that, at such a low nucleation temperature ($1130°K$), formation of amorphous phase from the melt is possible, too.

In the last drawing (fig.2c), melting-crystallization process of crystalline Si is shown. (This example presents a process of film treatment by most energetic pulse from sequence of pulses, when a-Si film has been already transformed into crystalline state.) The film is melted up to last Si sublayer followed by film regrowth (after a crossing of interface-line and T_{mc}-isotherm). Film regrowth is controlled by heat dissipation into substrate, and growth velocity is relatively low: it increases from 0 (in bottom of film) to $1.6m/s$ close to its surface.

7. Conclusions

The results of multipulse-scanning excimer-laser crystallization of amorphous Si films on glass substrates are presented. Relatively large-grained poly-Si films are obtained at energy density below threshold value E_m. Grain sizes are $50-80nm$ for a-Si:H films, and $200-500nm$ for a-Si ones.

Crystallization mechanisms of excimer-laser treatments of amorphous Si films are discussed. A main emphasis is made on problems of hydrogen content, of seeding control, and of melt undercooling. The influence of wetting condition on the survival of seeds-islands at film/substrate interface is considered. The issues of this discussions are confirmed by simulation results obtained. It is shown, that multipulse treatment of amorphous Si films provides most equilibrium conditions as far as it is possible during excimer processing.

8. Acknowledgments

The author would like to thank Prof. Givargizov E.I. for helpful discussions.

9. References

1. Usui, S., Sameshima, T. and Hara, M. (1989) The transformation of a-Si:H into polycrystalline silicon by excimer laser irradiation and its application to TFTs, *OPTOELECTRONICS-Devices and Technologies* **4**, 235-248.
2. Zhang, H., Kusumoto, N., Inushima, T., Yamazaki, S. (1992) KrF excimer laser annealed TFT with very high field-effect mobility of $329cm^2/Vs$, *IEEE Electron. Dev. Let.* **13**, 297-299.

3. Brotherton, S.D., McCulloch, D.J., and Edwards, M.J. (1994) Beam shape effects with excimer laser crystallization of plasma enhanced and low pressure chemical vapor deposited amorphous silicon, *Solid State Phenomena* **37-38**, 299-304.
4. Park, W.K., Chae, G.S., Givargizov, E.I., Limanov, A.B., and Kiselev, A.N. (1993) Crystallization of a-Si films on low-melting-point glass substrates, *Mat. Res. Soc. Symp. Proc.* **283**, 751-756.
5. Im, J.S., Kim, H.J. and Thompson, M.O. (1993) Phase transformation mechanisms involved in excimer laser crystallization of amorphous silicon films, *Appl. Phys. Lett.* **63**, 1969-1971.
6. Tsao, J.Y. and Peercy, P.S. (1987) Crystallization instability at the amorphous-silicon/liquid-silicon interface, *Phys. Rev. Lett* **54**, 2782-2785.
7. Thompson, M.O., Galvin, G.J., Mayer, J.W., Peercy, P.S., Poate, J.M., Jacobson, D.C., Cullis, A.G. and Chew, N.G. (1984) Melting temperature and explosive crystallization of amorphous silicon during pulsed laser irradiation, *Phys. Rev. Lett.* **52**, 2360-2363.
8. Stiffler, S.R., Thompson, M.O., and Peercy, P.S. (1991) Transient nucleation following pulsed-laser melting of thin silicon films, *Phys.Rev.B* **43**, 9851-9855.
9. Evans, P.V. and Stiffler, S.R. (1991) Interfacial atomic transport in the nucleation of crystalline silicon from the melt, *Acta. Metall. Mater.* **39**, 2727-2731.
10. Sameshima, T, and Usui, S. (1992) Pulsed laser-induced amorphization of silicon films, *Mat. Res. Soc. Symp. Proc.* **235**, 89-94.
11. Stiffler, S.R., Evans, P.V., and Greer, A.L. (1992) Interfacial transport kinetics during the solidification of silicon, *Acta. Metall. Mater.* **40**, 1617-1622.
12. De Unamuno, S., and Fogarassy, E. (1989) A thermal description of the melting of c- and a-silicon under pulsed excimer lasers, *Appl. Surface Science* **36**, 1-11.

MICROZONE LASER RECRYSTALLIZED POLYSILICON LAYERS ON INSULATOR

A.A. Druzhinin, V.G.Kostur, I.T.Kogut,
I.M.Pankevitch, Y.L. Deschinsky
"Lviv Polytechnika" State University,
"Polysensor" Research-Industry Company, Lviv, Ukraine

The increasing needs in polysilicon films, those are suitable to fabricate the low-cost microelectronic sensors and 3-D IC, stimulate the investigations in order to improve the properties of polysilicon. One of the ways to obtain the high-quality poly-Si layers is connected with the microzone laser recrystallization. This method allows to melt locally poly-Si films, to control both the processes of crystallites growth and defects formation, to change the structure and parameters of semiconductor.

The test patterns were fabricated by the standard IC techniques. They were developed on 0.5 μm thick poly-Si layers being obtained by LPCVD-method at $625^{\circ}C$ on thermally-oxidized p-type silicon plates with (100) orientation. The pattern's doping by boron have been carried out by the ion implantation method. To create the optimal temperature profile in the region of laser irradiation at the films' surface the combined capping layer was formed, that contained 0.5 μm thick SiO_2 layer and 0.15 μm thick Si_3N_4 strips. The recrystallization of polysilicon layers was performed in air by CW YAG laser irradiation ($\lambda=1.06$ μm). In order to optimize the technology of poly-Si on insulating wafers their treatment was carried out with following parameters:

diameter of the molten zone, μm	100 - 150
power, W	15 - 20
beams overlap, %	up to 40
scanning velocity, cm/s	15
substrate temperature, °C	up to 600

The treatment was performed in air by the laser beam with Gaussian distribution of intensity [1,2].

The features of structural changes at lateral epitaxial growth with a seed

The features of the structural changes in polysilicon layers and the seed effect on the lateral epitaxial growth of monocrystalline silicon on SiO_2 during the laser recrystallization have been studied. In our case the regions of poly-Si layers having different dimensions and configurations and being in direct contact with monocrystalline Si substrate were used as seeds.

The obtained results analysis showed that the seed region being affected by the scanning laser irradiation have been molten and epitaxially recrystallized. The effective lateral monocrystalline growth from the seed region to the region of Si-on-insulator substrate was shown to depend on the complanarity of SiO_2 layer as well as on monocrystalline Si wafer. A lot of defects arise at the edge of the seed during the laser recrystallization at the presence of the spikes at "polysilicon-on-monocrystal"-"polysilicon-on-SiO_2" interface region.

The investigation of laser irradiation influence on SOI-structure with a seed and without antireflective layer showed a presence of essential temperature difference between "polysilicon-on-insulator" and "polysilicon-on-monocrystalline-substrate" regions. It was found that irradiation power that provides a good recrystallization of poly-Si layers isn't enough for polysilicon melting through all the depth near the seed region. It's worth noting that laser irradiation power density, being optimized for recrystallization of poly-Si near the seed region, causes the destruction of poly-Si layer on insulator, forming separate agglomerates. The reason of this phenomenon is presence of the heat removal into the substrate near the seed region. In this case during the laser irradiation of the seed the recrystallization front moves from the seed to poly-Si on insulator, i.e. recrystallization is initiated by the seed. However, epitaxial recrystallization with total melting of poly-Si near the seed takes place without continuous formation of a monocrystalline layer. This process doesn't provide the lateral epitaxial growth of a large crystallite from the seed to poly-Si layer on SiO_2.

To carry out the lateral epitaxial growth the polysilicon layer should be melted both near the seed and on SiO_2 surface. However, this can be provided in the limited range of laser irradiation power. To obtain the uniform temperature profile in SOI-structure we used SiO_2/Si_3N_4 layer. It allows to achieve recrystallization conditions those promote melting of the whole poly-Si layer as at the seed regions as on SiO_2 for the same power densities of laser irradiation. It was established that the seed region has been recrystallized by forming the monocrystalline layer without grain boundaries. While the melting zone was moved from the seed to polysilicon-on-insulator a lateral epitaxial growth of crystallites with slope-oriented to the scanning direction has been observed. One should note that poly-Si recrystallization formed the large crystallites when the seeds were parallel to the scanning direction. In that condition the recrystallization front moved from the monocrystalline seed region having no undesirable nucleation centers to the center of the scanning strip. After the laser treatment of SOI structures with few seed regions we have obtained 500x800 μm^2 monocrystalline silicon regions on insulating substrate. Thus, at optimal location of the seed regions one can obtain polysilicon monocrystalline regions on insulating substrate those are suitable to fabricate IC elements and microelectronic sensors of physical values.

The structure influence of recrystallized polysilicon layer on its electrical properties

The electrical properties of poly-Si-on-insulator are defined by the structure of recrystallized material and by impurities and defects. Proceeding from that, an attention has been paid to the influence of these factors on the surface resistance of recrystallized layers [3] as well as on the phenomena in the boundary between recrystallized poly-Si and dielectric, on the transport phenomena in MOS-transistors on the basis of SOI-structures [4,5]. In consequence of the experiments carried out it was found that the electrical properties of recrystallized poly-Si layers and the devices on their basis depend not only on the defects number but mainly on their distribution. In particular, the electrical properties depend substantially on the grain boundaries orientation as related to current direction, on their localization when the recrystallization is carried out using antireflective layer.

The specimens being boron or phosphorous implanted have been investigated by measuring of surface resistance after the laser annealing. Fig.1,a shows the surface resistivity dependences on the power density of laser irradiation for solid phase recrystallization regime. The conductivity increase is caused not only by the crystal enlargement and electrical activation of the implanted impurities during the laser treatment but, probably, by reducing of the recombination centers amount caused by the density of the grain boundaries decrease.

Fig.1, b shows the measured surface resistivity of poly-Si layers as a function of laser irradiation power at liquid-phase recrystallization. In this case the resistance decreases with increasing of laser irradiation power. A sharp drop in resistivity is observed for the highest irradiation powers. One could explain it by the degradation of poly-Si layer that is confirmed by metallographic studies.

The properties of polysilicon layers essentially vary with the type of the grain boundaries, their interaction with background and dopant impurities. These differences are caused by the electrically active centers being created by some types of the grain boundaries. These centers form the localized states at the forbidden band and create the potential barriers those define the recombination, trapping and drift of the carriers. That's why it's interesting to measure the life time of the non-equilibrium carriers as a controlled parameter that is very sensitive to structural defects, including the grain boundaries in recrystallized poly-Si layer.

Recombination properties of both the initial and laser-recrystallized layers on insulating substrates have been investigated. The recombination was studied by the life time measurement using the (τ) phase method and 3 cm range microwave frequency equipment. An excitement of non-equilibrium conduction was made by sinusoidal-modulated red beam of He-Ne laser ($\lambda=0.64$ μm, output power 10 mW). The intensity of photoexcitement was changed by the neutral light filters.

Thus, it was established on the basis of investigations carried out that the LPCVD-method obtained initial poly-Si layers are the fine-grained ones having the life time of non-equilibrium carriers of 10 μs and relatively small barriers at the grain boundaries. After the laser recrystallization the life time decreases down to several μs and the potential barriers at the enlarged grains boundaries essentially

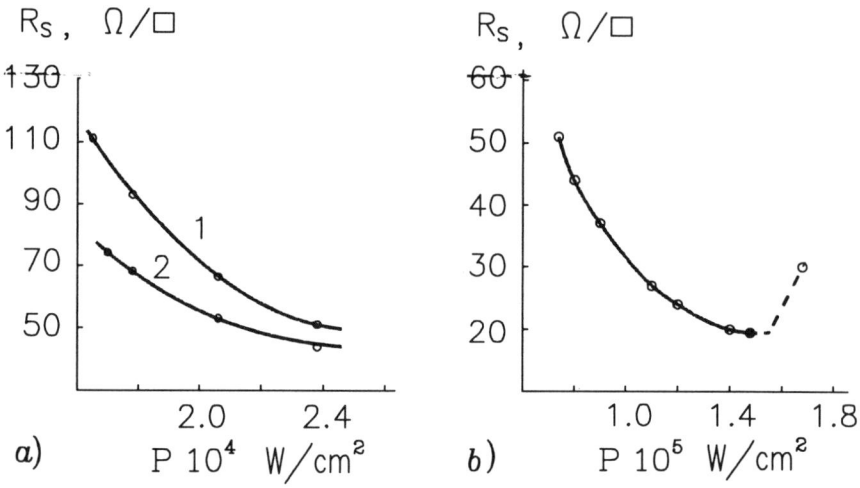

Fig.1. Polysilicon layer surface resistivity versus laser irradiation power density dependence for solid phase (a) [1-B^+,80keV, $7.5 \cdot 10^{15}$cm^{-2}; R_{S_0} =117.6Ω/□; 2-B^+,80keV, $1.1 \cdot 10^{16}$cm^{-2} ; R_{S_0} =79.5Ω/□] and liquid phase (b) [P^+,100keV; $1.5 \cdot 10^{16}$cm^{-2}; R_{S_0} =60 Ω/□] recrystallizations.

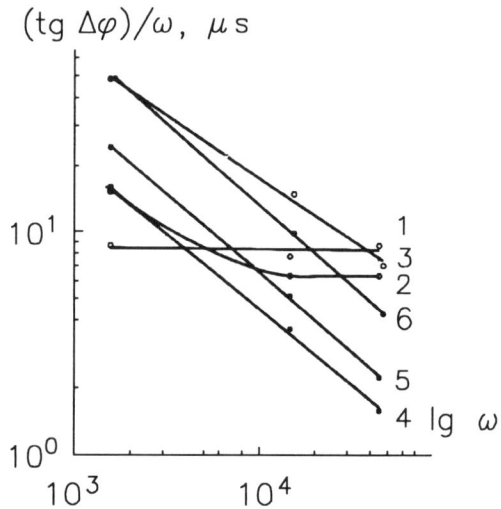

Fig.2. Frequency dependences of $(tg\,\Delta\varphi)/\omega$, at different levels of photoexcitation for initial polysilicon layer (1-3) and after laser recrystallization (4-6); 1,4-high excitation level; 2,5-intermediate level; 3,6-low level.

decrease. A dominant influence of the grain boundaries on non-equilibrium carriers recombination is confirmed by the tg($\Delta\phi/\omega$)=f(ω) dependences (Fig.2, curves 4-6). The life time reduction after recrystallization should be explained by the thermal stresses and the annealed defects appearence as well as by the electrically active grain boundaries formation during the laser irradiation. The fast cooling of the patterns after recrystallization could cause the new grain boundaries those should have the high densities of the boundary stated because of the non-equilibrium conditions of their formation. However, in the patterns having practically no the grain boundaries the effective life times of non-equilibrium carriers increase essentially.

The Hall-effect measurements showed the reduced carrier densities comparing to calculated values. This discrepancy should be explained by the partial trapping of the carriers at the defects of crystalline structure, at the grain boundaries mainly. In the annealed patterns the carrier densities increase. This effect is assumed to be due to the passivation and diminution of the trapping centers after the recrystallization.

To explain the carrier transport phenomena in polysilicon films the model has been proposed on the basis of τ-approach, taking into account the influence of crystallites size and layers thickness.

Therefore, the use of the laser recrystallization of poly-Si layers on insulating substrates to develop the IC elements and microelectronic sensors could be successfull in the case of essential reduction of the grain boundaries or of their defined localization in order to control their densities and distribution by the laser treatment.

The results of the investigations were applied to develop the microelectronic devices including those on the basis of SOI-structures.

References

1. Voronin V.A., Druzhinin A.A., Kostur V.G. et.al. (1986) Laser-induced epitaxial growth of silicon layers on insulating substrates, *Proc. 1st European Conf. on Crystal Growth, Budapest, Hungary, Apr. 1-7, 1990. Crystal Prop. Prep.* (1991), **32-34**, 83-88.
2. Druzhinin A.A., Kostur V.G., Lyba O.M. Structure changes and crystallization process peculiarities of polysilicon layers under laser-irradiation effect, *Proc. 3rd European Conf. on Crystal Growth, Budapest, Hungary, May 5-11, 1991. Crystal Prop. Prep. (1991),* **36-38**, p. 388-393.
3. Drushinin A.A., Kogut I.T., Kostur V.G. et al. Structural changes in capsulated polysilicon layers under influence of scanning laser irradiation, *Fizika i Himia Obrabotki Materialov.* (1992), Nr.3, 38-43.
4. Drushinin A.A., Kenio G.V., Kostur V.G. Polysilicon n-p-n-structures under influence of laser irradiation, *Electronnaya Promyshlennost* (1988), Nr.57, 21-22.
5. Druzhinin A.A., Kenio G.V., Kogut I.T. et al. Physical model of SOI MOS-transistor for non-equilibrium state, *Elektronnaya Technika, ser.3. Microelectronica* (1990), Nr.5 (139), 89-91.

Section 2:
SOI Materials Characterization Techniques

ELECTRICAL CHARACTERIZATION TECHNIQUES FOR SILICON ON INSULATOR MATERIALS AND DEVICES

SORIN CRISTOLOVEANU

Laboratoire de Physique des Composants à Semiconducteurs
(UA-CNRS 840 & INPG)
ENSERG, BP 257, 38016 Grenoble Cedex 1, France

1. Introduction

SOI technology has reached a stage of rapid and successful development. A major condition for competing with bulk silicon in commercial applications is the degree of understanding of material properties and device operation. This implies the use of adequate characterization techniques which overcome the difficulties induced by the thinness of the film and by the stacked interfaces. Since the quality of SOI structures has great impact on the performance of circuits integrated on them, device-based characterization is very suitable for exploring the properties of the starting material.

The discussion of electrical characterization methods will proceed systematically from the wafer to the device. The *pseudo*–MOS transistor is suitable for *in-situ* inspection of the quality of as-grown SOI wafers. More or less conventional transport measurements can be used for determining the carrier concentration and mobility in thin films. The MOS capacitance and conductance techniques are far more difficult to implement in SOI, where both the measurement and the parameter extraction are complicated by the contributions of several interfaces. Current-based techniques are more appropriate for thin films: the current is easily measurable even in small area devices, the experiment can be conducted separately from the front or the back interface, and simple single-interface models often are sufficient for data analysis. A brief survey of the experimental set-up, appropriate models, and parameter extraction will be made for MOS–Hall profiling, charge pumping, low-frequency noise spectroscopy, and Zerbst-like drain current transients.

2. Wafer Screening by Ψ–MOSFET

The pseudo–MOS transistor (Ψ–MOSFET) is a simple technique which takes advantage of the specific configuration of SOI and has the potential of being non-destructive. The Ψ–MOSFET is based on the upside-down MOS structure that is inherent in all SOI materials. Figure 1 shows that the bulk Si substrate acts as a gate terminal and can be biased, through the metal support, to induce a conduction channel at the interface. The buried oxide plays the role of a gate oxide and the Si film represents the transistor body. To operate *in situ* (without lithography and metallization) the Ψ–MOSFET, low pressure probes are placed on the film and form source and drain point contacts [1].

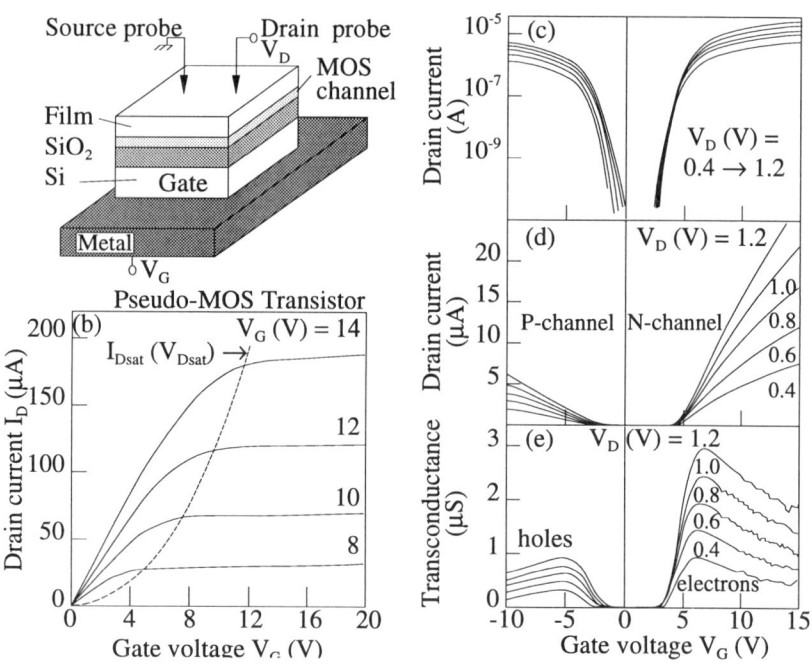

Figure 1. Schematic of the pseudo-MOS transistor and typical characteristics.

In spite of the device simplicity, very pure MOSFET–like characteristics are produced (Fig.1). The output $I_D(V_D)$ curves exhibit a nearly ideal behavior up to 20 V. According to the positive or negative bias applied on the gate, accumulation or inversion channels are activated at the interface. Typical subthreshold $I_D(V_G)$ characteristics for n- and p-channels, strong inversion and strong accumulation curves, as well as the corresponding transconductance characteristics are given in Fig.1. From these diagrams, many useful properties may be determined for electrons and holes.

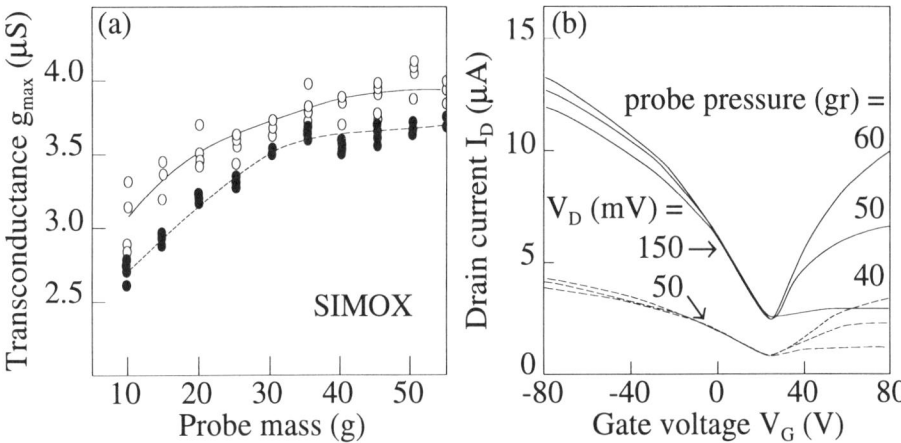

Figure 2. Influence of the probe mass (pressure) on (a) the transconductance peak, $g_{max} = f_g \mu_0 C_{ox} V_D$ and on (b) the $I_D(V_G)$ characteristics.

The basic setup is composed of any two-probe system, connected to a HP–4145 or equivalent pico-ameter [2]. However, for calibration purposes, it is more suitable to use a four-point prober which frequently serves for resistivity measurements [3]. Increasing the probe *pressure* gradually reduces the series resistances and provides more accurate carrier mobility values. In 200 nm thick fully depleted films, it was found that the transconductance improves by 35 % as the pressure increases from 10 g to 30 g, and then saturates (Fig.2a) [1].

Larger pressures are needed in thicker or more heavily doped films which cannot be totally depleted by applying a substrate bias. Figure 2b shows that the influence of the pressure is quite substantial in inversion ($V_G > 0$), where the tip contact to the n-channel is achieved through a p-doped overlay and a depletion zone. In contrast, the contact to the accumulation p-channel is direct and the pressure effect is much less relevant. The *spacing* between source and drain probes is not critical.

The accurate modeling of the Ψ–MOSFET basically requires the solutions of the Poisson and Gauss equations in the film and in the underlying substrate. In the simpler case of *fully depleted* films, the drain current is governed by either the inversion or the accumulation channel and the intercepts with V_G-axis roughly correspond to the threshold voltage V_T and flat-band voltage V_{FB}:

$$I_D = f_g C_{ox} \frac{\mu_0}{1 + \theta(V_G - V_{T,FB})}(V_G - V_{T,FB})V_D \qquad (1)$$

where μ_0 is the electron/hole mobility and $\theta = \theta^0 + f_g\mu_0 C_{ox} R_{SD} \gg \theta^0$ is the mobility attenuation factor. The ideal value, θ^0, is proportional to C_{ox} and can be very small ($\theta^0 \leq 0.01$ V^{-1}) for fully processed back channel SOI MOSFETs. In the Ψ–MOSFET, the Schottky nature of the probe contacts results in significant series resistances and larger θ values ($\theta \geq 0.05$ V^{-1}).

In a highly doped p-type film which is *partially depleted*, the total current is the sum of the inversion/accumulation channel current given by Eq.(1) and the volume current. As the depletion region shrinks, the volume current increases *linearly* with gate voltage

$$I_D = q f_g \mu_0 N_A V_D (t_{si} - w_d) = f_g \mu_0 C_{ox} V_D (V_0 - V_G) \qquad (2)$$

where w_d is the depletion width and V_0 is the fictive voltage that would lead to full depletion.

The *geometric coefficient* f_g accounts for non-parallel current lines since it is not possible to define the width and length of the Ψ–MOSFET. An empirical value, $f_g \simeq 0.75$, was determined by comparing Ψ–MOSFET data with four-point probe experiments performed with the same system [1].

The *threshold and flat-band* voltages are determined from $I_D(V_G)$ characteristics. The inversion region is discriminated from the accumulation region by a larger sensitivity to probe pressure. In order to cancel first order series resistance effects, the values of μ_0 and $V_{T,FB}$ are extracted by plotting the function $I_D/\sqrt{g_m}$ versus V_G. The slope, $\sqrt{f_g \mu_0 C_{ox} V_D}$, yields the mobility for electrons or holes, and the intercept gives V_T or V_{FB}.

A simple look taken at the Ψ–MOSFET $I_D(V_G)$ characteristics of Figs.1 and 2 is enough to detect whether the film behaves as fully or partially depleted. *Doping levels* above $5 \times 10^{15} cm^{-3}$ are quite precisely determined. The characterization of *fixed charges* in the buried oxide proceeds from the analysis of the flat-band voltage. The density of *interface traps* is determined from the subthreshold swing, $S = (kT/q)(1 + C_d/C_{ox} + qD_{it}/C_{ox})$, or from the difference between V_T and V_{FB}.

3. Transport Measurements

This section is dedicated to conventional measurements that provide basic transport parameters such as resistivity, carrier mobility and concentration, scattering process, etc. Although these methods are routinely used in bulk Si, they must carefully be reconsidered for application to thin SOI films. In response to intrinsic problems (large values of the sheet resistance, full depletion, in-depth inhomogeneity) and new opportunities (substrate bias influence), more refined experimental set-up and modeling are necessary.

3.1. FOUR-POINT PROBE

The *4–point probe* method is used for rapid inspection of wafer resistivity when the "volume" of the material is not accessible and only the surface may be probed. Two probes are dedicated to current injection and two other probes measure the voltage drop. This discrimination is necessitated by the existence of parasitic resistances at the metal–semiconductor contact.

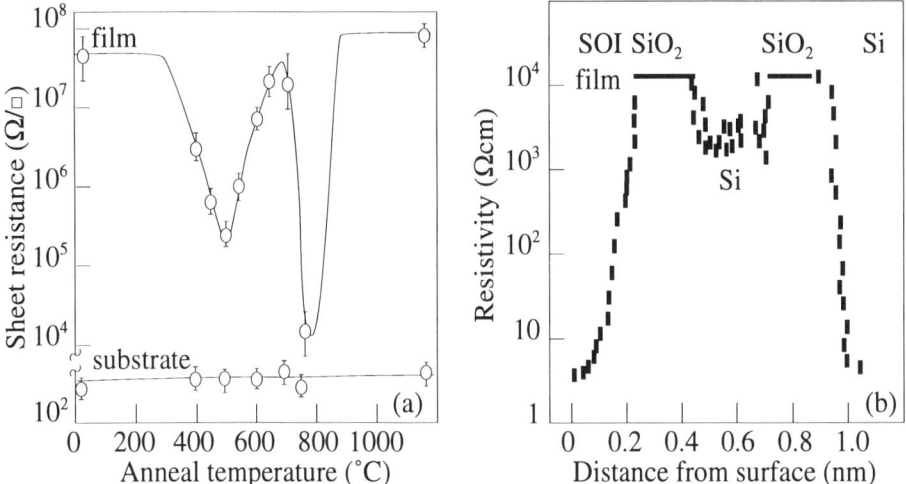

Figure 3. (a) Four-point probe sheet resistance variation in SIMOX film and substrate against anneal temperature, and (b) spreading resistance profile in a double SIMOX structure.

The most frequent set-up consists of 4 collinear and equidistant probes. The outer probes are carrying the current, whereas the voltage is sensed between the inner probes. The sheet resistance $R_\square$ and the *average* resistivity ρ are given by

$$R_\square = \frac{\rho}{t_{si}} = 4.53 \frac{V_{23}}{I_{14}} \qquad (3)$$

In undoped SOI films, the sheet resistance may be extremely large: $R_\square \geq 10^9 \, \Omega/\square$, in a 0.1 μm layer of resistivity 10^2–$10^4 \, \Omega$cm. The requirement for a high impedance detection system may be met by using a differential voltmeter connected to each terminal via Keithley electrometers.

Four-point probe measurements are currently performed on as-grown SOI structures to control the degree of contamination, the homogeneity across the wafer and the dopant activation. Figure 3a illustrates the generation of oxygen-related donors in early SIMOX material [4]. The sheet resistance of the film can drop by orders of magnitude after short anneals at 500 °C and 750 °C. This verifies the presence of many oxygen atoms in the Si film subsequent to the implantation process.

3.2. SPREADING RESISTANCE

In films that show in-depth inhomogeneities of doping and/or mobility, the resistivity profile is determined by spreading resistance. The basic idea is to scan laterally the resistance of a beveled sample and then to convert the data into a vertical profile.

A low angle ($< 15'$–$3°$) bevel is prepared by mounting the sample on a bevel block and lapping it with diamond paste. Two tungsten probes are moved along the bevel, by steps of a few micrometers, and the resistivity is measured at each location. A vertical resolution of about 35 nm is achieved by 2 μm steps along a 1° bevel. Higher resolutions are needed in very thin SOI films (50–100 nm). The spreading resistance R_{SR} is given by

$$R_{SR} = \frac{\rho}{\alpha r} \quad (4)$$

where r is the probe contact radius and α is a shape-dependent coefficient. The resistivity profile is deconvolved from the resistance values measured at each sample depth.

Figure 3b illustrates the resistivity profile in a double SIMOX structure formed by two successive oxygen implants at different energies. Two distinct oxides, separated by a Si region, are clearly visible. The sandwiched Si layer has poorer quality and higher resistivity than the top Si film. Spreading resistance measurements are currently being dedicated to the analysis of the activation and redistribution of implanted dopants in SOI structures.

3.3. VAN DER PAUW MEASUREMENTS

Van der Pauw has extended the Hall effect theory to samples of arbitrary shapes [5]. For *resistivity* measurements, the current is injected between adjacent contacts and the pseudo-resistances $R_{12,34} = V_{12}/I_{34}$ and $R_{23,41}$ are evaluated. Two measurements are performed with the roles of the contacts being shifted by a quarter of tour.

$$\rho_0 = \frac{\pi t_{si}}{\ln 2} \times \frac{R_{12,34} + R_{23,41}}{2} \times f \quad (5)$$

where f is a numerical coefficient related to the ratio $R_{12,34}/R_{23,41}$.

The *Hall coefficient* R_H is determined by injecting the current through diagonally opposite contacts and applying a magnetic field

$$R_H = \frac{t_{si}}{B} \times \frac{R_{13,24}(+B) - R_{13,24}(-B)}{2} \quad (6)$$

The Hall mobility is given by $\mu_H = R_H/\rho_B$. When the carrier energy distribution is taken into consideration, the relaxation time $\tau(\epsilon)$ between collisions must be averaged over the energy range, yielding

$$\mu_H = r_H \mu_0 \qquad R_H = \frac{r_H}{qn} \quad (7)$$

where the Hall scattering factor r_H is governed by collisions with acoustic phonons at high temperature ($r_H \simeq 1.18$) and ionized impurities at low temperature ($r_H \simeq 1.93$).

A major hypothesis of van der Pauw's theory is the presence of ideal point contacts, located at the sample circumference, over the whole thickness. If the size of the contacts and/or their distance to the sample edge is not negligible compared with the sample perimeter, corrections factors are needed [3].

The finality of Hall effect measurements is to monitor the optimization of SOI materials which is achieved by increasing the carrier mobility and suppressing contamination sources. Figure 4a shows the relation between mobility and doping observed in early SOS films [6]. A mobility degradation in thinner layers was systematically reported and attributed to numerous defects that subsist in the bottom of the film [7].

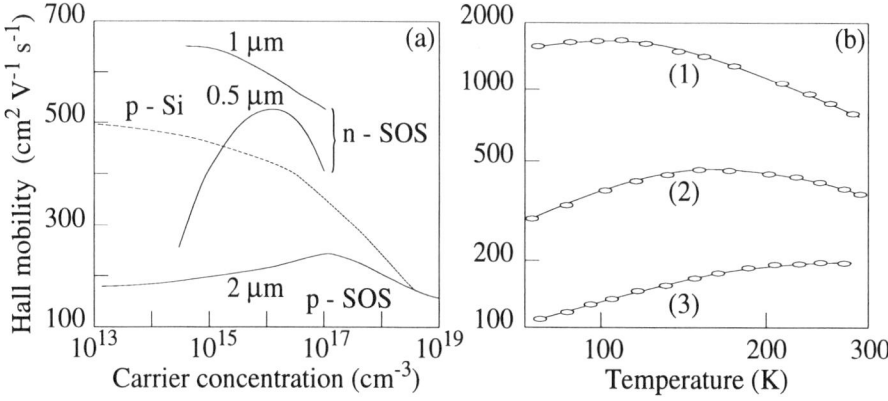

Figure 4. Hall effect mobility (a) versus doping in SOS films and (b) versus temperature in SIMOX films.

The activation energy of residual or dopant impurities is evaluated from Arrhenius plots, $\ln nT^{-1.5}$ vs. $1/T$. Hall effect "spectroscopy" may be used to enhance the resolution. It consists in taking the first or second order derivative of the carrier concentration with respect to the reciprocal temperature [9].

3.4. PHOTOCONDUCTIVITY

A variety of characterization methods are using light as an additional experimental parameter. The illumination wavelength is selected such that the photon energy exceeds the band-gap and causes carrier generation. In general, the absorption coefficient α is very large compared to the thickness of SOI films ($\alpha \gg 1/t_{si}$) and the sample is uniformly illuminated. The excess photoconductivity $\Delta\sigma_{pc}$ is expressed as

$$\Delta\sigma_{pc} \simeq \frac{q\alpha\phi_0}{t_{si}} \int_0^{t_{si}} (\tau_n\mu_n + \tau_p\mu_p)\,dz \qquad (8)$$

where $\tau_{n,p}$ are *recombination* lifetimes for electrons and holes in a n-type film and ϕ_0 is the photon flux. Only a small fraction of photons are absorbed. A global *photoconductivity lifetime* τ_{pc} is defined to overcome the difficulty raised by the independent variations of $\mu_{n,p}$ and $\tau_{n,p}$ with temperature:

$$\tau_{pc} = \frac{1}{q\alpha\phi_0} \times \frac{\Delta\sigma_{pc}}{\mu_H} \qquad (9)$$

where the temperature dependence of the Hall mobility is determined from van der Pauw measurements under darkness.

There are many other types of experiments based on photoconductivity principles.
 - The photoconductivity decay after exposure to a short laser pulse is monitored to determine the recombination lifetime.
 - The time-of-flight technique consists in applying a narrow light impulse near the terminal of a biased rectangular sample. The distribution of the photogenerated minority carriers is sensed by a p-n junction situated at the opposite terminal. The time delay between carrier injection and collection yields the mobility of minority carriers. The widened shape of the collected pulse is a measure of the diffusion coefficient, whereas the reduction of the pulse area accounts for the recombination lifetime.
 - The photo-magneto-electric (PME) effect combines light and magnetic field actions. This gives rise to a longitudinal short-circuit current or open-circuit voltage, from which the carrier lifetime can be calculated.

- The scope of Photo-Induced Current Transient Spectroscopy (PICTS) is to analyze the deep traps before any device processing [10].

4. Capacitance and Conductance Techniques

The capacitance and conductance techniques, frequently used for extracting the parameters of Si-SiO$_2$ interface in bulk silicon MOS systems, can still be applied to the SOI capacitor structures. The conventional MOS capacitor theory is modified to account for the top and bottom Si/SiO$_2$ interfaces of the buried oxide. The film is biased through a metal contact (gate) and the back of the wafer is grounded, The total capacitance C_T of this semiconductor-insulator-semiconductor (SIS) n-i-n capacitor is expressed by [11, 12]

$$\frac{1}{C_T} = \left(\frac{dQ_T}{dV_G}\right)^{-1} = \frac{1}{C_{s2} + C_{it2}} + \frac{1}{C_{ox2}} + \frac{1}{C_{s3} + C_{it3}} \tag{10}$$

where $C_s = -dQ_s/d\psi_s$ is the surface capacitance, $C_{it} = -dQ_{it}/d\psi_s$ is the static capacitance of the interface trap, and C_{ox2} is the buried oxide capacitance. We consider the case of a negative gate bias which tends to accumulate the upper interface and deplete the bottom interface. It is appropriate here to define a coupling factor K_2

$$K_2 \equiv \frac{d\psi_{s2}}{d\psi_{s3}} = -\frac{C_{s3} + C_{it3}}{C_{s2} + C_{it2}} \tag{11}$$

which yields

$$\frac{1}{C_T} = \frac{1}{C_{ox2}} + \frac{1 - K_2}{C_{s3} + C_{it3}} \tag{12}$$

According to Eq.(12), the equivalent circuit of a SIS capacitor reduces to a MOS-like equivalent circuit. For $K_2 = 0$, there is no charge coupling and the equivalent circuit for a SIS capacitor is identical to that of a conventional MOS capacitor. The interface trap capacitance is derived from "low-frequency" measurements

$$C_{it_3}(V_G) = (1 - K_2)\left(\frac{1}{C_{LF}} - \frac{1}{C_{ox_2}}\right)^{-1} - C_{s_3}(V_G) \tag{13}$$

The *conductance* method can also be employed for determining the interface trap density in a SIS capacitor operated in the depletion region [11]. The parallel branch of the equivalent circuit is converted into a frequency-dependent capacitance C_p and conductance G_p

$$C_p = (1 - K_2)^{-1}\left(C_{s3} + \frac{C_{it_3}}{1 + \omega^2\tau_{it}^2}\right) \qquad \frac{G_p}{\omega} = (1 - K_2)^{-1}\left(\frac{\omega\tau_{it}C_{it_3}}{1 + \omega^2\tau_{it}^2}\right) \tag{14}$$

The above expressions are for a single level interface trap and can be adapted to a continuum of traps throughout the band gap.

Prior to extracting the interface properties, we need to know the SIS capacitor parameters such as buried oxide thickness, doping concentration, and fixed oxide charge. The buried oxide thickness t_{ox_2} can be determined from the measured maximum high-frequency capacitance C_{HF}^{max} [2]. Then, the doping concentrations in film and substrate are calculated from the minimum high-frequency capacitance regardless of the interface trap density. The fixed oxide charge densities (Q_{f_2} and Q_{f_3}) are determined from the voltage shift of the measured high frequency C-V curve by comparing it with the ideal C-V curve. Applying a gate bias to accumulate one interface yields the fixed oxide charge at the other interface. The strechout and shift in the C-V curves, observed at positive gate biases, are attributed to an increase of interface trap density and fixed oxide charge density at the upper oxide interface.

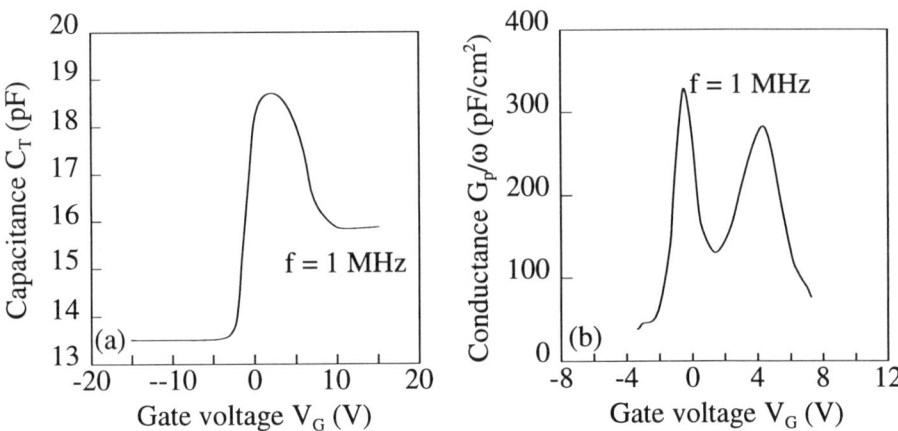

Figure 5. (a) High-frequency capacitance and conductance curves for a typical SIMOX SIS capacitor.

However, this method is less accurate if the film becomes fully depleted before reaching the minimum capacitance. In this case, the substrate doping density N_{D_3} can be deduced from

$$N_{D_3}(w_3) = -2\left\{q\epsilon_{si}\frac{d}{dV_G}\left[\frac{1}{(1-K_2)C_{HF}^2}\right]\right\}^{-1} \qquad (15)$$

where w_3 is the distance from the substrate/buried oxide interface. Since precise value of $(1 - K_2)$ requires the prior evaluation of the doping concentrations, an iteration procedure becomes necessary.

5. Profiling the Vertical Inhomogeneities

The properties of thin SOI films may vary with distance from the top surface to the buried insulator. Such a *vertical inhomogeneity* is often due to enhanced defect generation that takes place near the film–insulator interface during material synthesis. In non-uniform films, the average values of resistivity, mobility and autodoping are rather meaningless, since they may look very poor, even if the properties of the top portion of the layer reaches bulk Si standards. Moreover, contrasting and misleading average parameters may be obtained in SOI layers with identical quality but different thickness. A very efficient method, described hereafter, combines the MOS field effect with magnetoelectric measurements.

Figure 6a shows the configuration of an MOS-Hall device, which is nothing but a 7-terminal depletion-mode n-channel MOSFET. Besides the gate, source and drain, there are 4 small lateral N^+ contacts. The longitudinal voltage drop, measured between contacts $H_{1,2}$ (or $H_{3,4}$), gives the conductance and magnetoresistance values, free of the influence of source/drain junctions. Contacts $H_{2,4}$ (or $H_{1,3}$) serve for Hall voltage measurements.

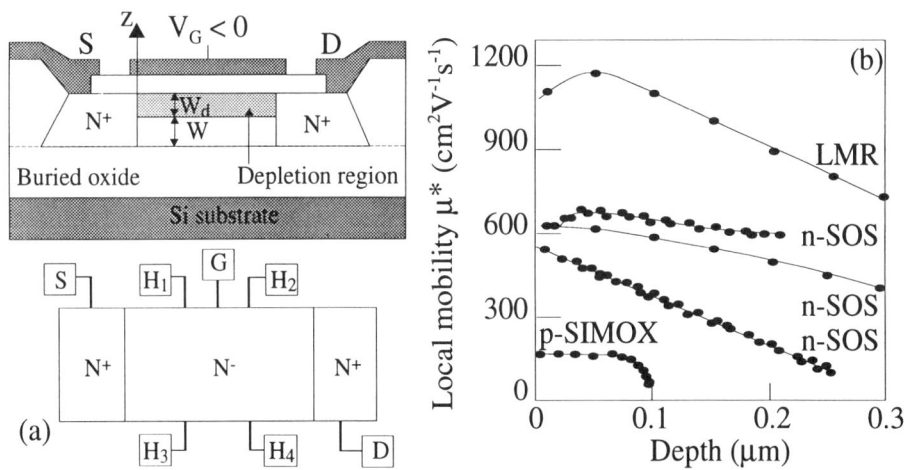

Figure 6. (a) Configuration of a seven-terminal MOS-Hall device and (b) carrier mobility profiles (after Lee *et al*).

Hall effect and magneto-conductance measurements are performed at low drain bias as a function of V_G. By gradually depleting the film, the thickness w of the conducting region is reduced and the measured (average) transport properties are modified. A procedure involving differentiation provides the contribution of the infinitesimal layer Δw, situated at the limit between depletion and "active" regions.

The gate voltage dependence of the depletion depth w_d is determined from small-signal $C(V_G)$ capacitance measurements. The vertical profile of a parameter A consists of successive local values $A^*(z)$ which are extracted from the experimental average values $\overline{A}(V_G)$. For example, the average resistance (or magneto-resistance) of a conductive sheet of thickness w is

$$\overline{R}(V_G) = \frac{l_x}{l_y} \left(\int_0^w \sigma^*(z) dz \right)^{-1} \tag{16}$$

Conversely, the local conductivity at $z = w$ is given by

$$\sigma^*(z = w) = \frac{l_x}{l_y} \frac{d\overline{R}^{-1}}{dV_G} \frac{dV_G}{dw} \tag{17}$$

The measured average Hall mobility, defined in Eq.(7), is expressed as [2]

$$\overline{\mu}_H(V_G) = \left(\int_0^w \mu_H^*(z) \sigma_0^*(z) dz \right) \left(\int_0^w \sigma_0^*(z) dz \right)^{-1} \tag{18}$$

where $\sigma_0^*(z)$ is the conductivity profile at zero magnetic field. Differentiating Eq.(18) and using Eq.(17) yields for the local Hall mobility

$$\mu_H^*(w) = \frac{l_x}{l_y} \frac{1}{\sigma_0^*} \frac{d\left(\overline{\mu}_H \overline{R}_0^{-1}\right)}{dV_G} \frac{dV_G}{dw} = \overline{\mu}_H - \overline{R}_0(V_G) \frac{d\overline{\mu}_H}{d\overline{R}_0} \tag{19}$$

The profiles of $\sigma^*(w)$ and $\mu_H^*(w)$ are free of any theoretical assumption except the validity of capacitance-depth curves.

Most experiments have been conducted on early SOS and SIMOX films, where the material homogeneity used to be a critical technology issue [13, 14]. Several results are collected in Fig.6b. The mobility decreases almost linearly with depth in SOS, whereas, in low temperature annealed SIMOX, it is flat in the upper half of the film and then degrades rapidly close to the buried interface.

6. Charge Pumping Technique

Charge pumping (CP) is a very sensitive method for the characterization of low concentrations of interface traps (10^9 cm^{-2}eV^{-1}) in short-channel MOS devices. The adaptation of CP to SOI transistors requires a contact to the Si film. Either 5-terminal MOSFET's or gate-controlled p-i-n diodes may be used [15]. The two terminals of the P$^+$P$^-$N$^+$ diode have different functions. The N$^+$ contact controls the charge of the inversion layer, whereas the P$^+$ terminal supplies the majority carriers.

The principle of conventional CP in enhancement-mode MOSFET's on bulk Si is to repeatedly switch the gate from inversion to accumulation and *vice versa*, while keeping the source and drain contacts grounded or slightly reverse biased. In inversion, some of the minority carriers provided by the source and drain reservoirs are trapped on the interface states. During the falling edge of the pulse, the mobile minority carriers are collected rapidly from the inversion layer by the source and drain, and then the trapped carriers recombine with majority carriers provided by the substrate. This recombination gives rise to an *average* charge pumping current I_{CP} in the substrate terminal, which is a frequency-amplified measure of the density of interface states.

The CP current is proportional to the average concentration of interface traps D_{it}, frequency f, and surface potential range $\Delta\Psi_s$ swept through during the pulse

$$I_{CP} = q^2 f W L D_{it} \Delta\Psi_s \qquad (20)$$

By including into the model the direct carrier emission to the conduction and valence bands, Groeseneken *et al* [16] found for trapezoidal pulses

$$\Delta\Psi_s = \frac{2kT}{q} \ln\left[v_{th} n_i \sqrt{\sigma_n \sigma_p} \sqrt{t_r t_f} \frac{|V_{FB} - V_T|}{|\Delta V_G|}\right] \qquad (21)$$

where v_{th} is the thermal velocity of carriers, $\sigma_{n,p}$ are the capture cross sections, and $t_{r,f}$ are the rise and fall times of the ΔV_G pulse.

A very meaningful CP curve (Fig.7a) is obtained by varying the bottom level V_{GL} of the trapezoidal pulse while keeping the pulse magnitude ΔV_G and frequency fixed. Three different regions can be distinguished on this *rectangle-like CP signature*: (i) as long as the top level of the pulse does not induce carrier inversion, there is no trapping and $I_{CP} \simeq 0$, (ii) a maximum CP current, given by Eqs.(20) and (21), is obtained when the top level exceeds V_T, and (iii) for higher V_{GL} values ($V_{GL} > V_{FB}$) the surface does not return in accumulation, hence the recombination rate and I_{CP} vanish. It follows that the left and right hand edges of the "rectangle" correspond to $(V_T - \Delta V_G)$ and V_{FB}, respectively, whereas the plateau level gives the trap concentration D_{it}.

We will focus on the specifics of the technique in SOI. In the inset of Fig.7a, the CP current was measured by applying constant pulses on the front gate and varying V_{G2}. A sharp peak is observed in the region where the back interface is *depleted*. This peak indicates that pulsing the front gate results in a scanning of the back surface potential which enables the pumping of some of the back interface traps. Another consequence of the interface coupling is that the values of the threshold voltage, flat band voltage and swept potential range $\Delta\Psi_{s1}$ depend on the back gate bias.

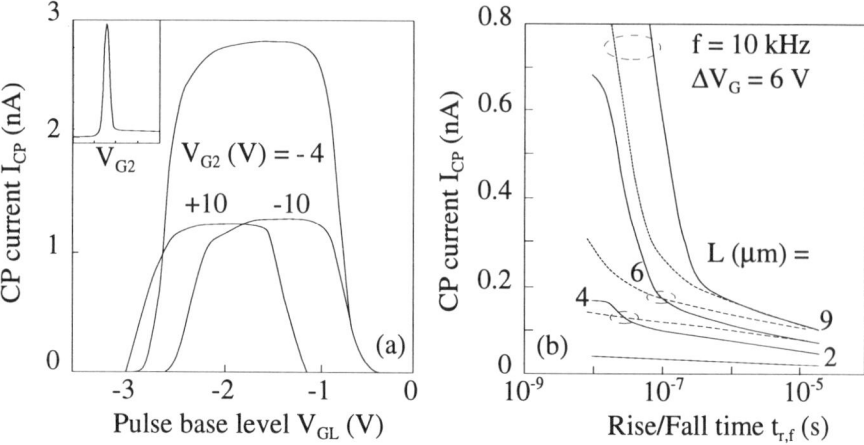

Figure 7. Front gate CP current versus (a) pulse base level for various back gate voltages, and (b) versus rise and fall times ($t_r = t_f$) for various diode lengths, with the back interface in inversion (——) or in accumulation (- - -) (after Ouisse).

This is the reason for the small difference existing between the CP currents measured for inversion and accumulation at the back interface.

Shown in Fig.7b is the modification of the "rectangle–like" CP curve as the back interface goes from inversion to depletion and accumulation. The shifts of the left and right edges are used to monitor the variations of V_{T1} and V_{FB1} versus V_{G2}. It is noted that for the front interface traps to be accurately characterized, the back gate has to be maintained in strong inversion or accumulation.

Charge pumping depends on several physical time constants (carrier transit times for the formation of the accumulation and inversion channels, carrier lifetime, trapping and emission times) as well as on experimental time constants (period, rise and fall times). The latter parameters have to be longer than the former ones, in order to prevent any source of parasitic recombination. For instance, if all mobile minority carriers are not removed rapidly enough (during the fall of the pulse) before the arrival of majority carriers, an undesirable recombination occurs. This parasitic component of I_{CP}, referred to as a "dimensional effect" [15], is responsible for an overestimation of D_{it}, It obviously happens in long devices (the time for channel formation being proportional to L^2) or if the rise/fall times are too short.

SOI structures are more subjected to dimensional effects because electron or hole reservoirs may exist at the back interface and interact with front channel carriers. The range of confidence for the CP parameters is established by drawing $I_{CP}(\ln t_{r,f})$ curves which, according to Eqs.(20) and (21), must be linear. The onset of a non-linearity is evocative for the onset of dimensional effects (Fig.7b). The most critical case, $V_{G2} > 0$, corresponds to a difficulty in evacuating the holes which are less mobile. However, Fig.7b shows that the rise/fall times may actually be small enough: 1 μs for $L = 10$ μm, 100 ns for $L = 4$ μm, 10 ns for $L \leq 2$ μm.

7. Low Frequency Noise

The analysis of the electrical noise is important in two respects: (i) it provides a practical limit for the operation of integrated circuits, and (ii) it represents a powerful tool for defect identification. The noise is defined as the power spectral density $S_x(f)$ of a time-dependent, random function $X(t)$, and it is given by the Fourier transform of the auto-correlation function of $X(t)$ [17]. In a semiconductor device, the noise reflects the current fluctuations brought about by random variations in the carrier mobility and/or concentration. Each physical mechanism involved in current fluctuations is expressed by a specific random distribution (Gaussian, Poisson, etc) and has, in general, a distinct noise spectrum.

There is strong evidence that the $1/f$ noise of MOS transistors originates from fluctuations in the carrier number. The basic mechanism is the carrier tunneling between the inversion channel and slow oxide traps N_t located at 1–2 nm from the Si–SiO$_2$ interface. The equivalent drain current noise is expressed as [18]

$$\frac{S_{I_D}}{I_D^2} = \frac{q^4 \lambda_{ox}}{\alpha^2 kTLW} \times \frac{N_t}{(C_{ox} + C_d + qD_{it} + C_{inv})^2} \times \frac{1}{f} \qquad (22)$$

where λ_{ox} is an average tunneling length. The concentrations of slow traps N_t and interface traps D_{it} may be correlated by a first order approximation $D_{it} \simeq \lambda_{ox} N_t$. The inversion layer capacitance is $C_{inv} = (q/\alpha kT)Q_{inv}$, with $\alpha = 1$ in weak inversion and $\alpha = 2$ in strong inversion. In practical cases, a $1/f^\gamma$ dependence, with $0.8 \leq \gamma \leq 1.2$, is observed instead of a pure $1/f$ noise.

The advantage of Eq.(22) is to illustrate in detail the influence of the most significant device parameters. In *weak inversion*, C_{inv} as well as the variation of C_d are negligible, so that the normalized noise factor S_{I_D}/I_D^2 is a constant (Fig.8a). It does not depend on either drain or gate voltages. In *strong inversion*, however, C_{inv} becomes rapidly prevailing in Eq.(22). Moreover, a slight increase of the noise factor with V_D is observed in Fig.8b due to the decrease in Q_{inv} along the channel. A typical noise curve, S_{I_D}/I_D^2 vs. V_G, is composed of a "plateau" in weak inversion, followed by a sharp decrease, proportional to $(V_G - V_T)^{-2}$ in strong inversion.

The *Random Telegraph Signal (RTS)* is attributed to individual carrier trapping at the Si–SiO$_2$ interface. In very small area devices, the trapping events are seldom and the trapping of one carrier can easily be detected in the time domain as a small pulse superimposed on the "normal" current value. The duration of the pulse is nothing but the time constant of the trap. These pulses have rather sharp transitions and a constant amplitude (Fig.9a). They correspond to the local modification of the flat-band voltage by one trapped carrier. It is shown that, for small enough fluctuations, the drain current pulse ΔI_D is independent of the trap area [19]

$$\frac{\Delta I_D}{I_D} = \frac{q}{LWC_{ox}} \frac{g_m}{I_D} \left(1 - \frac{d}{t_{ox}}\right) \quad (23)$$

where d is the distance of the trap from the interface. Not only, in very small devices, the trapping events become distinct but also does the amplitude of the fluctuation increase. The back channel of SOI MOSFET's is an ideal tool for studying the properties of RTS fluctuations because C_{ox} is very small and allows larger pulses to be obtained.

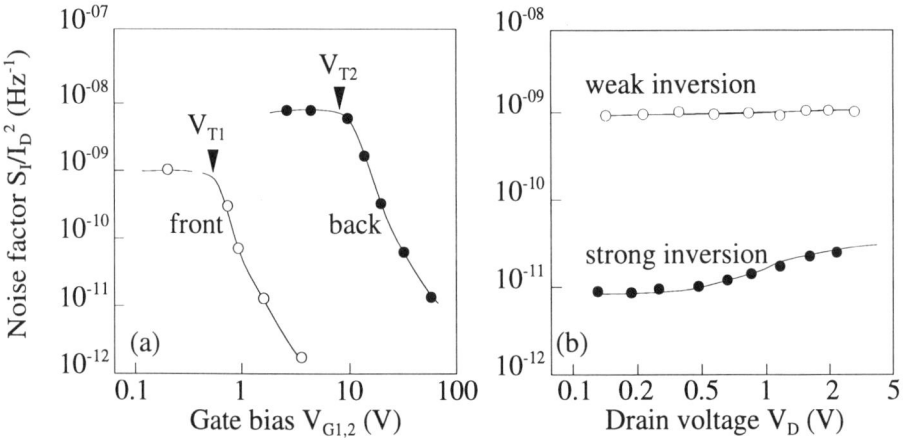

Figure 8. Normalized noise factor versus gate voltage (Fig.(a): oo front channel, •• back channel) and drain voltage (Fig.(b): oo weak inversion, •• strong inversion) in a SIMOX n–MOSFET.

The power spectral density of RTS signals, induced by a single type of traps, is similar to that of G–R noise: plateau at low frequency followed by an $1/f^2$ decrease. When many traps with different time constants coexist, the superposition of their RTS signals results in the standard $1/f$ noise.

In early SIMOX material, annealed at low temperatures ($< 1250°C$), the dominant noises were $1/f$ and Lorentzian (with $1/f^2$ tail) [20]. Figure

8 shows the typical variations of the normalized $1/f$ noise factor S_I/I_D^2 with the gate and drain voltages. The transition from the weak inversion "plateau" to the strong inversion slope (Fig.8a) well corresponds to the threshold voltage. The density of slow traps is calculated from the plateau level using Eq.(22). It is found that the buried oxide contains, in general, one order of magnitude more slow traps than the front thermal oxide where $N_{t1} \simeq 10^{18}\,\text{cm}^{-3}\text{eV}^{-1}$.

In SIMOX structures annealed at high temperatures ($\geq 1300\,°C$), the $1/f$ noise is reduced and can be masked by a Lorentzian noise. When probing the back channel, an additional source of G–R noise was sometimes observed for special values of the back gate voltage. It was attributed to the activation of traps localized near the buried oxide. Typical RTS signals have been detected in small-area SIMOX transistors (Fig.9a). The pure RTS noise had a Lorentzian spectrum. Once the RTS signal was subtracted from the total current fluctuations, the residual noise had a white or $1/f$ spectrum [19].

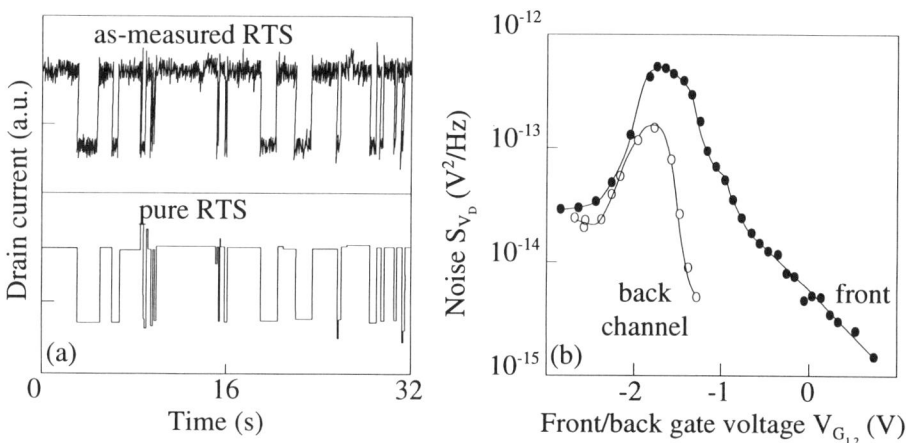

Figure 9. (a) As-measured and pure RTS signals in the time domain and (b) drain voltage noise (f=10 Hz, strong inversion) versus front and back gate voltages.

In *depletion-mode* SOI MOSFET's, the two interface noise sources are completed by the noise generated in the neutral region of the film [21]. The drain voltage fluctuations, $S_{V_D}^\star$, measured in a partially depleted SIMOX N$^+$NN$^+$ transistor are shown in Fig.9b. A very meaningful curve is obtained by varying the back gate bias while keeping the front interface in strong inversion. As long as the back interface is inverted too ($V_{G_2} < -20\,\text{V}$), the noise is constant and reflects the unique contribution of the neutral film whose thickness is minimum. When V_{G_2} scans the weak inversion and depletion regions, the noise increases sharply due to the interactions between free carriers and buried oxide/interface traps. The noise peaks at the flat-band voltage.

As V_{G_2} goes into accumulation, the interface traps are saturated and the measured noise decreases. A symmetrical behavior is obtained by varying V_{G_1} and maintaining the back interface strongly inverted. The interface coupling impacts on noise measurements in fully depleted SOI transistors in two main respects: *direct* noise generation at the opposite interface and *indirect* modification of the surface potential by the opposite gate bias.

In conclusion, the noise spectroscopy is an elegant method to discriminate and characterize the defects located in different regions of the transistor. This only requires to "decouple" a given region from its surroundings, by appropriately biasing the front and back gates, and measure the noise generated solely by that source.

8. Drain Current Transient Technique

The carrier generation-recombination properties govern the performance of MOS and bipolar devices (leakage current, refresh time of dynamic RAMs, floating body, etc). Moreover, the carrier lifetime depends on residual crystalline defects and stands as a figure of merit for the quality of SOI films.

The Zerbst method [22] consists in applying a pulse on the gate of a MOS capacitor, so that a deep-depletion region is formed. Equilibrium is reached through bulk and surface carrier generation processes. The monitoring of the capacitance transient provides the generation lifetime τ (which can substantially exceed the recombination lifetime) as well as the surface velocity S. Since capacitance measurements are not straightforward in thin SOI films, several modifications will be needed: (i) MOS transistors will serve as test devices, (ii) the drain current will be monitored, and (iii) the dual gate configuration will be used to form deep depletion regions and initiate carrier generation.

We first consider a N^+NN^+ transistor biased in the ohmic region. Applying, at $t = 0$, a large negative voltage on the gate results in a very fast build-up of positive charge (deep depletion region) to balance the gate charge. Then, the inversion charge starts forming. It is a much slower process since no reservoir of holes is available. The deep depletion region controls the thickness of the drain current path. The holes supplied by thermal generation allow the deep depletion region to shrink and the current $I_D(t)$ to increase towards an equilibrium value I_∞ (Fig.10) [23]. The transient variations of the inversion charge Q_{inv}, depletion charge Q_d, interface trap charge Q_{it}, and surface potential Ψ_s are governed by the charge balance

equation
$$Q_{inv} + Q_{it} = C_{ox}(V_G - V_{FB} - \Psi_s) - Q_d \qquad (24)$$

Expressing Q_d, Ψ_s, and I_D in terms of depletion depth $w_d(t)$ gives

$$Q_d = qw_d N_D \qquad \Psi_s = \frac{qw_d^2 N_D}{2\epsilon_{si}} \qquad (25)$$

$$I_D = \frac{q\mu N_D W V_D}{L}(t_{si} - w_d) = \frac{1}{K}(t_{si} - w_d) \qquad (26)$$

Combining Eqs.(24) and (25), yields the rate of change of inversion and interface trap charges which is perfectly balanced by the total carrier generation rate [24]

$$G = -\frac{N_D C_{ox}}{2\epsilon_{si}}\left(1 + \frac{w_d}{\epsilon_{si}}\right)\frac{dw_d}{dt} = \frac{n_i(w_d(t) - w_\infty)}{\tau} + n_i S \qquad (27)$$

The first term stands for volume generation which is usually assumed to occur in the excess depletion region $(w_d - w_\infty)$, where w_∞ is the depletion depth at equilibrium [25]. The second contribution comes from the Si-SiO$_2$ interface. Equation(27) can be rewritten in a generic form

$$LHS(t) = \frac{1}{\tau} \times RHS(t) + S \qquad (28)$$

where the left- and right-hand sides are readily expressed, using Eq.(26), in terms of drain current

$$LHS = -\frac{N_D C_{ox}}{2n_i \epsilon_s}\frac{d}{dt}\left[t_{si} + \frac{\epsilon_{si}}{C_{ox}} - KI_D\right]^2; \quad RHS = K(I_\infty - I_D) \qquad (29)$$

Instead of computing the current transient $I_D(t)$, Eq.(28) is an elegant way of presenting the experimental data, LHS(t) vs. RHS(t), on a straight line. The slope and intercept yield the values of τ and S, respectively.

By contrast to accumulation-mode transistors, in N$^+$PN$^+$ structures, the drain and source reservoirs do supply minority carriers to form quasi-instantaneously the inversion channel. Since there is no source of majority carriers (no P$^+$ contact), except the generation process, the experiment is designed to induce a deficit of holes. The front gate is biased in strong inversion, whereas the back gate is pulsed from depletion (or accumulation) to strong accumulation. The holes requested in the accumulation layer are released from the neutral region of the film. A deep depletion region forms under the gate and the film potential drops. To maintain front gate charge conservation (Eq.(24)), the excess depletion charge is compensated by a temporary drop in the inversion charge and drain current. The current relaxes back to equilibrium through carrier generation within the film and at the interfaces.

Since the front surface potential is almost constant in strong inversion ($\Psi_{s_1} \simeq 2\Phi_F$), it is the film potential $V_B(t)$ that governs the densities of inversion, depletion, and accumulation charges. The charge balance of majority carriers is fulfilled by equating the rate of change of depletion and accumulation charges with Eq.(27). This again yields Eq.(28), where the right-hand side RHS(t) is the same as in Eq.(29) and the left-hand side becomes

$$LHS(t) = -\frac{N_A C_{ox2}}{2 n_i \epsilon_{si}} \frac{d}{dt} \left[K(I_\infty - I_D) + \frac{\epsilon_{si}}{C_{ox2}} + \sqrt{\frac{4\epsilon_{si} \Phi_F}{q N_A}} \right]^2 \quad (30)$$

In some particular occasions [26], the variation of the accumulation charge may be neglected, and the current has a simple exponential transient [27]

$$I_\infty - I_D = I_\infty e^{-t/\tau_R} \quad (31)$$

where the recovery time is $\tau_R = \tau N_A / n_i$. This explains the difference by many orders of magnitude between the time constants of the generation process and of the drain current relaxation.

For large pulses, the deep depletion charge may be enough to temporarily suppress the drain current (for instance, when full depletion is achieved in accumulation-mode transistors). During a time t_{off}, freshly generated holes contribute to the relaxation of the front surface potential, which is swept from depletion and weak inversion to strong inversion [26]. Of course, the Zerbst analysis still holds for $t > t_{off}$. In films that are fully depleted at equilibrium, the concept of deep depletion and the expression $w_d(t)$ are meaningless. However, the dual gate pulsing technique can still be successful. Remark that as long as the two interfaces are depleted, they play equivalent roles in the generation process.

The experiment is performed in obscurity, with low drain voltage. Long channels are suitable for the extension of source and drain depletion regions to be less significant. Junction leakage or BOX leakage currents may jeopardize the accuracy of drain current transients.

The measurements are performed with a HP–4145 system and typical drain current transients are shown in Fig.10. As is usual for Zerbst plots, straight lines are obtained except at the beginning of the transient. This is due to the larger surface generation rate when the interface is depleted. The distinct features of front and back gate biases in partially depleted enhancement-mode transistors are illustrated in Fig.10. When the back gate voltage step is increased, the current variation range is more pronounced, the off-state period is longer, but the steady-state current does not change (Fig.10a).

Only in fully depleted MOSFETs, may I_∞ be modified by the pulse amplitude via interface coupling (Fig.10b); the bottom of the back gate pulse acts on the threshold voltage of the front channel. But, once the bottom of the pulse has reached strong accumulation, a further increase of the pulse will not modify any more V_{T_1} and I_∞ (Fig.10c).

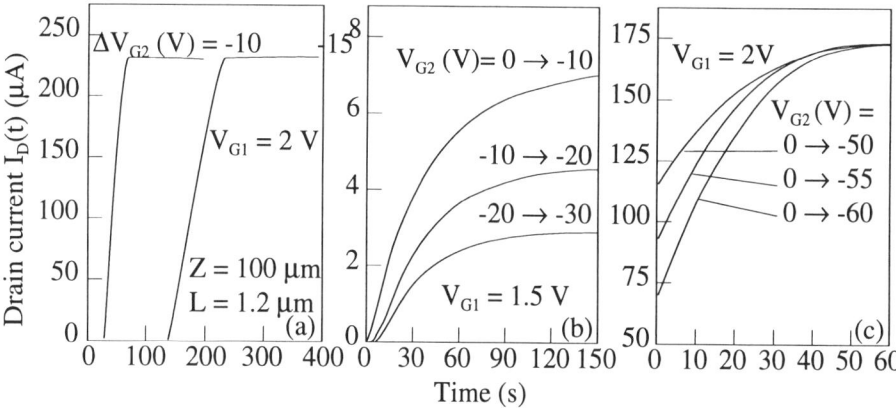

Figure 10. Current transients in N^+PN^+ transistors pulsed with various back gate steps at fixed front gate bias: (a) partially depleted SIMOX, (b) fully depleted SIMOX, (c) fully depleted wafer bonding MOSFETs.

Increasing the front gate bias, the steady-state current becomes larger and the off-state region may be suppressed. This cancels or reduces the transition through weak inversion regime and makes the strong inversion analysis with Eq.(30) more accurate. For intentionally selected low gate biases, a weak inversion model may successfully be applied, leading to same lifetime values [28].

Measurements performed on various SOI materials show that the generation lifetime is very sensitive to the crystalline quality of the film. In early SIMOX material annealed at low temperature, the carrier lifetime was very poor (10–100 ns). High temperature annealing above 1300 °C was very efficient and improved the lifetime by three orders of magnitude (10–100 μs). So far, lifetime in excess of 100 μs has only been achieved in multiple-implanted SIMOX [26].

The surface generation velocity, currently deduced from Zerbst transients, is extremely small (0.01–0.1 cm/s) and does not reflect the actual generation rate at a *depleted* interface. It was found from weak inversion analysis that the surface generation velocity decreases by 2–3 orders of magnitude from weak to strong inversion [28].

The roles of the two gates can of course be interchanged for probing different regions of the film. Note that the transient duration can be much smaller, owing to the difference in the capacitances of the gate oxide and buried oxide in Eqs.(29) and (30), without necessarily implying an inhomogeneity of the film quality.

The depth profiling of carrier lifetime is feasible by taking the derivative of the Zerbst curve with respect to the non-equilibrium depletion width w_d. The slopes of the tangents are simply the local values $\tau_{eff}^{-1}(w_d)$ at each point situated beyond w_∞.

In summary, the drain current transients are exclusive SOI techniques. They were imagined to exploit the special dual-gate configuration of SOI transistors, hence there is no chance for corresponding methods to exist in bulk Si.

9. Concluding Remarks

Several electrical characterization methods, especially conceived for SOI or imported from bulk silicon, have been revisited in the context of SOI materials and MOS transistors. Transport measurements in as-grown SOI wafers are rendered more difficult by the film thinness. By chance, the Ψ–MOSFET appears to be a very unique and attractive technique for in-situ material evaluation.

As far as the device-based methods are concerned, the adaptation of their experimental setup and basic theory to thin SOI transistors is straightforward only if the back interface is maintained in accumulation. Otherwise, unusual and more complex situations are reached, as compared to the conventional curves obtained in bulk MOSFETs. Full depletion is the condition for interface coupling effects to occur, in other words the apparent carrier transport and generation properties at one interface can be controlled by the opposite gate.

None of the above characterization techniques can negate the use of the others. Each of them is focused on a particular feature and presents distinct merits and weaknesses. Together rather than separated can they provide a complete enough electrical image of SOI structures.

References

1. S. Cristoloveanu and S. Williams, "Point-contact pseudo-MOSFET for in-situ characterization of as-grown silicon-on-insulator wafers," *IEEE Electron Device Lett.*, vol. 13, p. 102, 1992.
2. S. Cristoloveanu and S.S. Li, *"Electrical Characterization of SOI Materials and Devices,"* Kluwer, Norwell, 1995.
3. D.K. Schroder, *"Semiconductor Material and Device Characterization,"* J. Wiley & Sons, New York, 1990.

4. S. Cristoloveanu, J. Pumfrey, E. Scheid, P.L.F. Hemment, and R.P. Arrowsmith, "Thermal donor and new donor generation in SOI material formed by oxygen implantation," *Electron. Letts.*, vol. 21, p. 802, 1985.
5. Van der Pauw, "A method of measuring specific resistivity and Hall effect of discs of arbitrary shapes," *Philips Res. Repts.*, vol. 13, p. 1, 1958.
6. D.J. Dumin and P.H. Robinson, "Carrier transport in thin silicon films," *J. Appl. Phys.*, vol. 39, p. 2759, 1968.
7. S. Cristoloveanu, "Silicon films on sapphire," *Rep. Prog. Phys.*, vol. 50, p. 327, 1987.
8. S. Cristoloveanu, S. Gardner, C. Jaussaud, J. Margail, A.J. Auberton-Hervé, and M. Bruel, "Silicon on insulator material formed by oxygen implantation and high-temperature annealing: carrier transport, oxygen activity, and interface properties," *J. Appl. Phys.*, vol. 62, p. 2793, 1987.
9. B. Kleveland, S. Cristoloveanu, and J. Sicart, "Novel Hall effect spectroscopy of impurity levels in semiconductors," *Solid St. Electron.*, vol. 33, p. 743, 1990.
10. G. Papaioannou, V. Ioannou-Sougleridis, S. Cristoloveanu, and C. Jaussaud, "Photoinduced current transient spectroscopy in silicon-on-insulator films formed by oxygen implantation," *J. Appl. Phys.*, vol. 65, p. 3725, 1989.
11. H.S. Chen and S.S. Li, "A model for analyzing the interface properties of a semiconductor-insulator-semiconductor structure-I: Capacitance and conductance techniques," *IEEE Trans. Electron Dev.*, vol. 39, p. 1740, 1992.
12. F.T. Brady, S.S. Li, D.E. Burk, and W.A. Krull, "Determination of the fixed oxide charge and interface trap densities for buried oxide layers formed by oxygen implantation," *Appl. Phys. Lett.*, vol. 14, p. 886, 1988.
13. J-H. Lee, S. Cristoloveanu, and A. Chovet, "Non-homogeneous electrical transport through silicon-on-sapphire thin films: evidence of the internal stress influence," *Solid-State Electron.*, vol. 25, p. 947, 1982.
14. A.C. Ipri, "Electrical properties of silicon films on sapphire using the MOS Hall technique," *J. Appl. Phys.*, vol. 43, p. 2770, 1972.
15. T. Ouisse, S. Cristoloveanu, T. Elewa, H. Haddara, G. Borel, and D. Ioannou, "Adaptation of the charge pumping technique to gated p-i-n diodes fabricated on silicon on insulator," *IEEE Trans. Electron Devices*, vol. 38, p.1432, 1991.
16. G. Groeseneken, H.E. Maes, N. Beltran, and R.F. de Keersmaecker, "A reliable approach to charge pumping measurements in MOS transistors," *IEEE Trans. Electron Devices*, vol.ED-31, p. 42, 1984.
17. A. Chovet and P. Viktorovitch, "Le bruit électrique," *L'Onde électrique*, vol. 57, p. 699 and 773, 1977, and vol. 58, p. 69, 1978.
18. G. Reimbold, "Modified $1/f$ trapping noise theory and experiments in MOS transistors biased from weak to strong inversion. Influence of interface states," *IEEE Trans. Electron Devices*, vol. ED-31, p. 1190, 1984.
19. O. Roux dit Buisson, G. Ghibaudo, J. Brini, and T. Ouisse, "Measurements and analysis of random telegraph signals in small area SOI MOSFETs," in *Silicon-On-Insulator Technology and Devices* (W.E. Bailey ed.), The Electrochemical Soc., vol. 92-13, p. 191, 1992.
20. A. Chovet, B. Boukriss, T. Elewa, and S. Cristoloveanu, "Low frequency noise of front channel and back channel MOSFETs fabricated on silicon on insulator SIMOX substrates," in *Noise in Physical Systems* (C.M. Van Vliet ed.), World Scientific, Teaneck, New Jersey, p. 457, 1987.
21. T. Elewa, B. Boukriss, H.S. Haddara, A. Chovet, and S. Cristoloveanu, "Low-frequency noise in depletion-mode SIMOX MOS transistors," *IEEE Trans. Electron Devices*, vol. ED-38, p. 323, 1991.
22. M. Zerbst, "Relaxation effects at semiconductor-insulator interfaces," *Z. Angew. Phys.*, vol. 22, p. 30, 1966.
23. D.P. Vu and J.C. Pfister, "Determination of minority carrier generation lifetime in beam-recrystallized silicon-on-insulator by using a depletion-mode transistor," *Appl. Phys. Letts.*, vol. 47, p. 950, 1985.

24. T. Elewa, H. Haddara, and S. Cristoloveanu, "Interface properties and recombination mechanisms in SIMOX structures," in *The Physics and Technology of Amorphous SiO_2*, R.A.B. Devine ed, Plenum, New York, p. 553, 1988.
25. R.F. Pierret, "Rapid interpretation of the MOS–C C–t transient," *IEEE Trans. Electron Devices*, vol. 25, p. 1157, 1978.
26. D.E. Ioannou, S. Cristoloveanu, M. Mukherjee, and B. Mazhari, "Characterization of carrier generation in enhancement-mode SOI MOSFETs," *IEEE Electron Device Letts.*, vol. 11, p. 409, 1990.
27. P.W. Barth and J.B. Angell, "A dual-gate deep-depletion technique for generation lifetime measurements," *IEEE Trans. Electron Devices*, vol. 27, p. 2252, 1980.
28. A. Ionescu and S. Cristoloveanu, "Carrier generation in thin SIMOX films by deep-depletion pulsing of MOS transistors," *Nucl. Instr. and Methods in Phys. Res.*, vol. B84, p. 265, 1994.

THE DEFECT STRUCTURE OF BURIED OXIDE LAYERS IN SIMOX AND BESOI STRUCTURES

A. G. REVESZ   and   H. L. HUGHES
Revesz Associates   *Naval Research Laboratory*
Bethesda, MD 20817, USA   *Washington, DC 20375, USA*

1. INTRODUCTION

Both SIMOX [1] and BESOI [2] structures are potentially important for future SOI devices. A common feature of these structures is that two Si layers are separated by an SiO_2 layer. Thus, the oxide layer is buried or confined between the Si layers. As a result of this confinement, some of the properties of these two types of buried oxide (BOX) layers are similar, even though the preparation conditions are very different. Although both BOX layers are similar to the widely used thermally grown SiO_2 films in the sense that they are *noncrystalline,* some properties are quite different; this is particularly true for the SIMOX BOX layers. The primary reason for this difference is the difference in their defect structures. The defect structure of the BOX layers plays an important role in determining various properties of the completed SOI devices, particularly fully-depleted transistors, as well as the yield of the fabrication process and, very likely, the reliability of the devices. In this paper the properties of these BOX layers are reviewed with emphasis on their defect structure.

2. FORMATION OF BOX LAYERS

The formation of the BOX layers is quite different for SIMOX and BESOI structures; details have been reviewed [1, 2]. Only a brief summary is given here focussing on the effects of preparation conditions on the defect structure of the BOX layers.

2.1. FORMATION OF BOX LAYERS IN SIMOX STRUCTURES

The BOX layer is formed in the course of implanting O^+ ions into silicon and subsequent annealing. The energy and dose of the oxygen ions, beam current, and the temperature of the Si wafer during implantion are important parameters, but their exact role in determining the defect structure of the BOX layer is still not known. Another potentially important parameter is the orientation of the Si wafer with respect to the ion beam [3]. As will be discussed below, some features of the defect structure of the BOX apparently depend on the angle between the oxygen ion beam and the channeling directions in silicon.

Most of the studies reviewed here were performed on single-and triple-implanted samples obtained with 1.8×10^{18} O^+/cm^2 total dose at 200 KeV energy. In the case of triple-implanted samples each of the three implant processes was followed by annealing. A few samples have been prepared with 0.6×10^{18} O^+/cm^2 dose or even less. Since the molar volume of SiO_2 is 2.2 times greater than that of silicon (assuming that the density of the oxide is 2.20 g/cm^3), an "accomodation volume" must be provided for the oxide, otherwise very large stress would be generated. It is thought that some of this accomodation volume is provided by the generation of Si interstitials and their migration toward the Si substrate and top Si layer. Since the diffusivity of silicon in the oxide is extremely low, those Si interstitial atoms which cannot reach one of the Si/SiO_2 interfaces will represent excess silicon in the oxide.

Oxygen implantation is followed by high temperature annealing. Most of the results reviewed here were obtained with samples annealed at 1325°C for 5 hours in Ar-0.5% O_2. The BOX layer acts as an oxide precipitate of infinite radius attracting the smaller oxide precipitates; in this manner the BOX layer grows and the Si/SiO_2 interfaces sharpen. At the end of the annealing the nominal BOX thickness is 400 nm for 1.8×10^{18} O^+/cm^2 dose and the thickness of the top Si-layer is about 210 nm. The excess silicon, associated with the non-uniform distribution of oxygen in the unannealed BOX, largely disappears during annealing by gettering at the two Si surfaces. However, some silicon remains in the oxide where it forms small (probably less than 3 nm) amorphous (a) clusters and/or even relatively large ($> \sim 1 \mu m$) crystalline precipitates. Note that (a)-Si clusters are usually not detected by electron microscopy; the reason is that as they are imbedded in another noncrystalline solid (BOX), there is very little, if any, diffraction contrast. At the end of the annealing the BOX layer appears to be similar to a thermally grown oxide as it is noncrystalline and stoichiometric within a few atomic percent. However, as discussed below, there is plenty of evidence that the oxide contains excess silicon ranging from oxygen vacancies as network defects to crystalline Si platelets; the latter are usually found near the BOX/Si substrate interface in single-implanted samples. In addition to the implantation and annealing processes, confinement effects may also result in excess silicon in the BOX; this issue is discussed below. Excess silicon in the BOX layer is the cause of various effects which have never been observed with thermally grown oxides; its control appears to be a major problem in SIMOX technology.

In an attempt to oxidize the excess silicon some samples received supplemental oxygen implantation of 0.5×10^{18} O^+/cm^2 or less dose after annealing followed by heat treatment at 1000 C. [4]

2.2. FORMATION OF BOX LAYERS IN BESOI STRUCTURES

In contrast to the formation of BOX layers in SIMOX structures, the BOX layer in BESOI structures results from bonding together two Si wafers with thermally grown SiO_2 films [2]. In some cases, the oxide film on one of the Si wafers is the very thin ($< \sim 2$ nm) native oxide formed at room temperature; in other cases both wafers have been oxidized usually, but not necessarily exclusively, in steam. The total thickness of the BOX is usually larger than 100 nm. The critical aspects of this process are the flatness of the Si wafers, the nature of the oxide films, and particulate contamination. The flatness is very important to ensure that there will not be voids associated with the bonding interface. The nature of the oxide is intimately related to how the bonding takes place. The important points are: the hydrogen content of the oxide(s), especially at the bonding surface(s), as well as the temperature, ambient, and time of the bonding process. It is believed that these parameters control the formation of hydrogen bonds between the two oxide surfaces: SiOH...OSi where the O atom in the second oxide may be a bridging (Si-O-Si) or non-bridging oxygen. These parameters also control the transformation of the hydrogen bonds into bridging Si-O-Si bonds which provide the final bonding between the two wafers. The details of the bonding process are not completely understood but there is plenty of evidence, to be discussed below (Section 4.2), that most of the BESOI structures exhibit phenomena associated with hydrogen in the BOX. Like SIMOX structures, confinement effects are also important in BESOI structures; this is also discussed in Section 5.

3. STRUCTURE OF NONCRYSTALLINE SILICON DIOXIDE

Since the BOX layers are noncrystalline, the results of the observations will be discussed in terms of the structure of nc SiO_2 that is briefly described here. The essential feature of the the structure of nc-SiO_2 is that the practically invariant $SiO_{4/2}$ tetrahedra are linked together in a very flexible manner in the sense that both the angles of the Si-O-Si bridging bonds, ϕ, and the rotational (dihedral) angles, δ (characterizing the orientation of two tetrahedra relative to each other), exhibit wide ranges. Noncrystallinity, i.e. the lack of long range order, arises from the flexibility of the Si-O bond rather than bond breaking [5]. A manifestation of this bond flexibility is that the bond energy varies only by about 0.1 eV from its minimum value at $\phi = 145°$ within the range characteristic of nc-SiO_2, $\sim 110° < \phi < \sim 180°$. However, the bond energy rises very rapidly below 120°, for instance, it is about 1 eV higher at $\phi = 100°$ than its minimum value [6]. The variation of the bond energy with δ is about 0.05 eV within its range of 0 to 120°.

The relatively high energy bonds associated with $\phi < \sim 120°$ are responsible for interactions between nc SiO_2 and water or hydrogen, irradiation, etc. [6, 7]. The saturation concentration of defects (Si-OH groups and E'-centers [O_3Si-]) generated by these interactions is about $10^{18} - 10^{19}$ cm^{-3}, that is about the same as the concentration of Si-O bonds associated with $\phi < \sim 120°$. The ϕ distribution shifts to lower values for pressure-densified silica glass with a concomitant change in the δ distribution so that the O - 2nd O distances become smaller [8]. With the downward shift of ϕ distribution the proportion of high energy bonds increases so that the oxide becomes more reactive. This effect is shown, among others, by the increased generation rate of E'-centers in densified silica glass [9]. The main peak at 1090 cm^{-1} in the infrared absorption spectrum shifts to lower frequencies with increasing densification [8]. Ion bombardment also increases the proportion of high energy bonds and shifts the main IR absorption peak to lower frequency, but the increase in density is much less than expected if this shift were due to densification [8]. The high energy bonds can be considered as "strained" in the chemical sense. However, this does not necessarily mean that the oxide is strained in the sense of elastic deformation.

It is obvious from these considerations that, in contrast to crystalline solids, the medium-order structure of nc-SiO_2 is not uniquely defined. Hence, relatively small variations in the preparation conditions may have significant effects on those properties which are largely determined by the distributions of ϕ and δ. On the other hand, those properties which depend primarily on the short-range-order (i.e. the structure of $SiO_{4/2}$ tetrahedra) as, for instance, the band gap, depend very little on the preparation conditions as long as the overall integrity of the oxide network is maintained. The flexibility of the structure makes it possible that regions with various degrees of *localized* ordering may be present in the oxide which is, however, still noncrystalline, as seen, for example, by X-ray or other diffraction techniques. Examples of such a localized ordering are the *channels* in thermally grown oxides along which the O_2 molecules move during oxide growth and the dependence of various properties of the nc SiO_2 on the crystallographic orientation of the Si substrate [10]. In the case of SIMOX BOX layers these localized ordered regions may arise from interaction between the oxygen ion beam and silicon along the channeling directions in silicon (i.e. relatively easy paths for transport processes) resulting in localized conducting regions in the BOX (leakage paths) and localized interaction between the BOX and top Si layers resulting in defects in the top Si layer; these issues are discussed below (Sections 4.1.4 and 5).

4. PROPERTIES OF BOX LAYERS: RESULTS AND INTERPRETATION

In this section the results of various studies performed on BOX layers are reviewed and discussed. We attempt to interpret the results in a self-consistent manner in terms of the structure of nc SiO_2. Some of the properties reflect the nature of the oxide network while others are mostly determined by defects in the BOX. There is no sharp distinction between these two classes as the oxide network and defects are inter-related and some locally ordered regions can, in fact, be considered as defects in relationship to the overall randomness of the nc structure.

The most important feature of the BOX is that it is noncrystalline. It has been pointed out that the nc structure of thermally grown SiO_2 films is essential for achieving a low density of interface states needed for sufficient passivation of the Si surface and, hence, for the operation of MOS devices [5]. The reason for this is that, as a result of the structural flexibilty of nc SiO_2, a good structural match with silicon is possible without the strict epitaxy which would be necessary if the oxide were crystalline. As a result of the nc structure of the BOX layers in both SIMOX and BESOI structures, the density of interface states at the Si/BOX interfaces is usually of the order of 10^{10} cm^{-2}(eV)$^{-1}$.

Most of this section is related to BOX layers in SIMOX structures. The properties of BOX layers in BESOI structures have been studied less extensively. There are several reasons for this as, for example, that up to now the device research and development has been more oriented toward SIMOX than BESOI structures. Also, it was taken for granted that the BOX layers in BESOI structures are essentially identical to conventional thermally grown oxides. However, recent studies show that there are significant differences between conventional un-confined oxide and BESOI BOX layers which are confined.

4.1. SIMOX BOX LAYERS

The most important difference between SIMOX BOX layer and thermally grown SiO_2 films is that the former usually contains excess silicon in some form. In this sense the BOX in SIMOX structures is similar to <u>deposited</u> SiO_2 films with deliberately introduced excess silicon [11]. One of the manifestations of this similarity is the observation that BOX layers may contain amorphous (a) silicon dangling bond defects ("D-defect": Si_3Si-) [12]. This defect is also found in the deposited oxide films in which the excess silicon forms small ($< \sim 3$ nm) a-Si clusters. The excess silicon in SIMOX BOX may be present as network defects as, for instance, oxygen vacancy and/or O_3Si-SiO_3 bonds, a-Si clusters of various sizes, as well as crystalline Si precipitates. Practically all the properties discussed in this section depend on the nature and density of excess silicon.

4.1.1.Bulk Properties

Infrared spectroscopy showed that the structure of BOX layers, particularly the un-annealed ones, is strained in the sense that the distribution of ϕ is shifted to lower values. For un-annealed BOX layers the main IR absorption peak is around 1050 cm^{-1}; with increasing oxygen dose the frequency shifts upward [13]. Annealing also shifts the frequency upwards and after long annealing at sufficiently high temperature the position of the peak approaches (but usually does not quite reach) the range characteristic of silica glass (1090 cm^{-1}). Further evidence of the strained structure of BOX comes from etching studies that demonstrated that the etch rate in HF of BOX is about 20% less than that of thermal oxide [14]. It is well known that the sensitivity of Si-O bonds to HF decreases with ϕ.

Etching the top Si layer in hydrazine [15] or KOH [16] revealed the existence of columns (pipes) in annealed BOX. These pipes were dissolved during this etching. These pipes are probably involved in the defect conduction behavior of SIMOX structures discussed below (Section 4.1.4.). Most of the conducting defects in BOX are associated with light scattering centers present on the surface of the top Si layer [17]. These defects are protrusions and mostly contain oxygen. Exposure to HF results in etch pits in the silicon, some of which can penetrate the BOX. The density of the conducting defects depends on the orientation of the Si wafer with respect to the oxygen ion beam; the maximum density occurs when the ion beam is parallel with a channeling direction (e.g. [110]) in the silicon [3]. This observation clearly demonstrates that the structure of the nc BOX is not completely random, as some locally ordered regions (*channels*) are present. It is suggested, tentatively, that the energy of the Si-O bonds along such a channel is higher than average and, hence, more prone for some interaction with hydrogen or, perhaps, carbon. The interaction results in the breaking of these bonds. The transport of oxygen to the Si layer leaves behind a Si-rich region ("pipe"). A very small amount of hydrogen (or carbon) could act as a "catalyst" for this process. Excess silicon in SIMOX BOX may also be present as photo-active Si-clusters; see discussion below (Section 4.1.5). Confinement effects are, at least partially, responsible for excess silicon in BOX layers; this is also discussed below (Section 5.).

Both single-wavelength [18] and spectroscopic ellipsometric studies [15] indicated that the refractive index of BOX is usually higher than that of silica glass. This observation could result from a densification effect of the oxide and/or from excess silicon in the BOX. It should be realized that the evaluation of spectroscopic ellipsometry measurements does not give a unique result, as different assumptions may lead to the same result. In contrast, single-wavelength ellipsometric measurements performed on pseudo-SIMOX samples give an unambiguous result for the refractive index and thickness of the oxide, provided that the oxide is a perfect dielectric (i.e. no light absorption takes place) and the BOX/Si substrate interface is flat. These requirements are not necessarily met with every SIMOX sample.

Single-wavelength ellipsometric measurements performed on "pseudo-SIMOX" samples (top Si layer removed) showed refractive indices of BOX layers ranging up to 1.488. However, after heat treatment in oxygen at 1100°C the values have decreased to about 1.46 which is the value characteristic of silica glass [18]. Note that heat treatment in argon did not produce this effect; this observation indicates that excess silicon is probably responsible for the increased refractive indices. Considering the optical properties of BOX as determined by Si-O and Si-Si bonds, the 1.488 value corresponds to about 4.2×10^{20} Si-Si bonds/cm^3. However, some caution is needed even in this case as light absorption associated with the Si-Si bonds was not taken into account and the possibility, that the BOX/Si-substrate interface is not perfectly flat, could have influenced the results. Nevertheless, the density of excess silicon as estimated from the bulk conduction behavior of BOX is similar (see Section 4.1.4).

4.1.2. *Interaction between BOX and deuterium*

The interaction between BOX and deuterium reveals important structural features of the BOX layers. This interaction was studied by exposing the samples to D_2 gas at 700 - 900 C for up to 40 hours and determine the D concentration by nuclear

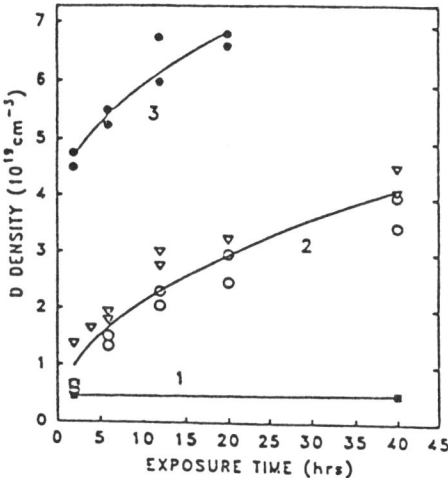

Figure 1. Concentration of D in BOX produced by exposure to D at 750 C and 650 Torr. Curve 1: 90 nm thick thermal oxide grown at 1325 C in Ar-0.5% O_2; Curve 2: annealed SIMOX, the two symbols refer to two set of samples; Curve 3: annealed SIMOX with epi-Si layer [18].

reaction technique [19]. The incorporation of D into the BOX is determined by the rate of reaction between D and SiO_2 resulting in the formation of D-containing defects (Si-OD, Si-D) that are stable up to ~1100°C. The reactions within the BOX exhibit two components: one is rapid and strongly depends on sample preparation, the other is much slower and not so sensitive to sample preparation; see Fig.1. Even at 40 hrs exposure to D_2 there is no sign of saturation; at this point the D-concentration approaches (and for un-annealed samples [not shown] even exceeds 10^{20} cm^{-3}). This value is much larger than the density of E'- centers (~ 10^{15} cm^{-3} for annealed BOX without irradiation). In contrast to the behavior of BOX, thermal oxide grown at 1325°C (SIMOX annealing temperature) shows the rapid component only with a saturation value of 5×10^{18} D/cm^3; the same behavior has been observed with silica glass [20]. It was suggested that in the case of silica glass and thermally grown oxide the reaction sites are high energy (strained) bonds associated with small bridging bond angles [17]. The larger concentration of reaction sites in BOX and their time-dependent reactivity are attributed to a larger proportion of the strained bonds and defects associated with excess silicon.

4.1.3. Electron-spin-resonance studies
Electron-spin-resonance (ESR) studies provide significant contributions to our understanding of the structure of BOX layers. The ESR-active defects observed in SIMOX samples are as follows:
1) a-Si dangling bond (D-defect), a-Si$_3$Si-, with a g-value of 2.0059. Their density in un-annealed BOX is about 10^{20} cm^{-3} and in annealed BOX is less than 2×10^{15} cm^{-3} (if present at all), assuming uniform concentration [12]. D-defects can also be generated by VUV illumination [21].

2) Crystalline (c) silicon dangling bond associated with Si/SiO$_2$ interfaces (P$_b$-center), c-Si$_3$Si-/SiO$_2$, with $2.0013 < g < 2.0088$; these have been observed in un-annealed samples only [22].

3) Si dangling bond in the oxide, O$_3$Si- (E'-center), which has two modifications: E'$_\gamma$ with $2.00023 < g < 2.00173$ and E'$_\delta$ with $2.00168 < g < 2.00209$. The E'$_\gamma$-centers have been observed in un-annealed samples but not in annealed ones above the sensitivity of the technique (10^{15} cm^3), unless they were irradiated. The generation rate of these centers at 10 Mrad dose in BOX is about the same as in silica glass densified by 10% which is about ten times higher than normal silica glass [12]. The E'$_\gamma$-centers are uniformly distributed in the BOX even when irradiated under bias [23]; this behavior is different from that of thermally grown oxides in which case the centers are mostly in the vicinity of the Si/SiO$_2$ interface. It was found that the generation rate of E'$_\gamma$-centers is increased by a factor of 25 when the top Si-layer was etched off [23]. Similarly, hydrogen exposure of SIMOX samples as, for

instance, during epi-Si deposition, resulted in tenfold increase in the density of these centers [24]. The suggested mechanism is as follows. Hydrogen reacts with the O_3Si-SiO_3 bond and forms O_3Si-H. This group dissociates upon irradiation to O_3Si- and H. The highly mobile H atoms can form H_2 molecules and so minimize the probability of recombination with a Si dangling bond. This process should be more efficient than the dissociation of O_3Si-SiO_3 to O_3Si-, $^+SiO_3$, and e^- because of the finite probability of the electron recombining with the positive charge and, thus, re-forming the Si-Si bond.

The E'_δ -center is also a Si dangling bond defect in nc SiO_2. This has been observed in *pseudo-SIMOX* samples (top Si layer removed) [25]. The defects were generated by impinging ions during Ar plasma exposure within a 10 nm thick layer of the BOX. The defect concentration profile was measured by repeated irradiation and etching. It was found that the profile essentially mirrors the profile of etch rate in dilute HF. This suggests that both effects are related to strained Si-O bonds associated with Si-Si bonds. The ESR signal of these centers indicates that the unpaired electron is delocalized among at least five Si atoms forming a cluster of Si-Si bonds. Heat treatment in oxygen at 950 C prior to irradiation apparently oxidized these bonds as no signal was observed.

In addition to these ESR centers in the BOX, there is also an ESR-active defect in the top Si layer and Si substrate which is unique to SIMOX and BESOI, in general, confined structures. This defect is a double-donor that becomes ESR-active when ionized by the presence of positive charge in the BOX or applying appropriate bias to the Si substrate [26]. This defect appears to be an oxygen interstitial and its density is roughly correlated with that of the E'_δ centers in BOX. The E'_δ precursors and the double donor defects appear to be present in confined structures only; see discussion below (Section 5.).

4.1.4. *Electrical conduction*

From the viewpoint of SIMOX devices the electrical properties of the BOX layers are very important. Electrical conduction in BOX has a bulk and localized ("defect") component; the first one is area-dependent while the second one is not. It appears that excess silicon in various forms plays an important role in both processes. The defect conduction can be a major factor in determining the yield of device fabrication.

Bulk Conduction. Buried oxide (BOX) layers in SIMOX structures exhibit well defined and reproducible bulk ("background") conduction which, in contrast to defect conduction, is area-dependent [27]. For annealed samples the bulk conduction is quasi-ohmic and time-dependent up to 80 V (for ~ 400 nm thick BOX); the temperature dependence indicates 0.3 eV as the activation energy associated with trapping effects.

At higher voltage the current-voltage characteristics are super-linear and resemble those observed for deposited SiO_2 films that contain excess silicon in the form of Si-clusters [11]; see fig.2. The high-field conduction depends very little on temperature, indicating that the conduction is controlled by tunneling between Si clusters. Fowler-Nordheim plots indicate an effective barrier height of 1.3 - 1.5 eV for annealed samples and 0.4 eV for un-annealed ones; this range is intermediate between 3.1 eV (thermal oxide) and 0.4 - 0.6 eV (Si-rich SiO_2). From the effective barrier heights of annealed samples the size of the clusters is estimated as 0.5 nm (three Si atoms) and their density as 2×10^{19} cm^{-3} so that the concentration of excess silicon is at least 6×10^{19} cm^{-3}. This value is comparable to that estimated from the refractive index, as mentioned above. For un-annealed BOX the order of magnitude of excess silicon concentration is estimated as 10^{21} cm^{-3}.

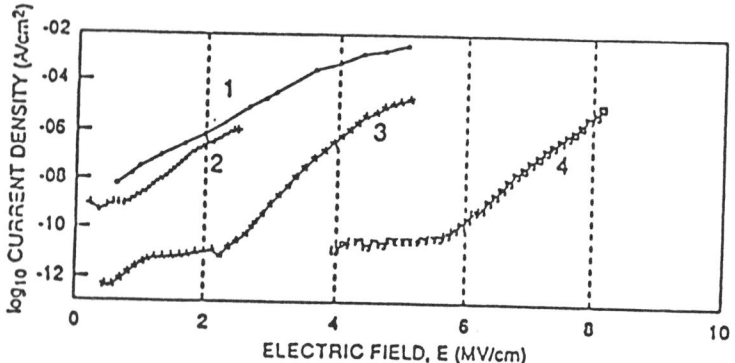

Figure 2. Logarithm of current density as a function of electric field.
Curve 1: Si-rich SiO_2 [11]; Curve 2: un-annealed BOX
Curve 3: annealed BOX; Curve 4: thermal oxide.[27]

The conduction depends on polarity; this effect is probably associated with the asymmetry in the distribution of excess silicon at the two interfaces. For un-annealed samples the current at a given voltage is several orders of magnitude higher than for the annealed ones and the onset of the super-linear regime is at lower voltage; this behavior is similar to the defect conduction exhibited by annealed SIMOX samples.

Supplemental oxygen implantation eliminates the trapping-controlled conduction at low field and renders the high field conduction somewhat similar to that characteristic of thermally grown oxide, but without becoming identical to it. All of the observations indicate that the BOX layers contain excess silicon; this may be in the form of O_3Si-SiO_3 bonds in the oxide and/or very small silicon (probably amorphous) clusters/filaments. The supplemental oxygen implantation apparently eliminates one kind of defects whereas the density of the other kind is only reduced; this effect is similar to that observed by photo-injection techniques to be discussed below.

Defect Conduction. The BOX layers also exhibit localized (defect) conduction superimposed on the background (bulk) conduction [28]. There are two types of conducting defects: Type I defect shows a pre-breakdown quasi-linear I-V characteristics with 10^{-6} A $<I< 10^{-3}$ A in the voltage range of 0.01 - 10 V. Type II defects exhibit a super-linear I-V behavior above 5 V and breakdown or current-saturation usually occurs at 10 - 50 V (for $\sim$400 nm thick BOX). A large number of samples prepared in various manners has been studied with automatic test equipment by which the number of Type I defects has been determined from several hundreds of capacitors on a given wafer. The defect density was calculated using the Poisson distribution; for annealed samples the values range, by and large, from 0.1 to 10 defects/cm^2, while for un-annealed samples the range is 40 - 120 defects/cm^2. The pre-breakdown I-V characteristics of Type II defects resemble the bulk conduction behavior of un-annealed BOX. These defects appear to be closely related to some fundamental aspects of the bulk conduction of BOX layers and/or confinement/contamination effects. Type II defects appear to be, at least partially, responsible for process-induced leakage effects in completed SIMOX IC's [29].

It has recently been discovered that the localized (defect) leakage conducting paths in BOX are associated with light scattering centers on the surface of the top Si-layer [17]. Some of these leakage paths are similar to Type I defects mentioned above in the sense that they are highly conducting, whereas some others become conducting as result of processing. The correlation between device process yield, as determined by leakage current in BOX, and the density of light scattering defects is shown in Fig. 3.

Most of the defects on the Si surface result in etch pits in HF, many of which penetrate even the BOX. Note that silicon cannot be etched in HF unless some oxidizing agent is present or the silicon is polarized anodically; the latter case is pit corrosion. It was found by X-ray technique that most of the defects are associated with oxygen but some of them contain nickel or chromium [17]. It is not clear at the present how these O-containing defects in the the top Si-layer are related to the conducting paths in the BOX and what is the mechanism of the leakage conduction. It is important to realize that the leakage conducting paths in the BOX are obviously heterogeneities in the oxide; photo-injection studies, discussed below (Section 4.1.5.), also indicate the presence of heterogeneities, amphoteric Si-clusters, in the BOX. Another point is that defects in the BOX (as observed by the copper plating technique) which are associated with channeling directions in the silicon are more conducting than those which are not [3]. The I-V characteristics of these defects resemble those of the Type II defects mentioned above.

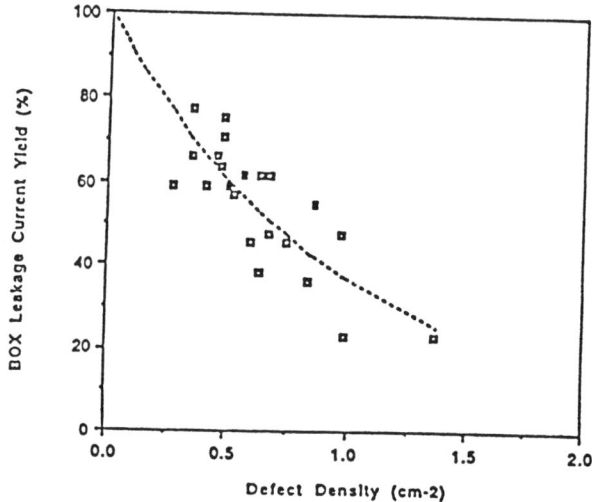

Fig. 3. Yield of device processing determined by BOX leakage current as a function of the density of light scattering defects [17].

These observations indicate that defect conduction in BOX is a more complex, perhaps even fundamental, aspect of SIMOX structures than being simply caused by particles on the surface of silicon which would locally mask the O-beam during implantation. Defect conduction in the BOX is presently a major problem in the technology of SIMOX devices as it can greatly reduce the yield of device processing and may even affect the reliability of the devices.

4.1.5. Photo-injection studies

The photo-injection studies were performed on pseudo-SIMOX samples where the top Si layer was replaced by Au electrodes [30]. The MOS structures were exposed to UV ($h\nu < 6$ eV) or VUV ($h\nu = 10$ eV) photons under positive or negative bias on the metal electrode. The charge accumulation in BOX was monitored by 1 MHz C-V measurements and electron photo-injection I-V characteristics. The processes occurring under these conditions are shown in Fig. 4. In the spectral range of $3.5 < h\nu < 4.3$ eV only electron injection from the metal electrode is possible (transition 1). If the photon energy is within 4.3 and 6 eV, photo-injection of electrons from the Si substrate (transition 2) and optical ionization of the photo-active defects in BOX (transition 3) become possible. Illumination with 10 eV photons generates electron-hole pairs in the oxide within 10 - 20 nm from the metal electrode (transition 4) and, depending on bias, electrons or holes can be injected into the BOX.

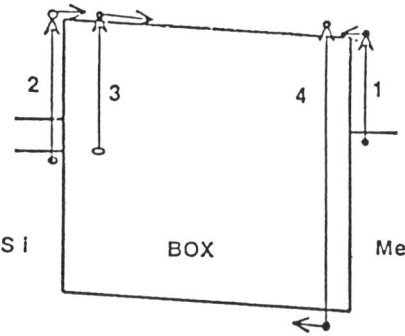

Figure 4. Processes occuring during photo-injection [31].

The unique result of these studies is that there are defects in the BOX which become positively charged by photo-ionization (transition 3) under relatively low electron injection level [30]; this effect is shown for various samples in Fig.5 [32].

This kind of positive charge generation has never been observed with thermally grown oxide films, including BOX in BESOI structures. The process is independent of the applied bias, indicating that it is not related to hole injection from the Si substrate. Notice in Fig. 5 the significant difference between the single-and triple-implanted samples.The high density of defects in the SI sample can be greatly reduced by supplemental O-implant, especially for defects with relatively small capture cross section. However, the density of high cross section defects is still large, even though less than without supplemental O-implant. This is a potentially important point as there is a tendency in the present technology to use SI rather than TI structures. If a standard thick BOX layer is etched down to 150 nm, its behavior is practically identical to that of the thin sample.

If electron injection continues after the saturation of the positive charge, the defects trap electrons so the oxide charge becomes negative. Clearly, the photo-active defects are amphoteric. The electron trapping behavior of the various samples is similar to their photo-ionization behavior, except that the saturation value of the negative charge is higher than that of the positive charge. The reason for this effect is that, in addition to the amphoteric defects, there are non-amphoteric ones as well. The minimum photo-excitation energy needed to generate the positive charge in BOX corresponds to the difference between the silicon valence band and oxide conduction band. The large capture cross section of the amphoteric defects (10^{-14} -10^{-13} cm^2) does not depend on the applied field. These observations indicate that these defects are not point (network) defects but heterogeneities in the oxide, very likely

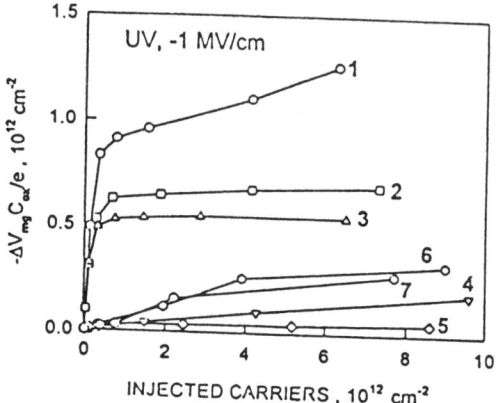

Figure 5. Generation of positive charge in BOX with photons of 4.3 eV or higher energy. Curve 1: standard (400 nm BOX) single-implanted (SI) sample; curve 2: standard (SI) sample with suppl. O-implant (10^{17} O/cm^2 dose); curve 3: standard (SI) sample with suppl. O-implant (5×10^{17} O/cm^2 dose); curve 4: standard triple-implant (TI) sample; curve 5: standard TI sample with suppl. O-implant (10^{17} O/cm^2 dose); curve 6: thin (150 nm) BOX; curve 7: thin BOX with suppl. O-implant (10^{17} O/cm^2 dose). [32]

Si-clusters. The behavior shown in Fig. 5 and the similar electron trapping effects show that the large clusters, which are predominant in the SI samples, are near the BOX/Si-substrate interface, whereas the small clusters are in the "bulk" of the oxide. The density of small clusters is reduced in the thin or thinned down samples as well as in the samples which received supplemental O-implant.

Heat treatment of un-confined pseudo-SIMOX samples in argon or oxygen, using resistance heated tube furnace, almost completely eliminates the amphoteric defects even in SI samples [32]. These defects have not been observed in the so-called equilibrium oxide either; this oxide is prepared by implanting such a high dose of oxygen that the top Si layer is totally consumed in forming the 550 nm thick oxide [33], hence, the oxide is un-confined.

These observations demonstrate that the formation of amphoteric Si-clusters is tied to oxygen implantation and the confined nature of the BOX. The removal of the large clusters by a post-anneal treatment is possible only in un-confined structures. Various aspects of confinement are discussed below (Section 5.).

As the Si-clusters were eliminated by heat treatment of pseudo-SIMOX samples, electron trapping in its "pure" form (i.e., not associated with the amphoteric defects) could be studied [34]. It was found that there are two, relatively deep, neutral traps with 3×10^{-16} and 2×10^{-17} cm^2 trapping cross sections. In addition there are shallow traps with very large cross section (of the order of 10^{-11} cm^2) near the Si-substrate/BOX interface; these traps are probably the large crystalline Si platelets that occur in single implant samples.

Comparison with results obtained by other techniques. The results obtained by the photo-injection technique can be compared with those of other, somewhat similar, techniques. Thus, for example, the electron trapping behavior of BOX layers with and without supplemental O-implant was studied by avalanche electron injection [35]. It was found that 5×10^{17} O$^+$/cm^2 dose of supplemetal O-implant reduced the density of traps from 3×10^{12} cm^{-2} to 2.1×10^{12} cm^{-2} and decreased their capture cross section from 6×10^{-14} to 10^{-14} cm^2; also, it introduced 10^{12}cm^{-2} new traps with 10^{-15} cm^2 cross section.

Electroluminescence (EL) studies showed the presence of 2.7 and 4.4 eV bands in the spectra obtained with BOX layers [36]. The defects responsible for these bands are associated with excess silicon (or oxygen vacancy). Similarly to the Si-clusters discussed above, the EL-active defects are located near the BOX/Si-substrate interface. It was also found that the 1.9 eV band, which is related to SiOH groups, is not present in the EL spectrum of SIMOX BOX layers.

4.1.6. Irradiation behavior

The irradiation behavior of SIMOX devices is similar to that of conventional MOS devices in the sense that ionizing radiation generates net positive charge in the oxide. However, there are important differences between BOX and thermal oxide:

1) Both hole and electron trapping occurs in BOX, the holes remain trapped in deep traps but a fraction of electrons is de-trapped from shallow traps leaving a net positive charge in the oxide; the electron-hole recombination is controlled by interactions between field and space charges during irradiation and the extent of electron de-trapping depends on temperature [37, 38]. No electron trapping associated with irradiation has been observed with thermal oxides.

2) The trapped holes are distributed uniformly in the BOX rather than being accumulated at the interface(s), as in the case of thermal oxides.

3) There is practically no hole transport, hence, the radiation-induced increase in the density of interface states is very low.

4) The density of radiation-induced E'_γ - centers is much higher than the net positive charge [23]. Accordingly, the model suggested for thermal oxides, that irradiation generates both E'_γ -centers and positive charge from precursors:

$$O_3Si...SiO_3 + h\nu \longrightarrow O_3Si^+ + {}^-SiO_3 + e^-, \qquad (4)$$

the electrons being swept away, does not hold for BOX. Here the bond between the Si atoms is not necessarily the same as the Si-Si bond referred to above or the mechanism of the generation of the E'_γ center and charge is different.

From field ionization (tunneling) it was determined that the shallow electron traps are distributed in the energy range of 0.3 to 1.5 eV with the peak at 1.0 eV and 10^{11} cm^{-2} (eV)$^{-1}$ density giving a total density of $\sim 10^{11}$ cm^{-2} [38]. This value is reasonably close to the barrier height determined from Fowler-Nordheim conduction, 1.3 - 1.5 eV, as discussed above. The shallow electron traps can be removed by heat treating pseudo-SIMOX samples [38]; this result is in agreement with the observation that such a heat treatment removes the shallow traps responsible for the time-dependent low field electric conduction of BOX (Section 4.1.4.)

Supplemental O-implant has also removed the shallow electron traps without affecting the deep electron and hole traps [39]; this result is similar to that found in the course of photo-injection studies (Section 4.1.5.).

Recent studies using very high radiation dose (up to 30 Mrad) showed that 4×10^{16} O$^+$/cm^2 supplemental O-implant followed by heat treatment at 1000 C did not affect the hole trapping behavior but had a slight effect on the electron trapping behavior [40]. Namely, it reduced somewhat the density of traps with 2.2×10^{-13} cm^2 trapping cross section from 2.5×10^{12} to 1.5×10^{12} cm^{-2} and introduced 2.5×10^{11} cm^{-2} new traps with a 10^{-15} cm^2 cross section. Thus, the total density of traps remained unchanged. If the heat treatment after the supplemental O-implant was performed at 1320 C, the effect of the supplemental O-implant vanished. This result is very likely related to confinement effects which take place at the higher but not at the lower temperature (see Section 5.)

An interesting feature of the irradiation behavior of BOX/Si interfaces is that there is practically no increase in the density of interface states. This may be due to the lack of hole transport, the stability of the interface formed at high temperature [5], or the undetectable amount of hydrogen ($<\sim 10^{17}$ cm^{-3}) or to a combination of these factors.

4.2 BESOI BOX LAYERS

As the BOX layers in BESOI structures are thermally grown or one of them is the native oxide grown at room temperature on the Si surface, they are more similar (but not identical) to conventional SiO$_2$ films than SIMOX BOX layers. The similarity between SIMOX and BESOI BOX layers is due to the fact that they are confined. The main features of the defect structure of BESOI BOX layers are that voids may be present in the oxide; also, there are SiOH and SiH groups, an H-related precursor of an ESR-active center, and the precursor of the ESR-active E'$_\delta$-center. Accompanying the E'$_\delta$ precursors, there are double-donor oxygen interstitials in the silicon; these two defects result form the confined nature of the BOX layer (see Section 5.).

4.2.1. Voids in the BOX layer

Voids result from insufficient wafer flatness and surface finish as well as from surface contamination and particulates [41]. The voids at the bonding interface contain air pockets which are under pressure, resulting in localized stress in the Si wafer. The stress may result in plastic deformation during subsequent thermal processing. Heat treatment at 1050 C generated dislocations that were not confined to the voids but extended into the surrounding area. Pinholes are created in the Si layer where voids are present in the BOX and etch pits form in HF [42].

4.2.2. Hydrogen in the BOX

As the first step in the bonding process is the formation of hydrogen bonds, the bonding process usually requires the presence of water [2]. SIMS analysis indicated large amount of hydrogen at the bonding interface and, in the case of SiO$_2$ - SiO$_2$ bonding (as opposed to Si(native oxide) - SiO$_2$ bonding), in the oxide as well [43]. Infrared analysis showed that after bonding hydrophylic wafers there are 10^{15} cm^{-2} H-bonded H$_2$O molecules present. [44] Their IR signature gradually disappears above 150 C. The signature of H-bonded SiOH groups appears at 300 C; these groups become isolated (not H-bonded) at 600 C. In addition, there are two types of SiH groups present: one is associated with the bonding interface, it disappears at 700 C; at this temperature the other one in the oxide appears. The concentration of the latter decreases with increasing temperature of heat treatment and becomes nil at 1100 C.

Fowler-Nordheim injection of electrons into the oxide layer of pseudo-BESOI [Si(native oxide-SiO$_2$] structure results in charging the oxide and increasing the density of interface states at the bonding interface [43].

In another study Si/BOX/Al structures were exposed to electrical stress at room temperature [45]. Positive ionic mode instability was observed to develop when the BOX layer was exposed to H$_2$O. An isotopic effect was observed when H was replaced by D indicating H/D transport through the BOX. Electron injection and UV illumination were found to enhance the charge instability. Electron traps with 10^{-17} cm^2 cross section (most likely SiOH groups) are present in much larger concentration in the BOX layer than in non-bonded thermal oxide. These features suggest structural changes in the thermal SiO_2 associated with the bonding process used for preparing BESOI structures which leads to H-enrichment.

Another aspect of hydrogen in BESOI BOX is the presence of a new ESR-active defect with delocalized spin observed after VUV illumination; this defect is labelled as the EH center [21]. The hyperfine coupling of this defect indicates that it is associated with hydrogen. Most of the EH and other ESR centers are near the bonding interface. These observations are in agreement with the results obtained with photo-injection measurements.

4.2.3. Irradiation behavior

The irradiation behavior of BESOI structures bonded at 900 C is similar to that of conventional MOS structures with thermally grown oxide. For instance, in contrast to SIMOX BOX, hole transport takes place but strong trapping effects may be associated with the bonding interface [46]. If the bonding is performed at higher temperature or the sample is heat treated, then with increasing temperature its behavior is becoming increasingly similar to that of SIMOX samples; heat treatment at 1325 C renders the BESOI samples indistinguishable from SIMOX samples [38]. This is strong manifestation of the confinement effect.

5. CONFINEMENT EFFECTS

The common feature of SIMOX and BESOI BOX layers, which sets them apart from conventional thermally grown SiO_2 films, is that they are confined (contained) between two silicon layers. As a result of the confinement, SIMOX BOX may not fully relax the strain generated by the transformation of silicon into the oxide associated with a large increase in the molar volume (see Section 2.1.). Hence, the SIMOX BOX is very likely in some densified form in which the distribution of the Si-O-Si bridging bond angles is shifted to lower values and the distribution of the rotational (dihedral) angles is such that the O - 2nd O (or Si - 2nd Si) distances are shorter than in a fully relaxed oxide (see Section 3.) The lack of relaxation may promote the persistence of medium range structural features which are influenced by the structure of the Si substrate as, for instance, channels as well as Si-clusters, especially near the Si/BOX interface (see Section 4.1.5.). It is possible that these structural features are responsible for the effect of Si crystallographic orientation on the density of conducting defects in SIMOX BOX. The densified (strained) structure also renders the oxide more reactive which is manifested in its reaction with hydrogen, increased generation rate of ESR centers, etc. In this sense the effect of confinement is a fundamental aspect of the behavior of SIMOX BOX layers.

Since the oxide films in BESOI structures are un-confined before bonding, they are not as strained (if at all) as the SIMOX BOX layers. However, the BESOI BOX layer resulting from bonding is confined and some of its properties are very similar to those of SIMOX BOX layers. These properties are the presence of defects in the oxide which upon irradiation, VUV illumination, or hole injection become the E'_δ ESR - centers [21, 23]. The density of these centers is practically identical (usually of the order of 10^{12} cm^{-2}) to that of the O-interstitial double-donor defects in the Si layers which become ESR - active if positive charge is present in the BOX or the sample is appropriately biased [26]. As the E'_δ defect is an O-vacancy in the oxide, the strong coupling between the two defects indicates that oxygen atoms from the BOX got transferred into the Si layer(s). Apparently, this process occurs in confined structures only as these defects have not been observed in un-confined structures. It is curious that heat treatment of pseudo-SIMOX sample in oxygen removed both defects even though one would think that such a treatment would increase the concentration of O-interstitials in silicon. It is difficult to reconcile these observations with the suggestion that interaction between BOX and silicon is a simple transfer of O atoms from the oxide (and generating O-vacancies there) to silicon where O-interstitials form [47].

One might think that the E'_δ precursors are the shallow traps observed in SIMOX and in those BESOI structures which experienced heat treatment above 900 C. However, these centers are present even in those BESOI samples whch were bonded at 900 C and no further heat treatment took place [23]; these samples did not show shallow trapping effects. Hence, the correlation between E'_δ precursors and shallow electron trap is not good. Also, there is no correlation between E'_δ defects and Si-clusters as the latter are not present in BESOI BOX. Apparently, there are various forms of excess silicon (oxygen vacancies) in the BOX layers and only some (or just one) of them can interact with the Si layer(s). The nature of this interaction between BOX and silicon is not clear at the present. One can speculate that the excess silicon in confined structures is generated by a process which is completely different from that characteristic of un-confined structures. It is possible that the generation of O interstitials in silicon (double-donor defects) in confined structures is associated with injection of Si interstitials into the BOX where they become part of the oxide network, perhaps, by forming bonds with dissolved oxygen that remained in the oxide at the end of thermal oxidation or oxygen implantation. Another possibility is that Si atoms injected into the confined oxide form aggregates. The generation of excess silicon in this manner cannot occur in un-confined structures as the Si atoms in the oxide would be immediately oxidized by oxygen entering the oxide through its surface.

It is very clear that confinement effects represent a major issue in both SIMOX and BESOI technologies as they generate defects in both the BOX and silicon which have not been encountered with conventional un-confined structures and, hence, have not been studied in such a great detail as the defects there.

6. CONCLUSION

The defect structure of BOX layers in both SIMOX and BESOI structures is significantly different from that of conventional thermally grown SiO_2 films even though all these oxides are noncrystalline. This difference is primarily due to the presence of various forms of defects associated with excess silicon or oxygen vacancy in the BOX layers which result from the confined nature of the BOX layers and, hence, are not present in conventional, un-confined structures. These defects are the precursors of the ESR-active E'_δ centers that are linked somehow to double-donor oxygen interstitials in the silicon layer(s). Another form of excess silicon is the Si-cluster in SIMOX BOX layers; oxygen ion implantation and confinement effects are apparently responsible for their formation. Voids and hydrogen in various forms are important defects in BESOI BOX layers. The defects in the BOX layers strongly affect the electrical properties; and may be the cause of various problems in the yield of device processing and the reliability of the devices.

ACKNOWLEDGEMENTS

The contribution of A. G. Revesz was sponsored by the Naval Research Laboratory, Washington DC under Contract N00014-92-C-2264 supported by the Defense Nuclear Agency, Washington, DC. The authors are indebted to A. Stesmans for helpful discussions.

REFERENCES

1. Colinge, J. P. (1991) *Silicon-on- Insulator Technology*, Kluver Academic Publishers, Dordrecht.
2. Maszara, W. P., Goetz, G., Caviglia, A., and McKitterick, J. B. (1998) Bonding of silicon wafers for silicon-on-insulator, *J. Appl. Phys.* **64**, 4943.
3. Anc, M. and Krull, W. A. (1994) Effects of implant angle on SIMOX structures and properties of buried oxide, *1994 IEEE International SOI Conference Proc.*, pp.79-80.
4. Lawrence, R. K., Hughes, H. L., and Revesz, A. G. (1992) Photocurrent measurements of electron traps in SIMOX, *1992 IEEE International SOI Conference Proc.* pp. 106-107.
5. Revesz, A. G. (1973) Noncrystalline silicon dioxide films on silicon: a review, *J. Non-Cryst. Solids* **11**, 309-330.
6. Revesz, A. G. and Gibbs, G. V. (1980) Structural and bond flexibility of vitreous SiO_2 films, in G. Lucovsky, S. T. Pantelides, and F. L. Galeener (eds.), *The Physics of MOS Insulatorrs*, Pergamon, New York, pp. 92-96.
7. Revesz, A. G. and Walrafen, G. (1983) Structural interpretations for some Raman lines from vitreous silica, *J. Non-Cryst. Solids* **54**, 323-333.
8. Devine, R. A. B. (1993) Ion implanation and radiation induced structural modifications in amorphous SiO_2, *J. Non-Cryst. Solids* **152**, 50.
9. Devine, R. A. B. and Arndt, J. (1989) Correlated defect creation and dose-dependent radiation sensitivity in amorphous SiO_2, *Phys. Rev.* **B39**, 5132.
10. Revesz, A. G., Mrstik, B. J., and Hughes, H. L. (1988) Thermal oxidation of silicon, in R. A. B. Devine (ed.), *The Physics and technology of Amorphous SiO_2*, Plenum, New York, pp.297-306.
11. DiMaria, D. J., Wong, D. W., Falcony, C., Theis, T. N., Kirtley, J. R., Tsang, J. C., Young, D. R., and Pesavento, F. L. (1983) Charge transport and trapping phenomena in off-stoichiometric silicon dioxide films, *J. Appl. Phys.* **54**, 5801-5827.
12. Stesmans, A., Devine, R. A. B., Revesz, A. G., and Hughes, H. L. (1990) Irradiation induced ESR active defects in SIMOX sructures, *IEEE Trans.* **NS-37**, 2008-20012.
13. Stahlbush, R. E., Carlos, W. E., and Prokes, S. M. (1987) Radiation induced effects in SIMOX: a spectroscopic study, *IEEE Trans.* **NS-34**, 1680-1685.
14. Stesmans, A. and Vanheusden, K. (1991) Chemical etch rates in HF solutions as a function of thickness of thermal SiO_2 and buried SiO_2 formed by oxygen ion implantation, *J. Appl. Phys.* **69**, 6656.

15. McMarr, P. J., Mrstik, B. J., Barger, M. S. Bowden, G., and Blanco, J. R. (1990) A study of Si implanted with oxygen using spectroscopic ellipsometry, *J. Appl. Phys.* **67**, 7211-7222.
16. Roitman, P., Edelstein, M., Kraus, S., and Kisitserntrukal (1990) Residual defects in SIMOX: threading dislocations and pipes, in *1990 IEEE SOI Proc.*, pp.104-105.
17. Yue, J., Liu, S. T., Fahner, P., Gardner, G., Witcraft, W., and Finn, C. (1994) An effective method to screen SOI wafers for mass production, *1994 IEEE International SOI Conference Proc.* pp. 113-114.
18. Revesz, A. G., Brown, G. A., and Hughes, H. L. (1993) in J. Kanicki, W. L. Warren, R. A. B. Devine, and M. Matsumara (eds.), *Mat. Res. Soc. Symp. Proc.*, Vol. 284, pp. 555-566.
19. Myers, S. M., Brown, G. A., Revesz, A. G., and Hughes, H. L. (1993) Deuterium interaction with ion-implanted layers in silicon, *J. Appl. Phys.* **73**, 2196-2206.
20. Shelby, J. E. (1980) Reaction of hydrogen with hydroxyl-free vitreous silica, *J. Appl. Phys.* **51**, 2589-2593.
21. Warren, W. L., Shaneyfelt, M. R., Schwank, J. R., Fleetwood, D. M., and Winokur, P. S. (1993) Paramagnetic defect centers in BESOI and SIMOX buried oxides, *IEEE Trans.* **NS-40**, 1755-1764.
22. Barklie, R. C., Hobbs, A., Hemment, P. L. F., and Reason, K. (1986), *J. Phys. C.* **19**, 6417.
23. Hervé, D., Leray, J. L., and Devine, R. A. B. (1992) Comparative study of radiation-induced electrical and spin-active defects in buried SiO_2 layers, *J. Appl. Phys.* **72**, 3634.
24. Zvanut, M. E., Stahlbush, R. E., Carlos, W. E., Lawrence, R. K., Hevey, R., and Brown, G. A. (1991) SIMOX with epitaxial silicon: point defects and positive charge, *IEEE Trans.* **NS-38**, 1253.
25. Stesmans, A. and Vanheusden, K. (1993) Depth profiling of oxygen vacancy generation in buried SiO_2 thin films, in J. Kanicki, W. L. Warren, R. A. B. Devine, and M. Matsumura (eds.), *Mat. Res. Soc. Symp. Proc.* Vol. 284, p. 299.
26. Stesmans, A., Revesz, A. G., and Hughes, H. L. (1991) Electron spin resonance of defects in silicon-on-insulator structures formed by oxygen implantation: influence of γ-irradiation, *J. Appl. Phys.* **69**, 175-181.
27. Revesz, A. G., Brown, G.A., and Hughes, H. L. (1993) Bulk electrical conduction in buried oxide of SIMOX structures, *J. Electrochem. Soc.* **140**, 3222-3229.
28. Brown, G. A. and Revesz, A. G. (1993) Defect electric conduction in SIMOX buried oxides, *IEEE Trans.* **ED-40**, 1700-1705.

29. Hosack, H. H., Joyner, K. A., El-Ghor, M. K., Hollingsworth, J., Brown, G. A., and Pollack, G. P. (1992) Particle effects on buried oxide leakage in SIMOX materials, *1992 IEEE International SOI Conference Proc.*, pp. 98-99.
30. Fedosenko, S. I., Adamchuk, V. K., and Afanas'ev, V. V. (1993) Silicon clusters as photo-active traps in buried oxide layers of SIMOX, *Microelectronics Eng.* **22**, 367.
31. Fedosenko, S. I., Afanas'ev, V. V., and Revesz, A. G. (1994) Charge trapping in BOX layers of SIMOX structures covered with epitaxial silicon, in S. Cristoloveanu (ed.), *Proc. Sixth International Symposium on Silicon-on-Insulator Technology and Devices,* The Electrochem. Soc. Pennington, NJ, pp.253-258.
32. Afanas'ev, V. V., Revesz, A. G., and Hughes, H. L. (1994) Photo-injection study of SIMOX structures with supplemental oxygen implant, *1994 IEEE International SOI Conference Proc.*, pp. 87-88.
33. Hosack, H. H., Hollingsworth, J., El-Ghor, M. K., and Joyner, K. A. (1992) "Equilibrium oxide" fetures of the SIMOX process, *Mat. Res. Soc. Sump. Proc.* **235**, pp.159-164.
34. Afanas'ev, V. V., Revesz, A. G., Brown, G. A., and Hughes, H. L. (1994) Deep and shallow electron trapping in the buried oxide layers of SIMOX structures, *J. Electrochem. Soc.* **141**, 2801-2804.
35. Bhar, T. N., Lambert, R. J. and Hughes, H. L. (1994) Decrease in electron capture cross section in SIMOX with supplemental implant, *1994 IEEE International SOI Conference Proc.*, pp. 115-116.
36. Bota, S., Perez-Rodriguez, A., Morante, J. R., Baraban, A., and Konorov, P. P. (1994) Electroluminescence analysis of the screen oxide in SIMOX structure, in S. Cristoloveanu (ed.), *Proc. Sixth International Symposium on Silicon-on Insulator Technology and Devices,* The Electrochemical Society, Pennington, NJ
37. Boesch, H. E., Taylor, T. L., Hite, I. R., and Bailey, W. E. (1990) Time-dependent hole and electron trapping effects in SIMOX buried oxide, *IEEE Trans.* **NS-37**, 1982.
38. Stahlbush, R. E., Campisi, G. J., McKitterrick, J. B., Maszara, W., Roitman, P., and Brown, G. A. (1992) Electron and hole trapping in iradiated SIMOX, ZMR, and BESOI buried oxides, *IEEE Trans.* **NS-39**, 2086.
39. Stahlbush, R. E., Hughes, H. L., and Krull, W. (1993) Reduction of charge trapping and electron tunneling in SIMOX by supplemental implantation of oxygen, *IEEE Trans.* **NS-40**, 1740-1754.
40. Flament, O., Paillet, P., Leray, J. L., Asper, B., Griffard, B., and Auberton-Hervé, A. J. (1994) Effect of supplemental dose in SIMOX on very high ionizing dose response, in S. Cristoloveanu (ed.), *Proc. Sixth International Symposium on Silicon-on-Insulator Technology and Devices,* The Electrochemical Societ, Pennington, NJ, pp. 381-389.

41. Baumgast, H., Pinker, R. D., Steigmeier, E. F., Anderset, H. and deKroch, A. J. R. (1989) Impact of interface preparation on defect generation during wafer bonding, *1989 IEEE SOS/SOI Conference Proc.*, pp. 95-96.
42. Sadana, D. K., Lasky, J., Hovel, H. J., Petrillo, K., and Roitman, P. (1994) Nano-defects in commercial bonded SOI and SIMOX, *1994 IEEE International SOI Conference Proceedings*, pp. 111-112.
43. Bengtsson, S., Ericson, P., Mitani, K., and Abe, T. (1994) Charge carrier injection into the buried oxide of wafer bonded silicon-on-insulator materials, in S. Cristoloveanu (ed.), *Proc. Sixth International Symposium on Silicon-on - InsulatorTechnology and Devices*, The Electrochemical Society, Pennington, NJ, pp. 245-252.
44. Feijoo, D., Chabal, Y. J., and Christman, S. B. (1994) Multiple internal reflection infrared absorption analysis of bonded silicon wafers, *1994 IEEE International Conference Proc.*, pp. 89-90.
45. Afanas'ev, V. V., Revesz, A. G., Brown, G. A., and Hughes, H. L. (1994) Hydrogen-induced charge instability of buried oxide of BESOI structures, submitted to the *J. Electrochem. Soc.*
46. Pennise, C. A., Boesch, H. E., Goetz, G., and McKitterick (1993) Radiation-induced charge effects in buried oxides with different processing treatments, *IEEE Trans.* **NS-40**, 1765-1773.
47. Devine, R. A. B., Mathiot, D., Warren, W. L., Fleetwood, D. M., and Aspar, B. (1993) Point defect generation and oxide degradation during annealing of the Si-SiO_2 Interface, *Appl. Phys. Lett.* **63**, 2926.

IR STUDY OF BURIED LAYER STRUCTURE ON DIFFERENT STAGES OF TECHNOLOGY

LITOVCHENKO V.G., LISOVSKII I.P., LOZINSKII V.B.,
ROMANYUK B.N., MELNIK V.P.

*Institute of Semiconductor Physics Academy of Sciences,
Kiev, 252028, Ukraine*

1. Introduction.

SOI systems show promise for creation of reliable, temperature-stable and radiation hardened microelectronic devices, and, hence, have been intensively studied during the last decade. A series of theoretical and experimental results concerning the mechanism of producing SOI structures, the structural and electrophysical properties of the buried oxide layer has been obtained (see, for example, ref.1-3). These investigations made it possible to elaborate new approaches in the field of physics, processing and characterization techniques for SOI systems.

Among the available experimental techniques for study of the buried layer structure IR-spectroscopy is used. Its capability, however, is quite limited because of the traditional approach of spectra analysis: the maximum position of the absorption band is usually used to draw conclusions on the buried layer composition and its structure. Recently we have shown [4-5] that the characteristics of the short-range and local order in silica films can be determined using IR-spectroscopy with subsequent analysis of the absorption band shape. Since the silicon-oxygen phase in SOI systems should possess the main structural features typical for SiO_2 and SiO_x ($x<2$) (bond length and bond angles) such a technique be also used in this case. In the present work IR study of the silica layer in SOI systems has been carried out, which enabled us to follow its development during different technological treatments.

2. Experimental techniques.

Samples about 450 μm thick were cut from a (100)-oriented P-doped ($\rho \approx 4.5 \Omega$ cm) Cz-Si wafer; both sides of the samples were polished. The initial concentration of interstitial oxygen and carbon was about 8×10^{17} and less than 2×10^{16} cm^{-3}, respectively. The region of silica phase was produced by combined implantation of oxygen (150

KeV, 2 - 4x10^{-2} C cm^{-2}) and carbon (100 KeV, 0.2, 1 and 5x10^{-2} C cm^{-2}) ions, and by subsequent heat treatment (650, 1150, 650+1150° C during 2 hours) in inert (Ar) ambient. Specification of the samples under investigation is presented in Table 1.

Table 1. Specification of the samples

Sample	Regime of treatments	C_i (10^{17} cm^{-3})	C_p (10^{17} cm^{-3})	d (nm)
A1	initial	8.4	5.7	-
A2	Implantation of O$^+$, 1..2x10^{17} cm^{-2} ;C$^+$, 1.2x10^{15} cm^{-2}	9.1	11.0	-
A3	Postannealing at 650° C, 2 h	7.7	2.8	-
A4	Postannealing at 1150° C, 2 h	9.1	14.5	-
A5	Postannealing at 650° C, 2 h + 1150° C, 2h	8.5	layer	21
B1	initial	8.1	5.7	-
B2	Implantation of O$^+$, 1..2x10^{17} cm^{-2} ;C$^+$, 3.5x10^{15} cm^{-2}	8.4	6.7	-
B3	Postannealing at 650° C, 2 h	8.0	4.5	-
B4	Postannealing at 1150° C, 2 h	8.5	layer	20
B5	Postannealing at 650° C, 2 h + 1150° C, 2h	9.6	layer	27

IR transmission spectra were measured in the wavenumber range 950<v<1300cm^{-1} where the Si-O stretching band is located. The wafer of a float-zoned Si was used as a reference sample. The analysis of the shape of the absorption band was carried out using a method of computer deconvolution of the absorbance curves into gaussian profiles. As a result their main parameters (maximum position W, full width H at half maximum, intensity I) were determined. The details of this method were described earlier [4,5]. To make this procedure more unambigous the parameters of the gaussian profile connected with interstitial oxygen atoms were fixed at W_1 =1107.5±0.5 cm^{-1} and H_1 =33±1cm^{-1} [6,7]. The deconvolution accuracy was characterized by the r.m.s. deviation of the summed gaussian profiles from the experimental spectrum and did not exceed ~2x10^{-2}. To determine the structural components of the silica phase, the results of deconvolution were interpreted in the framework of the random bonding mo-del (RBM). The value of interstitial oxygen concentration (C_1) was determined from the intensity of the 1107.5 cm^{-1} band using a calibration factor of 3.0x10^{17} at./cm^{-3}; the concentration of the oxygen atoms incorporated in silicon-oxygen phase (C_p) was estimated from the intensity of silica absorption band according to [7]. In both cases the effect of multiple reflections was taken into account.

Changes in chemical composition with the depth of the samples on the different stages of technological treatments were studied by means of the Auger electron spectroscopy; the spectra were measured between successive steps of sample sputtering with Ar$^+$ (2.8 KeV) ions.

3. Results and discussion.

Figure 1 shows changes in IR absorption spectra due to technological treatments of the samples "A". It is seen that spectrum of the initial sample was characterized by a rather symmetrical absorption band at 1107cm^{-1} which is known to be inherent in interstitial oxygen atoms. Ion implantation and subsequent low-temperature heat treatment slightly influenced the spectral curve. In the case of high temperature heat treatment of the implanted sample (especially in the combination with low temperature pre-anne-

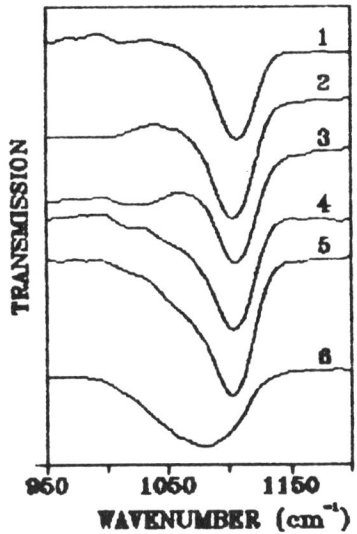

Figure 1. IR transmission spectra measured after different stages of technological treatments of the Si samples. Curves 1-5 correspond to samp-les A1-A5. Curve 6 represents a differential spectrum (i.e. refferred to that of the initial samp-le) measured on the sample A5.

aling) IR spectra changed drastically: the maximum position shifted towards smaller wavenumbers to reach ~1100 cm^{-1}. The spectra broadened in the range of 1000-1100 cm^{-1} and became asymmetrical. The differential spectrum of such a sample (i.e. referred to that of the initial sample) has the maximum position at ~1080 cm^{-1} which is known to be inherent in bridging oxygen atoms in silica. Deconvolution of the Si-O stretching band into gaussian profiles is presented in Fig.2. It is seen that in the case of the initial sample (a) Si-O band can be described mainly with elementary band of 1107.5 cm^{-1} ; two other bands (1065 and 1085 cm^{-1}) have the vaues H characteristic for oxygen precipitate inclusions (~22 cm^{-1}) [7]. Analysis of their maximum positions in the framework of RBM statistics enables to conclude that these precipitates consist of the mixture of SiO$_4$ tetrahedra and Si-O$_3$-Si clusters. In the case of the sample B5 (b) the absorption band con-sists of three powerful contributions; one of them (1107.5 cm^{-1}) is connected to intersti-tial oxygen, two other have the same parameters (W- ~1050 and ~1085cm^{-1} ,H - ~60 and 45 cm^{-1} , respectively) as in the case of the spectra of silicon dioxide films ther-mally grown on silicon [4,8]. The same results were obtained also for the samples A5 and B4. This fact makes it possible to conclude that in these samples the buried silica layer exists and its structure is very similar to that of the thermal SiO$_2$ films, i.e. it may be represented as mixture of interconnecting 4- and 6-fold SiO$_4$ rings.

Figure 3 shows the data of layer-by-layer Auger analysis. Oxygen distribution pro-files for the most of samples practically coincide with that for the initial one both for depth and for the meaning of the straggling ($2\Delta R_p \approx 200$ nm). For the sample B5 the oxygen distribution profile changes significantly: the implanted oxygen is tries to po-sition itself in the narrow region situated between R_{pC+} and R_{pO+} . This result indicates the formation of a buried layer.

Using the intensity value of the absorption band created by elementary profiles (~1050 and ~1085cm^{-1}) and taking into account that the absorption coefficient for silicon dioxide is 3.4×10^4cm^{-1} [9] the thickness of the buried layer was estimated for the studied samples. This value amounts to ~20-27nm depending on the dose of carbon im-plantation and the regime of heat treatments. Auger spectroscopy gave a value of the buried layer thickness (the etching rate of silica in our case was ~14 nm s^{-1}) of about

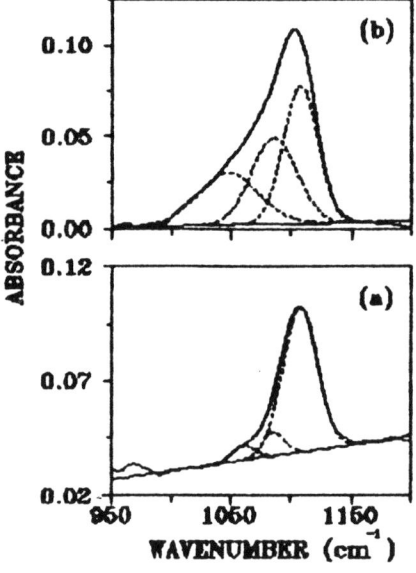 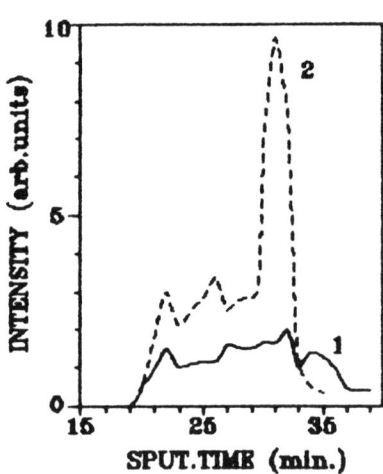

Figure 2. Examples of the deconvolution of the absorbance spectra into gaussian profiles for the samples B1 (a) and B5 (b).

Figure 3. Oxygen distribution measured by means of Auger electron spectroscopy in the samples B1 (1) and B5 (2).

22 nm. On the other hand, if one assumes that all implanted oxygen ions (the dose is 1.2×10^{17} cm^{-2})are collected, due to the high-temperature treatment, in a silica layer having a density of ~2.2 g cm^{-3} , the thickness of this layer is ~27 nm. The agreement is rather good.

The results obtained make it possible to follow the evolution of the structure of the buried silica layer as a function of implantation and annealing conditions (see Table 1). Implantation of carbon and oxygen ions leads to some increase of the oxygen content in interstitials and in the precipitate phase. Subsequent low-temperature annealing contrary decreases the concentration of briging oxygen atoms that agrees with the known fact [10]. The most substantial changes take place during high-temperature annealing. The efficiency of such annealings is strongly dependent on the dose of implanted carbon. In the case of a rather small implanted dose of C (samples A) heat treatment at 1150^0 C alone leads only to substantial growth of precipitate phase, but no buried layer is formed. To produce it, low-temperature preannealing has to be carried out. If the dose of the carbon implantation is sufficiently high (samples B) to produce the silica layer, high-temperature treatment alone is enough. If a two-stage heat-treatment is used, the efficiency of the buried layer formation is higher - the thickness of this layer is maximized under such condition. Thus, carbon implantation enhance the process of silicon-oxygen phase creation due to the formation of preferable conditions for oxygen accumulation in a rather narrow region and its subsequent interaction with the silicon lattice.

The authors wish to thank A.G.Revesz for fruitful discussions. This work was supported, in part, by a Soros Foundation Grant awarded by the American Physical Society.

4. References

1. Crowder, S.W., Hsieh, C.J., Griffin, P.B., Plummer, J.D. (1994) Effect of buried Si-SiO$_2$ interfaces on oxidation and implant-enhanced dopant diffusion in thin silicon-on-insulator films, *J.Appl. Phys.* **76**, 2756-2764.
2. Revesz, A.G., Brown, G.A., Hughes, H.L. (1993) Bulk electrical conduction in the buried oxide of SIMOX structures, *J.Electrochem.Soc.* **140**, 3222-3229.
3. Litovchenko, V.G., Romanyuk, B.N., Melnik, V.P. (1994) The effect of mechanical stresses on oxygen precipitation during formation of SOI structures, in S.Cristoloveanu (ed.), *Silicon on insulator technology and devices*, The Electrochem.Soc.Inc., San-Francisco, pp.104-110.
4. Lisovskii, I.P., Litovchenko, V.G, Lozinskii, V.B., Steblovskii, G.I. (1992) IR spectroscopic investigation of SiO$_2$ film structure, *Thin Solid Films* **213**, 164-169.
5. Lisovskii, I.P., Litovchenko, V.G., Lozinskii, V.B., Melnik, V.P., Frolov, S.I. (1994) Structure of the modified surface layer formed by ion bombardment of SiO$_2$ films, *Thin Solid Films* **247**, 264-270.
6. Pajot,B., Stein, H.J., Cales, B., Naud, C. (1985) Quantitative spectroscopy of interstitial oxygen in silicon, *J.Electrochem.Soc.* **132**, 3034-3037.
7. Lisovskii, I.P. (1993) IR study of the structural arrangement and chemical composition of vitreous films, *Optoelectronika i poluprovodnikovaya tekhnika* **26**, 93-113.
8. Boyd, I.W. (1987) Deconvolution of the infrared absorption peak of the vibrational stretching mode of silicon dioxide: evidence for structural order?, *Appl.Phys.Lett.* **51**, 418-420.
9. Boyd, I.W., Wilson, J.B. (1982) A study of thin silicon dioxide films using infrared absorption technique, *J Appl.Phys.* **53**, 4166-4172.
10. Gaworzevski, P., Hild, E., Kirscht, F.G., Vescem, L. (1984) Infrared spectroscopical and TEM I investigations of oxygen precipitation in silicon crystals with medium and high oxygen concentrations, *Phys. Stat. Sol.* **A85**, 133-147.

OPTICAL INVESTIGATION OF SILICON IMPLANTED WITH HIGH DOSES OF OXYGEN AND HYDROGEN IONS

P.A. ALEKSANDROV, E.K. BARANOVA, I.V. BARANOVA, V.V. BUDARAGIN AND V.L. LITVINOV
Institute of Information Technologies,
Russian Research Center "Kurchatov' Institute"
Kurchatov sq., 1, Moscow, 123181, Russia

Abstract. Silicon layers implanted with high doses of oxygen and hydrogen ions (separately and consequently) were investigated by IR-spectroscopy technique. The absorption bands centered at 615, 630, 890, and 2110 cm^{-1} were observed in the case of only hydrogen implantation. The bands centered at 460, 800, and 1060 cm^{-1} were detected in the case of irradiation with O^+ ions. Transmission spectra show the interaction between oxygen and hydrogen implanted both to oxygen bedding region and "behind" of it. The interaction follows from the facts: intensity of the band centered at 890 cm^{-1} increases, while the 615 cm^{-1} band disappears, and intensity of the 630 cm^{-1} band decreases. Measurements of the transmission spectra during thin removal of the implanted layers permit to draw and to analyze oxygen and hydrogen states distribution profiles before and after interaction. The result of interaction is different in the cases of hydrogen implantation to the oxygen bedding region and "behind" of it, but in both cases the oxygen removal from the surface to the deeper layers occurs. The observed changes of the bands centered at 615, 630 and 890 cm^{-1} are caused probably by the formation of defect complexes similar to E'H center ($O_3 - Si - H$). The obtained profiles of the oxygen and hydrogen states before and after interaction can be used for simulation of the processes taking place during Si oxide profile transformation by following hydrogen implantation.

It is well known, that there is an active interaction between hydrogen, defects, and impurities in SiO_2 [?].

The effect of transformation of the profile of previously implanted oxygen as a result of following implantation of H^+ ions into oxide layer and low temperature annealing at (1150 °C) have been shown. Mechanism of the profile reconstruction is not clear yet.

The purpose of the work is the investigation of the as-implanted Si layers, obtained by implantation of H^+ and O^+ ions separately and consequently. Hydrogen was implanted after oxygen for the two cases: into the profile of implanted oxygen and "behind" of it.

Oxygen and hydrogen implantation was carried out on ion beam accelerator **ILU-4/17** into p-type (111)-Si with $\rho = 10 \, \Omega \cdot cm$. Target temperature was (320 ± 30) °C, ion dose was $8 \cdot 10^{17}$ $ions/cm^2$, oxygen ions energy was 40 keV, hydrogen ions energies were 6.5 and 20 keV. In the case of subsequent irradiation there were two cases: 1) the calculated hydrogen profile is superimposed on that of oxygen (at H^+ ions energy of 6.5 keV), and 2) the hydrogen profile is positioned deeper than that of oxygen (at H^+ ions energy of 20 keV).

Transmission spectra of the implanted Si were received at room temperature by using of spectral photometer **IRS-29**. At the same time, the removing of thin Si layers was carried out by chemical and electrochemical etching methods. To control of the removed layers thickness, the weighting technique was employed.

Defect complexes and chemical bonds displayed on the transmission spectra, see Figure 1, agree with those published in [5,7].

Irradiation by H^+ ions gives broad intensive band centered at 630 cm^{-1} and two bands of less intensity, centered at 890 and 2100 cm^{-1}, Figure 1(a),(b). The band at 630 cm^{-1} is complex. It can be decomposed on Gaussians, Figure 2(a),(b). The main Gaussians are centered at 630 and 615 cm^{-1}. They correspond to the deformation oscillations of the bonds attributed to $Si-H$ and $Si-H_2$ respectively. The band centered at 2110 cm^{-1} corresponds to the stretching mode of these bonds [5,7]. The band centered at 890 cm^{-1} is not obviously identified – in this part of spectrum the deformation oscillation of the $Si-H$, $Si-H_3$, $Si-H_2$ bonds on the pore surface, and also **E'H** (O_3-Si-H) defect center [8] are situated.

As a result of the implantation by O^+ ions simple bands centered at 450 and 800 cm^{-1} (corresponding to deformation oscillations of $Si-O-Si$ bonds in SiO_2) and complex band centered at 1060 cm^{-1} appear in transmission spectra, Figure 1(c). To the last one the oscillations of bonds of O and Si in precipitates ($\nu_{min} = 1060 \, cm^{-1}$ [9]), and in SiO_x network ($x \leq 2$) ($\nu_{min} = 1060 \, cm^{-1}$ [10]) make contribution. Estimation of the x [10] gives the value of ≈ 1.75.

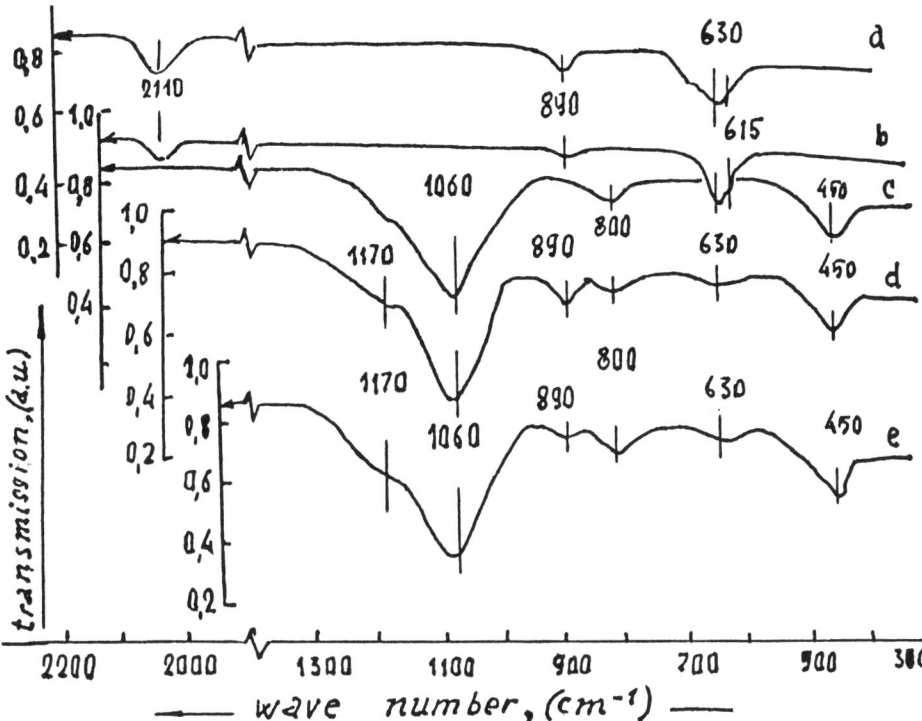

Figure 1. Infrared transmission spectra of Si implanted by oxygen and hydrogen ions with the following implantation parameters: a) H^+ 6.5 keV, b) H^+ 20 keV, c) O^+ 40 keV, d) H^+ 6.5 keV – after O^+ 40 keV, e) H^+ 20 keV – after O^+ 40 keV. Ion doses were of $8 \cdot 10^{17}$ $ions/cm^2$, in all cases.

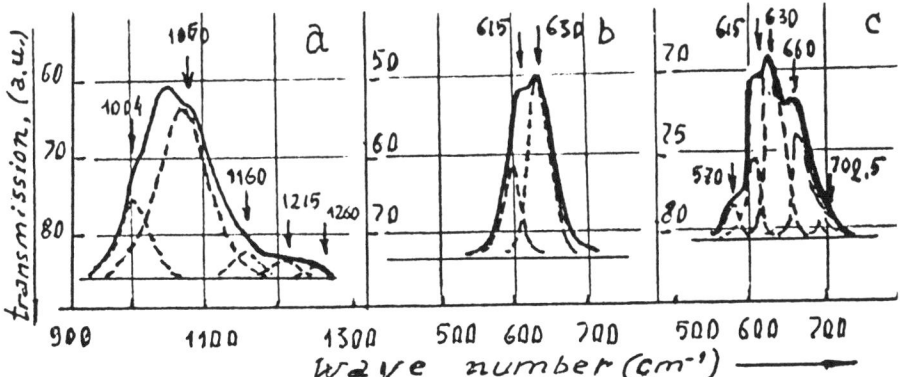

Figure 2. The representation of the complex shape transmission band as set of simple Gauss-shaped bands. Implantations were performed with the following parameters: a) O^+ 40 keV, b) H^+ 20 keV, c) H^+ 6.5 keV. Dose= $8 \cdot 10^{17}$ $ions/cm^2$

Difference between the precipitates shape leads to expansion of the transmission band towards to the high wave numbers ($1230\ cm^{-1}$) [9].

Implantation of the H^+ ions after O^+ ions results to the decrease of the concentration of centers similar to the $Si-H$, $Si-H$ ($630\ cm^{-1}$),see Figure 1 (d),(e), and quantity of the oxygen states. At the same time, hydrogen implantation creates centers with transmission band at $890\ cm^{-1}$.

Intensity of the each simple transmission peak (T) (or the component of the complex band) is proportional to the concentration of the correspondent states. Measurement of the spectra in conjunction with thin layer removing gives the T dependence on the removed layer thickness y, and $\frac{\partial T}{\partial y} = f(y)$ -is the profile of each type of bond, Figure 3 (a)–(d). The areas under the curves are proportional to these bonds concentration.

Differences in the H-centers distribution for H^+ implantation at 6.5 and 20 keV ions are consistent with the diffusion redistribution of hydrogen in Si. In the case of small quantity of radiation defects (at $6,5\ keV$ for H^+ ion implantation), the implanted hydrogen diffuses until capturing by impurities or another defects in Si lattice. In the case of 20 keV H^+ ion implantation the quantity of created radiations defects is higher (accordingly to the calculated elastic energy losses) and H is captured in the region of radiation defects, Figure 3 (a),(b). The significant part of the implanted hydrogen is trapped on the pores surface, the latter were observed by optical microscope.

The implantation of hydrogen into the oxygen implanted region and "behind" of it leads to redistribution of the oxygen bonds profiles. The redistribution effect is different for cases of implantation of H^+ with energy 6,5 and 20 keV, Figure 3 (d),(e). In the both cases implanted hydrogen removes oxygen from the surface to the deeper layer. In the region near R_p^0, see Figure 3 (e),the minimum in the distribution of $1060\ cm^{-1}$ band appears, with increasing of the band centered at $890\ cm^{-1}$ situated in the same region. Taking into account that the appearance of the $890\ cm^{-1}$ band is accompanied by the decreasing of the concentration of $Si-H$ state ($630\ cm^{-1}$) and the disappearance of $Si-H$ state, one can believe that in our case $890\ cm^{-1}$ band is associated with **E'H**-center ($O_3 - Si - H$) [8]. It should be noticed that interaction in this case is connected with SiO_x network and oxygen state in precipitates ($1170\ cm^{-1}$) are also changed.

As a result of this work, the following conclusions should be emphasized:

1. Interaction of implanted hydrogen with oxygen and radiation defects have an effect on transmission spectra obtained immediately after implantation.
2. The combination of optical measurements and thin layer removal technique permits to identify the defect states and to obtain their depth distributions.

167

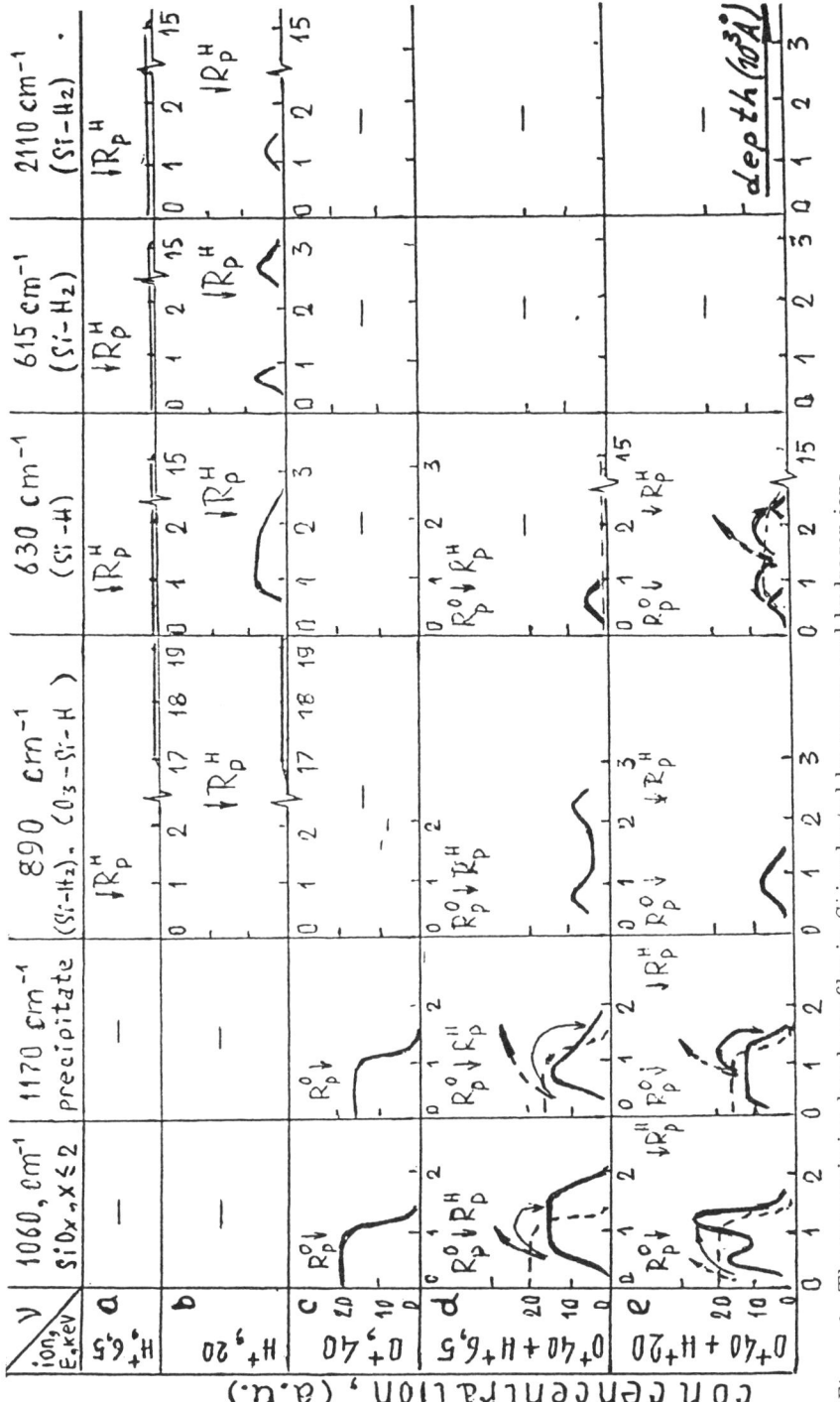

Figure 3. The transmission bands profiles in Si implanted by oxygen and hydrogen ions. Implantation conditions are the same as in Figure 1.

3. The profile of the $Si-H$ states is essentially dependent on H^+ energy, and consequently on the concentration of radiation defects. The shape of the profile may be explained by hydrogen diffusion in Si affected by the defects [1].
4. Oxygen states distribution is table-shaped in the case of the only O^+ ion implantation. In the case of subsequent implantation, the influence of the hydrogen on oxygen profile is dependent on mutual position of hydrogen and oxygen profiles. The only space redistribution is observed for oxygen in SiO_x network ($\nu_{min} = 1060\ cm^{-1}$). The partial departure of oxygen from precipitates occurs (probably to **E'H** centers ($O_3 - Si - H$), $\nu_{min} = 890 cm^{-1}$). Our conclusions are only qualitative at this point.
5. Thus, the obtained data are important for understanding of the processes, which take place during the transformation of the implanted oxide profile in silicon due to the subsequent H^+ implantation.

References

1. Pearton S.J., Corbett J.W., Stavola M. (1992) *Hydrogen in crystalline semiconductors*, Springer Verlag, Berlin.
2. Aleksandrov P.A., Baranova E.K., Budaragin V.V., Shemardov S.G. (1991) The influence of the secondary implantation of hydrogen on the oxygen profile change in Si *Proceedings of the All-Union conference on Physics of Interaction of Charged Particles with Crystals* 27–29 May 1991, Moscow, Moscow State University Publishers, pp. 82–83.
3. Aleksandrov P.A., Baranova E.K., Baranova I.V., Budaragin V.V., Litvinov V.L., Shemardov S.G., Ushkova E.B. (in press) The investigations of the oxygen and hydrogen interaction at the dielectric layers production in the Si by ion implantation *Proceeding of the scientific-technology seminar: Physical-technology problems of SOI structures and electronic elements on their base*, 21–23 October 1993, L'vov, Ukraine, report
4. Burenkov A.F., Komarov F.F., Kumachov M.A., Temkin M.M. (1985) *Distributions of energy deposed in the cascade of atomic collisions in solids*, Atomizdat Publishers, Moscow
5. Stein H.J. (1975) Bonding and thermal stability of implanted hydrogen in silicon, *Journal Electronic Materials*Vol. 4 no. 1, pp.159–174
6. Gerasimenko N.N., Rolle M., Chug I.J., Lee Y.H., Corelli J.C.,Corbett J.W. (1978) *Phys. State Sol.(b)*Vol. 90, pp. 689–695.
7. Mukashev B.N., Tamendarov M.F., Tokmoldin S.ZH., Frolov V.V. (1985) *Phys. State.Sol.(a)* Vol 91, pp. 509–522
8. Stahlbush R.E., Griscom A.N., Mrstic B.J.(1993), Post- irradiation cracking of H and formation of interface states in irradiated MOS field effect transistors, *J.Appl.Phys.*Vol. 73, pp. 658–667
9. Hu S.M. (1980) IR absorption spectra of SiO_2 precipitates of various shapes in Si: calculated and experimental, *J.Appl.Phys.*Vol. 51, pp. 5945–5948
10. Lisovskii I.P., Lozinskii V.B., Frolov S.I. (1993) Study of oxygen structural state in SiO_x films using the method of IR -spectroscopy, *Ukr.Phys.Journ.*Vol. 38 no. 5, pp. 745–751

ELECTRICAL PROPERTIES OF ZMR SOI STRUCTURES: CHARACTERIZATION TECHNIQUES AND EXPERIMENTAL RESULTS

T.E. RUDENKO, A.N. RUDENKO, V.S. LYSENKO
Institute of Semiconductor Physics, National Academy of Sciences of Ukraine
252650, Kiev-28, Prospect Nauki, 45, Ukraine

1. Introduction

The advantages of silicon on insulator (SOI) technology have been demonstrated in a wide range of applications, which include high-speed and radiation-tolerant CMOS devices, three-dimensional integration and very short channel VLSI circuits [1-3]. However, design and modeling of any new device require a detailed knowledge of electrical properties of used material. It is well known, that conventional characterization techniques used for bulk Si structures can not be directly applied in the case of SOI. A number of electrical characterization techniques suitable for SOI has been reported in the literature [4-6]. In this paper, characterization methods based on the use of SOI gated diode and a combined depletion-mode MOSFET and gated diode structure are presented. Proposed methods are applied to characterize both thick- and thin-film laser ZMR SOI structures.

2. Device Fabrication

The measurements reported in this study were performed on laser ZMR SOI films containing subgrain boundaries. The details of the fabrication ZMR SOI wafers are described in [7]. Test structures of various geometries were fabricated by the conventional polysilicon gate CMOS process. The thickness of the silicon film was 350 nm. The thicknesses of the gate oxide and buried oxide were 65nm and 1000nm, respectively. For comparison simultaneously with SOI-based test structures silicon on sapphire (SOS) and bulk Si-based structures were prepared in the same technological processes.

3. Capacitance-Voltage Measurements of SOI Gated Diodes

It has been known that capacitance methods are not very suitable for SOI because of the high series resistance of the Si film and the presence of parasitic capacitances [8]. However, these problems can be overcome by the use of proper measuring conditions.

The influence of the series resistance can be avoided by decreasing of measuring frequency, which depends on the device geometry, doping level and the Si film thickness. The parasitic capacitances of the conducting leads relative to the substrate can be excluded by an appropriate bridge measuring circuit. In this case, the capacitance-voltage measurements of SOI MOS structures allow one to determine a number of SOI parameters.

Below specific features of C-V-characteristics of SOI gated diodes are discussed. Fig.1 demonstrates capacitance versus front-gate voltage characteristics of the SOI gated diode with an undoped (or "natural") ZMR SOI film, which usually exhibits low n$^-$-type residual doping. Capacitance have been measured between the front gate and body with p-n-junction reverse biased and for different back-gate biases.

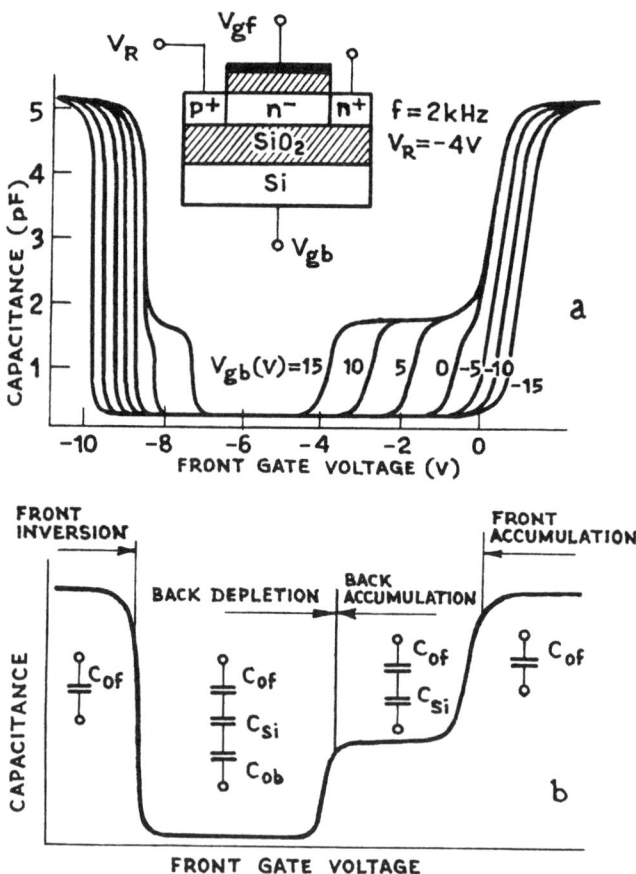

Figure 1. (a) The experimental capacitance versus front-gate voltage characteristics of a SOI gated diode with "natural" ZMR film, measured for different back-gate voltages and with p-n-junction reverse biased. (b) Simplified equivalent capacitor schemes in different regions of measurements.

In gated diodes the minority carrier response is controlled by p-n-junction, which acts as a source and a sink for minority carriers of inversion layer, and not by generation-recombination processes as in the case of MOS capacitors. Therefore, the inversion layer can follow a.c. gate voltage and contribute to measured high-frequency capacitance. As a result the capacitance sharply rises up to the front-gate oxide capacitance when the inversion layer is formed at the front interface. It takes place at $\varphi_{sf}=2\varphi_F+V_R$, where φ_F is Fermi potential, V_R is the reverse bias of p-n-junction. A portion of nearly constant capacitance, which corresponds to the series combination of the front-gate oxide capacitance and depleted silicon film capacitance, is observed as the Si film is depleted and the back interface is accumulated or inverted. When the back interface conducting channel disappears, the capacitance drops due to the series connection of the low substrate capacitance. This is just the flat-band condition at the back interface. Thus, there are clearly-defined points on C-V-characteristics, corresponding to the front surface inversion and back surface flat-band conditions. Linear shifts of these points with the back-gate voltage are described by the thin-film SOI theory [9] and therefore, can be used for rapid evaluation of the Si film and buried oxide thicknesses as well as of front and back fixed oxide charge densities. It is worth noting, that the reverse bias of gated p-n-junction extends the linear part of interface coupling dependence (for $(2\varphi_F+V_R)C_{Si}/C_{of}$ along front-gate voltage axis and for $(2\varphi_F+V_R)(1+C_{Si}/C_{ob})$ along back-gate voltage axis), which is convenient for characterization purposes.

4. Transconductance Measurements

The results similar to those available from above C-V-measurements can be obtained by transconductance measurements of SOI MOSFET with biased body contact. For characterization purposes it is suitable to use the combined structure of depletion-mode and enhancement-mode MOSFETs. A schematic top view of this structure, which can be considered also as a gated diode, is shown in Fig.2a. Plotted in Fig.2b are static front-gate transconductance characteristics of the depletion-mode MOSFET, fabricated in "natural" (n-) ZMR SOI film. These characteristics have been measured with the reverse bias of lateral p-n-junction, overlapped by the gate, and for different back-gate biases. A region of almost constant transconductance, varying with the back-gate voltage, is caused by an accumulation channel at the back interface. One can see, that transconductance characteristics in Fig.2b look similar to discussed above C-V-curves and can be used for extracting the same parameters of the SOI structure. In addition, the transconductance measurements in the considered combined structure allow one to evaluate front and back surface channel mobilities of both electrons and holes in the same silicon film. In undoped SOI films, having typically residual donor concentration of $(2-4) \times 10^{15}$ cm^{-3}, they were about 800 and 400 cm^{2}V^{-1}s^{-1} for electrons and holes, respectively. It worth noting, that as a rule back surface mobilities in the investigated ZMR SOI films are close to front surface values, which is in a general agreement with relatively low back surface state density (approximately 10^{11}cm^{-2} or less) obtained by charge pumping technigue.

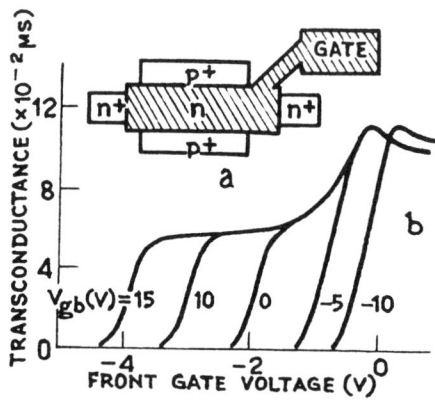

Figure 2.(a) Schematic top view of the combined depletion-mode and enhancement-mode MOSFET structure. (b) Transconductance versus front-gate voltage, measured in a depletion-mode MOSFET with "natural" ZMR SOI film with p-n-junction reverse biased (V_R=-4V) and for different back-gate biases (W/L=20μm/500μm, V_D=-0.1V).

5. Determination of Impurity Concentration and Carrier Mobility Profiles in SOI MOS Structures

If the silicon film thickness is larger than a few Debye lengths, it is possible to determine doping concentration and drift carrier mobility profiles using C-V and I-V-measurements in the above mentioned combined depletion-mode MOSFET and gated diode structure. Indeed, reverse biased gated p-n-junction removes the restriction on the widening of depleted region under the gate and thus makes possible full depletion of thick SOI films. In this case non-equilibrium C-V-characteristics can be measured and used for extraction of the impurity concentration (or, more precisely, free carrier concentration) profile in the silicon film [10].

On the other hand, the drain current in depletion-mode MOSFET is determined by the conduction in an undepleted part of the silicon film, which is modulated by the gate voltage through the vertical variation of the depletion layer width. For low drain voltages and very long channels we can neglect the variaton of the depletion layer width along the channel. In this case the drain current variation, corresponding to the vertical variation of the depletion layer width dx , is given by:

$$dI_D = q \frac{W}{L} V_D \mu(x) n(x) dx , \qquad (1)$$

where W and L are channel width and length, respectively, n(x) is the free carrier concentration and μ(x) is drift majority carrier mobility at the front-gate depletion layer edge.
Using the depletion approximation and applying Gauss' law at the front interface, we can obtain the variation in front-gate voltage, associated with dx:

$$dV_{gf} = \frac{q}{\varepsilon_s} x N(x) dx + \frac{q}{C_{of}} N(x) dx , \qquad (2)$$

where x is the front-gate depletion layer width, N(x) is the ionized impurity concentration at the front-gate depletion layer edge.
Assuming that N(x)≅n(x) and combining (1) and (2), we obtain:

$$\mu(x) = \frac{dI_D}{dV_{gf}} \frac{x/\varepsilon_s + 1/C_{of}}{V_D W/L} \qquad (3)$$

Taking into account that $x = \varepsilon_s(1/C + 1/C_{of})$ equation (3) can be rewritten:

$$\mu(x) = \frac{dI_D}{dV_{gf}} \frac{1}{CV_D W/L} \qquad (4)$$

where C is front-gate-to-body capacitance.

Carrier mobility profiles have been determined in SOI films with different doping profiles and compared with those in SOS films. Fig.3a gives an example of C-V-characteristics (solid line) measured in depletion-mode SOI MOSFET with p-type Si film doped by deep boron implantation to prevent creation of the inversion channel at the back interface. Reverse biasing of lateral gated n-p-junction provides non-equilibrium full depletion of electrically thick silicon film. The corresponding I-V-curve is shown by dashed line. A sharp drop in the capacitance is due to pinch of the conducting channel in the Si film, which is equivalent to the series connection of the substrate capacitance. As can be seen from the figure, the current cutoff is observed at the same gate voltage. Fig.3b shows the calculated acceptor concentration and hole mobility profiles. It is clearly seen, that non-uniformly doped SOI film exhibit also non-uniform mobility profile. In uniformly doped ZMR SOI films the carrier mobility is nearly constant across the silicon film thickness, in contrast to SOS films, in which it drops abruptly in the vicinity of the silicon-sapphire interface. This is readily apparent from the fact, that as a rule in SOS structures the current cutoff takes place long before full depletion of the Si film determined from C-V-curve (Fig.4).

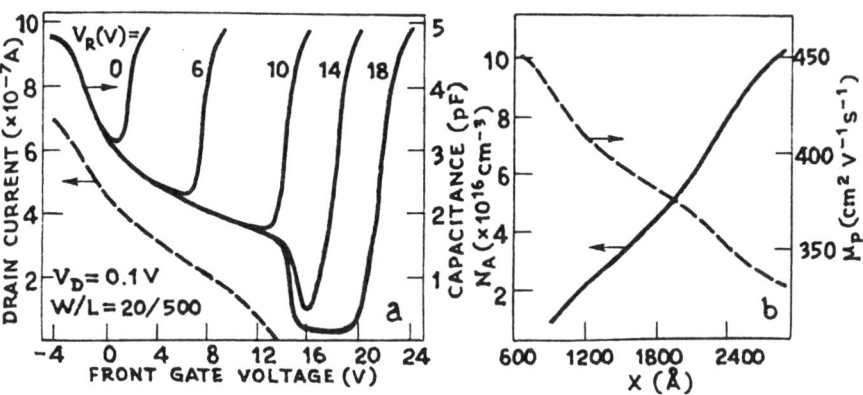

Figure 3. (a) C-V-characteristics (solid line) and I-V-curve (dashed line) of a depletion-mode MOSFET with p-type Si film doped by deep boron implantation (D=3.75x10^{12} cm^{-2}, E=100 keV). Capacitance was measured with a 10-kHz signal and for different reverse biases of n-p-junction. The substrate was held grounded. I-V-measurements were performed with V$_R$=18V. (b) The calculated acceptor concentration profile (solid line) and the hole mobility profile (dashed line).

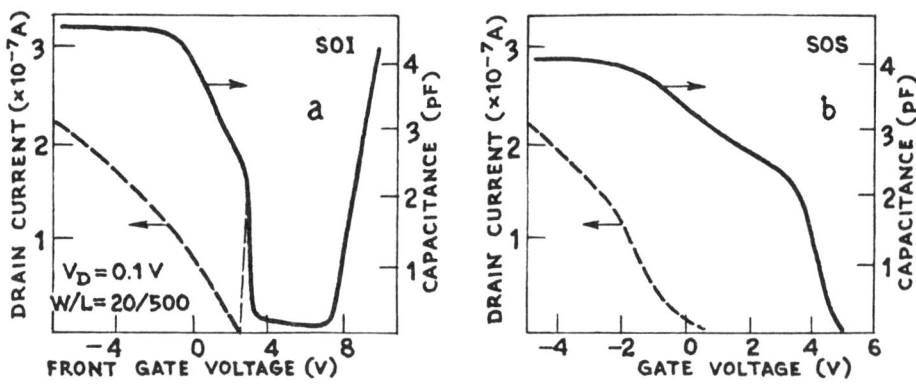

Figure 4. C-V-measurements (solid line) and I-V-measurements (dashed line) in SOI (a) and SOS (b) structures. Note that in contrast to SOS, in SOI-based structure a drop in the capacitance and current cutoff take place at the same V$_{gf}$.

In Fig.5 an average electron mobility is plotted as a function of the donor concentration in SOI and SOS films. It is evident, that an electron mobility in the investigated ZMR SOI films considerably exceeds that in SOS films and, in spite of the presence of subgrain boundaries, is closely approaching to corresponding values in bulk Si with the same doping levels. Actually, it is well-established, that subgrain boundaries do not affect the majority carrier transport.

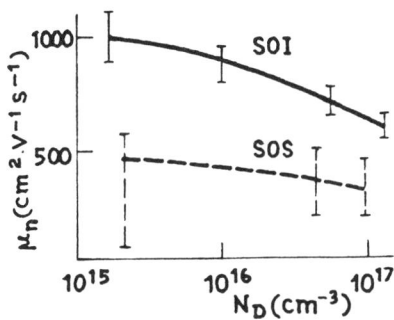

Figure 5. Drift electron mobility as a function of the donor concentration in SOI and SOS films.

6. Generation and Recombination Currents in SOI Gated Diodes

Because generation processes are responsible for the junction leakage, they are of great importance for SOI, main application of which is CMOS techology. On the other hand, the generation carrier lifetime reflects the material quality, as it is known to be extremely sensitive to the material crystalline defects and purity. An understanding of the recombination current behaviour in thin-film SOI MOS structures may be useful when analyzing parasitic bipolar effects.

In this paper particular attention is given to an unusual behaviour of generation and recombination currents in thin-film SOI MOS structures associated with interface coupling and specific free carrier distribution in the Si film. It can be readily illustrated by the example of the gated diode technique. The gated diode technique is widely used for bulk Si structures, because it enables one to separate easily the surface and volume generation and recombination components [11]. But in thin-film SOI gated diodes generation (reverse) and recombination (forward) current characteristics substantially differ from those of bulk Si structures.

Fig.6a shows the reverse current as a function of the front-gate voltage measured for different back-gate biases in p^+-n^--n^+ gated diode made in "natural" ZMR SOI film. It is convenient to consider these results in conjunction with C-V-measurements discussed above. The characteristics on the left-hand side of Fig.6a, measured at positive back-gate biases, in general, look similar to those of bulk Si structures. The rise in the reverse current corresponding to "switching-on" generation under the gate, takes place when $\varphi_{sf} \cong \varphi_F$. The current rise is stopped as the Si film and back interface become depleted. The step down in the reverse current is observed at the onset of inversion at the front interface ($\varphi_{sf} \cong 2\varphi_F + V_R$). In the case of bulk Si structures these steps down are usually attributed to "switching-off" surface generation under the gate, when the inversion layer is formed. However, as can be seen from Fig.6a, in the case of SOI the height of these current steps is not constant, but essentially depends on the back-gate voltage. At negative back-gate voltages the reverse current drops near to zero when the front surface becomes inverted, and characteristics are entirely distorted. It is explained by the fact, that only silicon regions with free carrier concentration lower than intrinsic concentration contribute to the generation current. Due to interface coupling in fully depleted SOI MOS structures the free carrier distribution and the thicknesses of inverted and depleted parts of the Si film (and consequently the actual generation volume) depend significantly on the back surface potential.

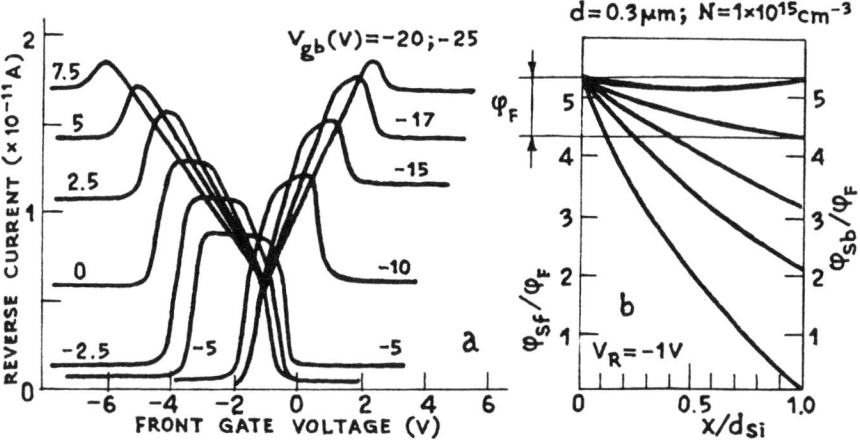

Figure 6.(a) Reverse current versus front-gate voltage, measured in a SOI gated diode with n⁻-type silicon film for different back-gate biases. (b) The potential distribution in the Si film for different back surface potentials with front surface inverted.

Fig.6b shows the potential distribution in the silicon film ($N=1 \times 10^{15}$ cm^{-3}, $d_{Si}=300$nm), calculated for the front surface inversion and for different back surface potentials. The inversion layer thickness can be evaluated as the distance from the surface where the potential reduces by φ_F. As can be seen from the figure, the inversion layer thickness is minimum for accumulated back surface. If the back surface potential differs from front one by Fermi potential or less, the inversion layer occupies the whole film. Therefore, the steps down observed in the reverse current at the onset of inversion are caused by "switching off" not only surface generation under the gate, but also generation in the significant part of the Si film, where minority carrier concentration exceeds intrinsic concentration. When the whole film under the gate becomes inverted, the generation current falls to zero. If the back interface is initially inverted or close to inversion (negative back-gate voltages in Fig.6a), an increasing of negative front-gate voltage leads to spreading the back surface inversion layer into the Si film, thereby reducing the generation volume and the generation current.

From the above it follows, that in determination of generation parameters in SOI MOS structures the free carrier distribution, dramatically changing the actual generation volume, should be taken into account. This is difficult to perform in transient techniques (DLTS [12], Zerbst-like technique [13]), but it is quite possible in steady-state non-equilibrium techniques as junction leakage or gated diode current measurements.

In a similar manner, the behaviour of the forward (recombination) current in thin-film SOI gated diodes greatly depends on back surface conditions. In general case the forward current in SOI diode consists of several components, including the diffusion current, the bulk recombination current and surface recombination at front and back silicon film surfaces. All of them depend on the free carrier concentration and are decreased with increasing n and p. Since the variation in the back-gate voltage changes free carrier distribution in thin-film SOI MOS structures, it changes the forward current also. As the whole silicon film is accumulated or inverted, the forward current is minimum. Under intermediate conditions different shapes of the characteristics can be observed, making difficult extracting recombination parameters.

Unique δ-shaped current-gate voltage characteristics are observed in double-gate regime at low forward biases in p-i-n SOI gated diodes (Fig.7). It is explained as follows. In double-gate regime in thin (low doped) SOI film the potential is nearly constant across the film thickness, as shown in Fig.7. As a consequence, at positive gate voltages the entire silicon film under the gate is enriched by electrons, so that i-region transforms into n^+, reducing forward current components. Similarly, at negative gate voltages i-region transforms into p^+, resulting in the negligible forward current. The forward current is maximum, when the Si film and its both interfaces are depleted (more exactly, when $\varphi \cong (\varphi_n + \varphi_p)/2$ with φ_n and φ_p being the corresponding quasi Fermi levels). The current maximum rises exponentially with forward bias. At low forward biases an ideality index equals 2. It means that the recombination component is dominant in the observed forward current. In this case it is possible to evaluate the effective recombination lifetime, which includes the influence of surface recombination. In the example presented in Fig.7, it is about 90 ns.

The generation carrier lifetime obtained by the gated diode technique in the investigated ZMR SOI films lies in the range $(0.4-1.2) \times 10^{-6}$ s, which is two orders higher than that in their SOS counterparts, but nearly two orders lower than in bulk Si structures. It is probable, that the reason for lower generation lifetimes in ZMR SOI films is the presence of subgrain boundaries.

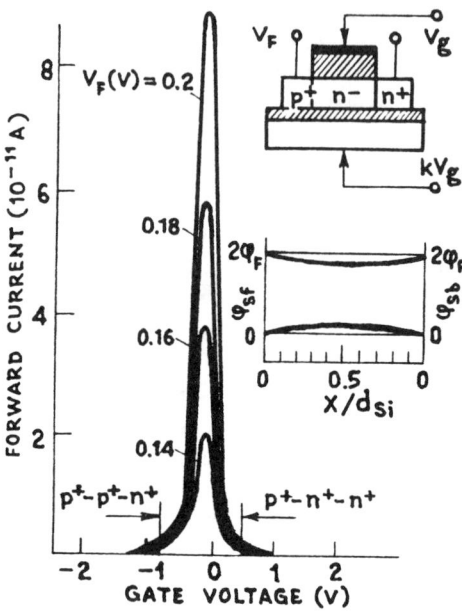

Figure 7. Forward current versus gate voltage, measured in p^+-n^--n^+ SOI gated diode in double-gate regime for different forward biases of p-n-junction. The insert shows the experimental setup and the potential distribution in the Si film (N=2×10^{15} cm^{-3}, d_{Si}=300nm).

7. Conclusion

In this paper, electrical characterization methods based on the use of a SOI gated diode and a combined depletion-mode MOSFET and gated diode structure have been presented and applied to investigate laser ZMR SOI structures.

It is shown that capacitance-voltage measurements of SOI gated diodes under proper measuring conditions allow one to evaluate a number of parameters in both thick- and thin-film SOI structures. A simple method to determine doping concentration and carrier mobility profiles in the Si film is proposed, based on depletion approximation and using capacitance-voltage and current-voltage measurements in the combined test structure under consideraton. Proposed method is demonstrated on SOI and SOS films with different doping profiles.

An unusual behaviour of the reverse and forward currents in SOI gated diodes, related to interface coupling and specific free carrier distribution in thin SOI films, has been analyzed. It is shown, that the back interface conditions strongly affect the observed generation and recombination currents as a result of the variation in the thickness of the inverted and depleted parts of the Si film. δ-shaped current-gate voltage characteristics observed in double-gate regime in forward biased p^+-i-n^+ SOI gated diode are presented. It is explained by sharp decreasing of the recombination current when the whole silicon film under the gate becomes accumulated or inverted.

Acknowledgements

The authors would like to thank Prof. Givargizov E.I. and Dr. Limanov A.B. for helping us in the SOI wafer fabrication.

9. References

1. Colinge, J-P. (1988) Silicon-on-insulator MOS devices for integrated circuit application, Hewlett-Packard J.2, 87-93.
2. Auberton-Hervé, A.J. (1989) CMOS SOI technologies for high speed and radiation hard circuits, in A.Heuberg, H.Ryssel, P.Lang (eds.), 19th ESSDERC Conference, Berlin, pp. 740-748.
3. Akaska, Y. (1991) 3D-IC technologies and possible application, IEICE Transactions 74, 328-336.
4. Vu, D.P., Pfister, I.C. (1988) Characterization of beam-recrystallized Si film and their Si/SiO$_2$ interfaces in silicon-on insulator structures: The very thin-film case, Appl. Phys. Lett. 52,45-48.
5. Cristoloveanu, S. (1989) Advanced silicon on insulator materials: processing, characterization and devices, in M.Schulz and C.Hardeke (eds.), Silicon Material Science and Technology, Berlin, Germany: Springler-Verlag, pp.223-248.
6. Ouiss, T., Cristoloveanu, S., Elewa, T., Haddara, H., Borrel, G., Ioannou, D.E. (1991) Adaptation of the charge pumping technique to gated p-i-n diodes fabricated on silicon on insulator, IEEE Trans. Electron Dev. 38, 1432-1443.
7. Limanov, A.B., Givargizov, E.I. (1983) Control of the structure in zone-melted silicon films on amorphous substrates, Mater. Lett. 2, 93-96.
8. McDaid, L.J., Hall, S., Eccleston, W., Alderman, J.C. (1989) Interpretation of capacitance-voltage characteristics of SOI capacitors, Sol. State Electron. 32, 65-68.
9. Lim, H.K., Fossum, J.G. (1983) Thershold voltage of thin-film silicon-on-insulator (SOI) MOSFET's, IEEE Trans. Electron Dev. 30,1244-1251
10. Rudenko, T.E., Rudenko, A.N., Lysenko, V.S., Limanov, A.B., Givargizov, E.I. (1993) Investigation of impurity concentration and carrier mobility profiles in ZMR SOI structures, Mikroelectronika 22,3-13 (in Russian).
11. Grove, A.S., Fitzgerald, D.J. (1966) Surface effect on p-n junction characteristics of surface space charge regions under nonequilibrium conditions, Sol. State Electron. 9, 783-806.
12. McLarty, P.K., Ioannou, D.E. (1990) DLTS analysis of generation transients in thin SOI MOSFET's, IEEE Trans. Electron Dev. 37, 262-266.
13. Ioannou, D.E., Cristoloveanu, S., Mukherjee, M. and Mazhari, B. (1990) Characterization of carrier generation in enhancemant-mode SOI MOSFET's, IEEE Electron Dev. Lett. 11,409-411.

Section 3:
SOI Devices

FABRICATION AND CHARACTERISATION OF POLY-SI TFTS ON GLASS

S D Brotherton, J R Ayres, D J McCulloch, N D Young

Philips Research Laboratories
Cross Oak Lane
Redhill
Surrey

ABSTRACT

In this paper we review the physics and technology of poly-Si TFTs fabricated with a glass substrate compatible technology, and, where appropriate, compare them with SOI devices. The topics covered include the formation and crystallisation of poly-Si, the trapping state distribution and leakage current and hot carrier instability phenomena.

1. Introduction

Poly-Si thin film transistors on glass are commanding increasing attention because of their application to active matrix flat panel displays (AMLCD). As their name indicates, the active material is poly-crystalline, and therefore the performance achieved from these devices is inferior to their single crystal SOI counterparts. However, as the material improves, it is likely that some of the issues common to SOI devices will be observed in poly-Si devices. In spite of their limited performance compared with SOI devices, it is with a-Si:H TFTs that the comparison should be made in context of the AMLCD application. The electron field effect mobility in a-Si:H is $< 1 cm^2/Vs$, whereas values of $> 100 cm^2/Vs$ have been reported for poly-Si [1,2]. It is this considerably higher mobility which is the key feature in this application, because it enables various addressing circuits to be fabricated on the AMLCD plate.

Detailed discussion of the AMLCD application is beyond the scope of this paper and the following material will focus on reviewing the present status of the physics and technology of poly-Si TFTs on glass. Topics covered include deposition and crystallisation of poly-Si, the influence of the trapping state density on device properties and a brief discussion of leakage current and hot carrier instability effects. Finally, a more conventional SOI technology has been reported for the fabrication of small projection displays on glass [3] and quartz [4] and these approaches will be briefly reviewed and compared with poly-Si.

2. TFT Structure

The two basic processes used for low temperature poly-Si TFT fabrication employ either furnace or laser crystallisation of amorphous pre-cursor films.

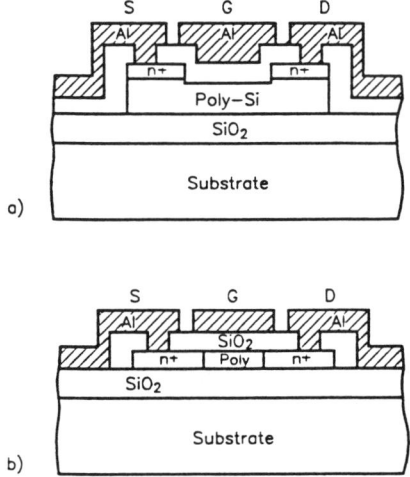

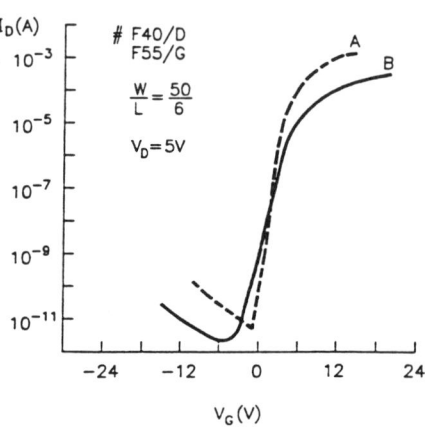

Fig. 1. Cross-sectional diagrams of poly-Si TFT architectures a) channel etched coplanar, b) ion implanted coplanar.

Fig. 2. Transfer characteristics of poly Si TFTs: A- laser crystallised, B- solid phase crystallised.

The most commonly used structure, especially for furnace processed devices, is a poly-Si gated auto-registered device which is similar in design to standard single crystal MOS and SOI devices. Ion implantation is used to dope the auto-registered source, drain and gate regions, and low temperature oxide deposition is used for the gate dielectric. A variety of procedures, including APCVD, PECVD and LPCVD, have been successfully used for this layer. After annealing at temperatures of ~600°C plus an MOS wet bake, interface state densities and fixed charge densities of ~5×10^{10}cm^{-2}eV^{-1} and $2-3 \times 10^{11}$ cm^{-2}, respectively, have been obtained [5].

A wider variety of device designs has been used for the laser crystallised devices, largely because this technology was initially seen as a process addition to a-Si:H TFTs. In many instances this has resulted in the use of non-implanted structures [6-8], in which the source and the drain regions have been formed from doped deposited layers. The architectures have included top gated staggered [6], channel etched coplanar [7] and inverted staggered [8]. In our own work we have used both the channel etched, non- implanted structure shown in figure 1a as well as the implanted structure in figure 1b. An important difference between these structures and the auto-registered device is the presence of gate-drain overlap and the associated extra capacitance which will deleteriously affect circuit performance.

The device characteristics obtainable with the different structures and processes are qualitatively similar as shown by the transfer characteristics in figure 2. A principal difference between the two processes is the larger on-current with the laser crystallised device due to its higher electron field effect mobility. Aspects of the device characteristics, are discussed in the following sections.

3. Poly-Si Deposition and Crystallisation

The principal technique for the deposition of poly-Si is by low pressure chemical vapour deposition (LPCVD) using silane. Practical deposition rates of 1 - 10 nm/min are obtained over the temperature range 540-620°C. For typical silane pressures of 100-200 mtorr, films will be deposited at the upper temperature in a fine grain columnar form. The grain size is typically 100nm and with a preferred <110> orientation. At the lower temperatures the films are more likely to be amorphous, but the precise structure of the film depends upon both pressure and temperature as shown by the work of Joubert et al [9]. In particular, it was shown that at the lower temperatures, as the silane partial pressure reduced, the films were more likely to be deposited in a poly-crystalline rather than an amorphous form. This work was extended by Voutsas and Hatalis [10], who demonstrated that the variation of crystallinity with system pressure and silane partial pressure could be correlated with the film deposition rate. This was explained in terms of the silicon adatom diffusion length during the time taken to grow a further mono-layer, which then immobilised the adatom.

Although early work used directly deposited LPCVD poly-Si at 620°C, the fine grain nature of this material resulted in electron mobilities of $<10 cm^2/Vs$. More recent work in this area includes the innovative use of plasma deposition to realise higher mobility material at much lower temperatures [11].

Films deposited in the amorphous state, and then thermally crystallised into poly-Si, have been shown to have higher carrier mobilities [12,13], due to the larger grain size compared with films deposited in the poly-crystalline state. The use of amorphous films, as pre-cursor material, has meant that large area plasma enhanced (PE)CVD is also used [5]. Crystallisation of the amorphous material is achieved either by solid phase crystallisation (SPC) at temperatures of ~600°C or by the use of a laser. The important features of these different approaches are summarised below.

3.1. SOLID PHASE CRYSTALLISED (SPC) AMORPHOUS SILICON

Amorphous films of silicon can be readily deposited at ~550°C by LPCVD with a silane pressure of 200 mtorr. At this temperature and pressure useful growth rates of ~2nm/min can be achieved. Amorphous silicon films deposited at ~250°C using plasma enhanced (PE)CVD can also be used as pre-cursor material. This exploits the use of large area deposition equipment available for a-Si:H TFTs and also benefits from the higher deposition rates achievable with this technique (~25nm/min). As the crystallisation of all these films is not directly seeded, but relies upon random nucleation from regions of incipient micro-crystallinity, crystallisation times of several hours are needed at ~600°C. Typical crystallisation results obtained from LPCVD and PECVD films are shown by the surface reflectance data [14] in figure 3. The process is characterised by a transient nucleation period followed by grain growth from these centres leading to complete crystallisation of the film. A detailed description of the time dependence of the process is given by the Avrami-Johnson-Mehl equation [13]. The PECVD films are generally found to have a longer nucleation period than the LPCVD films which is attributed to a lower degree of inherent order in the films (due to the lower deposition temperature). It is generally expected that the inherent order, or incipient crystallinity, within the pre-cursor film will decrease with reducing deposition temperature and with increasing deposition rates [13]. Hence, by optimising these conditions, larger grain films may be obtained.

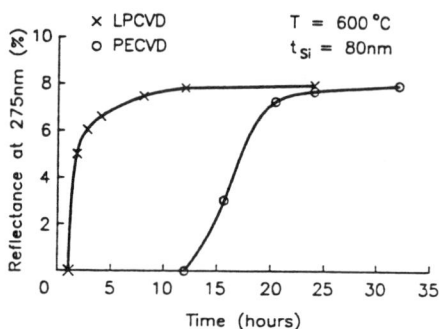

Fig. 3. Time dependence for SPC of LPCVD and PECVD amorphous silicon films.

The grains resulting from this process are generally elliptical in shape due to preferential growth in the <112> direction [15], and dendritic due to the formation of twins along [111] boundaries. The highly faulted nature of these grains is believed to limit device performance compared with the crystallographically more perfect grains formed by laser crystallisation (as discussed below). The influence of grain size on device behaviour is shown by the results in figure 4 in which the electron field effect mobility is shown to increase with film thickness for LPCVD material [16]. However, as also shown in this figure, the variation with film thickness is due to a comparable variation in grain size with film thickness. Transmission electron microscopy has shown that grain nucleation in these films occurs at the rear Si/SiO_2 interface, and suggests that, for the thinner films, the inter-nuclei separation is greater than the film thickness, such that unimpeded grain growth can occur until the top surface is reached.

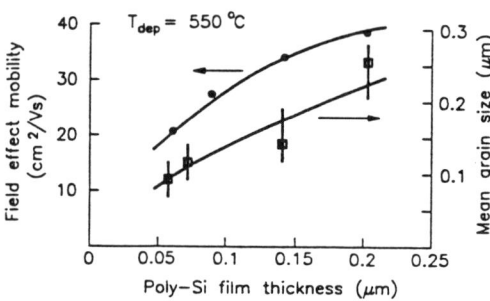

Fig. 4. Variation of field effect mobility and mean grain size with film thickness after SPC of a-LPCVD Si.

3.2 LASER CRYSTALLISED POLY-SI

The most widely reported laser crystallisation technique of a-Si on glass has been with rare-gas halogen excimer lasers [6-8, 17-19]. The standard gas mixtures and output wavelengths are ArF (193nm), KrF (248nm) and XeCl (308nm). These are all short duration, pulsed lasers (10-30ns) operating in the ultra-violet wave band and have been found to be well suited to the crystallisation of silicon on glass. This is because the optical absorption depth in amorphous silicon is about 6nm [20] at these wavelengths, such that the radiation is strongly absorbed in the silicon surface and can readily cause melting, whilst the short pulse duration results in a correspondingly small heat diffusion length of ~100nm in the silicon itself [20] and ~200nm in films of SiO_2 which can be used to cap the glass substrate and thereby protect it from excessive temperature excursions. Measurements [21] and calculations [21,22] have demonstrated that with suitable SiO_2 film capping, the temperature of the underlying glass surface can be kept below ~400°C.

The crystallisation system used in the work reported below has been previously described [7] and consists of a KrF laser delivering 30ns duration pulses at a typical frequency of ~10Hz. A semi-gaussian beam was used for the irradiations with stripes of crystallised material being formed by sweeping the sample through the beam in the direction of the gaussian energy distribution. The energy densities quoted are those measured at the peak of the gaussian. In order to avoid uncontrolled hydrogen release from hydrogen-rich PECVD films, the swept semi-gaussian beam is preferable to a stepped homogenised beam [7], unless multiple low energy passes are used with the latter to de-hydrogenate the silicon before the final high energy crystallisation [8].

Figure 5 shows the typical grain structure observed in cross section transmission electron microscopy after laser crystallisation of a 150nm thick a-Si:H precursor [1] layer. The film is seen to be stratified into a large grain surface layer

Fig. 5. Cross-sectional TEM micrograph of laser crystallised PECVD Si.

and a fine grain underlying region. Comparable stratification has also been reported by other workers for a-Si:H [23,24] as well for low hydrogen content LPCVD a-Si [24]. The depth of the large grain layer is related to the primary melt depth produced by the laser and the fine grain material is believed, by analogy with results on crystalline silicon [25], to result from the rapid propagation of a buried molten layer through the film. The relative depths of these two layers are a function of incident laser energy as shown by the results in figure 6. It will be seen from this figure that the large grain layer achieves a limiting maximum thickness of ~50-60nm, although the average lateral size of the grains continues to grow with increasing energy after the depth has saturated. As found with the SPC material, and as discussed below, the grain size influences the device behaviour. By extrapolating the grain depth data in figure 6 back to zero, an approximate value of 95 mJ/cm^2/pulse was obtained for the melt threshold energy of the 150nm thick a-Si:H layer.

The threshold energy for melting the film, E_T, can be obtained from the following approximate analytical solution of the heat diffusion equation [26], (assuming that the material constants are independent of temperature and that the thermal diffusion length, $\sqrt{Dt}$, is greater than the optical absorption depth):

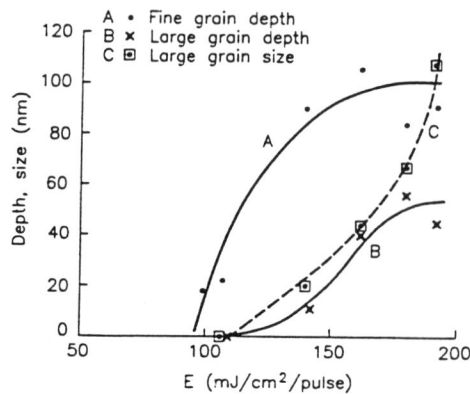

Fig. 6. Energy dependence of the stratified film features shown in figure 5.

$$E_T = \frac{(T_m - T_o)\sqrt{\pi}\varrho C_p \sqrt{D\tau}}{2(1-R)} \qquad (3.1)$$

where τ is the pulse duration, ϱ is the density, C_p is the specific heat, R is the surface reflectivity, T_m is the melting temperature and T_o is the initial temperature. Equation 3.1 applies if the film thickness, W, is greater than the thermal diffusion length $\sqrt{D\tau}$. If this condition is not satisfied, ie the sample is thinner than the diffusion length, then $\sqrt{D\tau}$ is replaced by W.

Using quoted optical and thermal constants for a-Si [20] in equation 3.1, the energy threshold for melting is calculated to be 82mJ/cm² for a 30ns pulse. This value is comparable to numerical simulations of Unamuno and Fogarassy [27]. In view of the simplifications involved in deriving the analytical expression, the agreement with the experimentally measured threshold energy, of ~95mJ/cm² is quite good. Moreover, other published data fall within the range of 100 ± 30mJ/cm² [17,23,28]. Although there is a clustering of the experimental results, they do cover an appreciable range of energies. There are several reasons for this: firstly, as shown by equation 3.1, the threshold energy is a function of pulse duration and this will vary with different gas mixtures and is rarely quoted. Secondly, the hydrogen content of the material affects the threshold energy [1] as

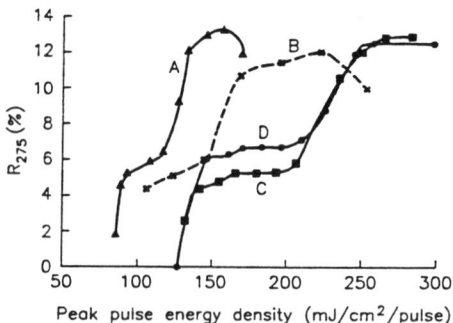

Fig. 7. Variation of laser crystallisation behaviour with hydrogen content (see table I) of a-Si films.

shown by the 275nm surface reflectance [14] data in figure 7. The step in the curves has been correlated with the stratified nature of the films. These results show how the melt threshold changes from ~80mJ/cm²/pulse to ~130mJ/cm²/pulse as the hydrogen content of the films is reduced by changing the preparation technique. The hydrogen content of these films, as measured by SIMS, as given in table I

Table I

#	Film	T_{dep} (°C)	Post Anneal	H (cm^{-3})
A	PECVD	200		$8 \pm 2 \times 10^{21}$
B	PECVD	250		$4.5 \pm 1.5 \times 10^{21}$
C	PECVD	250	1h @ 500°C	1.5×10^{20}
D	LPCVD	540		2×10^{19}

It will be noticed that the thermally de-hydrogenated PECVD sample behaves similarly to the low hydrogen content LPCVD sample, confirming that the significant variable between these films is the hydrogen content. The explanation of the role of hydrogen can be obtained by reference to equation 3.1, from which it can be concluded that increased hydogen is either reducing the melt temperature or the thermal diffusivity of the films. From the observation that the melt threshold energy increases more strongly with film thickness for LPCVD material compared with PECVD material, we conclude that the major effect is upon the diffusivity. Another factor contributing to the spread in reported melt thresholds is that the value has been found to vary with both pulse shape and scanning mode and the total number of pulses [24].

The electron field effect mobility is generally found to increase with laser energy as shown in figure 8. This is partly an artefact of the field effect mobility measurement and is initially due to the increased depth of the large grain surface layer to a value such that the band bending necessary for surface inversion can be accommodated within this layer [1]; beyond this point, the increase is due to the increase in lateral grain size shown in figure 6. The results shown in figure 8 were obtained from a 150nm thick film, but at the higher energies, at which the maximum depth of large grain crystallisation is achieved, there is also a film thickness dependence to the mobility. This is shown in figure 9 in which there is an appreciable increase in electron mobility as the film thickness is reduced from 80nm to 40nm. TEM examination of these samples has shown a 3-fold increase in grain size in the thinner films. This change in grain structure can be

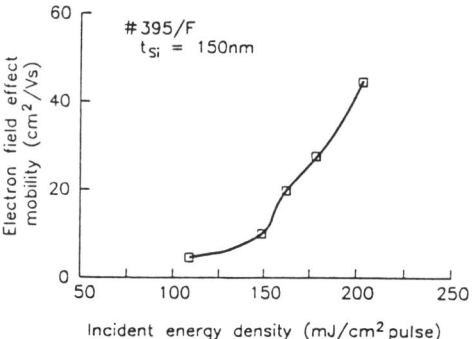

Fig. 8. Variation of electron field effect mobility with incident laser energy density.

correlated with the complete melting of the thinner films with a consequent modification to the crystallisation process.

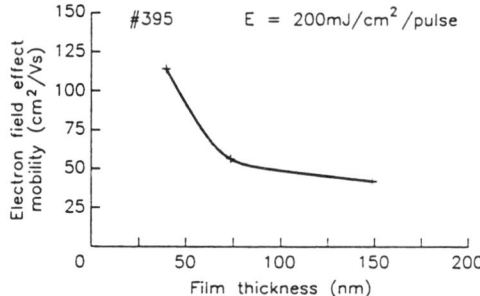

Fig. 9. Variation of electron field effect mobility with film thickness following laser crystallisation at 200mJ/cm²/pulse.

By and large, the high field effect mobilities ($>100 cm^2/Vs$) [2,6,28,29] are greater than those obtained by solid phase crystallisation of a-Si, discussed in section 3.1, in spite of the grain size being comparable or smaller ($\sim 100\text{-}300 nm$). The apparent inconsistency results from the difference in grain quality between the two processes. The laser crystallised material has few line or plane defects within the grains, whereas the SPC material consists of large dendritic grains which are rich in intra-grain defects such as twins, micro-twins, stacking faults etc. Apart from these potential scattering centres, the grain boundaries themselves may be different, but detailed high resolution electron microscopy is needed to clarify these differences.

4. Trapping State Distributions

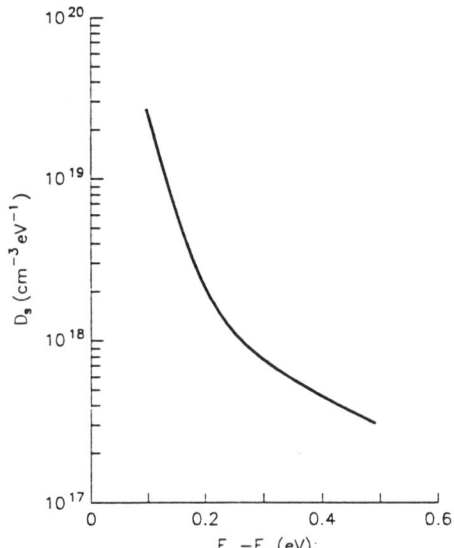

Fig. 10. Poly-Si trap state density measured across the upper half of the band gap by DLTS.

The TFT transfer characteristics shown in Figure 2 have sub-threshold slopes of ~ 0.6 -1.0 V/dec and these values are a direct indication of the fixed charge density in the surface space charge layer as the energy bands are bent towards electron accumulation. Using the standard single crystal MOSFET analysis, these values would be indicative of space charge densities of $10^{17}\text{-}10^{18} cm^{-3}$. The high space charge density arises from trapping states at grain boundaries and at intra-grain defects. The arguments above are merely illustrative of the order of magnitude effects observed in poly-Si, and are not a rigorous analysis. To do the latter, the energy distribution of deep lying trapping states has to be known and appropiately allowed for in the solution of Poissons' equation.

The trapping state distribution, obtained from DLTS measurements on a poly-Si TFT [30], is shown in figure 10 (the absolute accuracy of the measurement is about a factor of 2); the TFT used for this measurement had a sub-threshold slope of ~1.14V/dec. The distribution is continuous across the upper half of the band gap and rising towards the conduction band edge, with volume trap densities of 10^{18}-10^{19}cm^{-3}eV^{-1}. Qualitatively similar distributions have been obtained by other techniques such as the field effect analysis of the transfer characteristics [31] and numerical simulation of the TFT characteristic [32].

In a number of analyses [31,32] if has been found convenient to treat the trap distribution as an exponential or the sum of two exponentials:

$$N_T(E) = N_{T1}\exp-\frac{E}{E_1} + N_{T2}\exp-\frac{E}{E_2} \quad (4.1)$$

where E is measured from the nearest band edge, E_1 and E_2 are the characteristic energy widths of the distributions and N_{T1} and N_{T2} are the trap concentrations per unit volume per eV measured at the band edges. For instance, the trap distribution extracted from the DLTS measurement can be represented by

$$N_T(E) = 2.8x10^{18}\exp-\frac{E}{0.23} + 3.2x10^{20}\exp-\frac{E}{3.8x10^{-2}} cm^{-3}eV^{-1} \quad (4.2)$$

In analogy with a-Si analyses [33], the two distributions have been ascribed to deep states and band tail states. The former are associated with dangling bonds at grain boundaries and the latter with disorder induced states arising from features such as micro-twins and weak Si-Si bonds.

Large volume densities of trapping states distributed through the band gap can have a significant effect upon both threshold voltage and field effect mobility [33], as well as their dependence on V_G [34]. In these respects, the poly-Si TFT is different in its behaviour from its single crystal counterpart. The effect of the continuous distribution of states can be readily appreciated by noting that, in contrast to a single crystal or SOI device in which the surface charge induced by the gate voltage beyond threshold consists mainly of free carriers, in the case of poly-Si there may be a continued partition of induced charge between trapping states and free carriers. In extracting a field effect mobility, the implicit assumption is that all induced charge is free charge; if, however, a fraction of it is trapped then the carrier mobility will be underestimated. This has been demonstrated in a series of illustrative calculations by Migliorato and Quinn [34]. Clearly the magnitude and significance of the effect will be determined by the detailed distribution of trapping states and an order of magnitude appreciation of this can be obtained by considering the rate of change, with Fermi level position, of the free carrier (n_f) and trapped carrier (n_t) volume densities at the surface:

$$\frac{dn_f}{dE_F} = \frac{-N_c}{kT}\exp-\frac{E_F}{kT} \simeq 1.7x10^{21}\exp-\frac{E_F}{kT} \quad (4.3)$$

and, applying the zero Kelvin occupancy approximation to an exponential trap distribution,

$$\frac{dn_t}{dE_F} = -N_T \exp\frac{-E_F}{E_1} \qquad (4.4)$$

(It should be noted that the zero Kelvin approximation will underestimate the trapped charge density, but it has the merit of yielding a simple analytical expression).

Unless the value of equation 4.3 is appreciably greater than that of equation 4.4, carrier trapping effects can be significant. For instance, the values of equations 4.3 and 4.4, with E_F positioned at E_C-0.1eV ($n_f=9.1\times10^{17}$cm^{-3}), are 3.5×10^{19}cm^{-3}ev^{-1} and 2.2×10^{19}cm^{-3}eV^{-1}, respectively, using the values of N_T and E_T obtained from the DLTS measurement. This would indicate a field effect mobility of no more than 61% of the free carrier mobility. This approximate indication has been confirmed by using the trap distribution obtained from the DLTS measurement in a 2-D device simulator to calculate the TFT transfer characteristic: this had a sub-threshold slope of 0.94V/dec and a field effect mobility of 53% of the input free carrier mobility. However, the effect of the distribution of trapping states in reducing the field effect mobility by 47% is likely to be an upper estimate in well engineered TFTs. This is because the sub-threshold slope of the device used was 1.14V/dec and values down to 0.5V/dec have been obtained in higher quality TFTs, fabricated by, for instance, laser crystallisation. However, even in these devices, the values of sub-threshold slope and threshold voltage are still larger than in single crystal or SOI devices and are directly detemined by the trapping state distribution. This is illustrated by the results in figure 11 in which we show the calculated dependence of the sub-threshold slope on the value of N_{T1} from equation 4.1, in which the other parameters are the DLTS values quoted in equation 4.2.

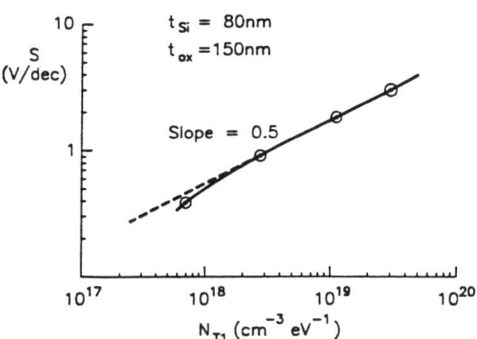

Fig. 11. Calculated variation of TFT sub-threshold slope with N_{T1} (from equation 4.2).

5. TFT LEAKAGE CURRENTS

The attainment of low, well controlled off-state leakage currents is a prime requirement for pixel TFTs in AMLCDs. This has been one driving factor in the investigation of the leakage current phenomena, the other has been to identify the mechanism which results in the widely reported exponential dependence of the leakage current on gate and drain bias. These bias dependences are shown in figures 2 and 12a, respectively. The current is independent of channel length and results from a generation process at the drain junction [35]. In view of the strong bias dependence, the generation mechanism cannot be pure thermal generation, and has been associated with a field enhanced process. Similar phenomena have been reported in single crystal Si devices [36] and in SOI devices [37] under high field conditions.

Various models have been put forward to characterise the mechanism, including trap to band tunnelling [38], the Poole-Frenkel effect [39] and phonon assisted tunnelling from traps [35,40-42]. For pure tunnelling, the non-zero activation energy, shown in figure 12b, would not be expected; the Poole-Frenkel effect does not predict the correct form of the exponential current dependence on bias, nor would it give the observed dependence of activation energy on bias [43]. Phonon assisted tunnelling offers the most satisfactory explanation for the majority of the published data, although the detailed explanations have varied between mid-gap trap to band transitions [35], inter-trap hopping [42] and band tail state to band transitions [41]. The expression given by Vincent et al [44] for phonon assisted tunnelling from traps in the drain space charge region has been incorporated into a simple junction generation current model and fitted to the experimental data shown in figures 12a and b. To achieve these fits it was necessary to use a carrier effective mass of $0.2m_o$ (transverse electron effective mass) and a drain space charge density of $1 \times 10^{17} cm^{-3}$ to

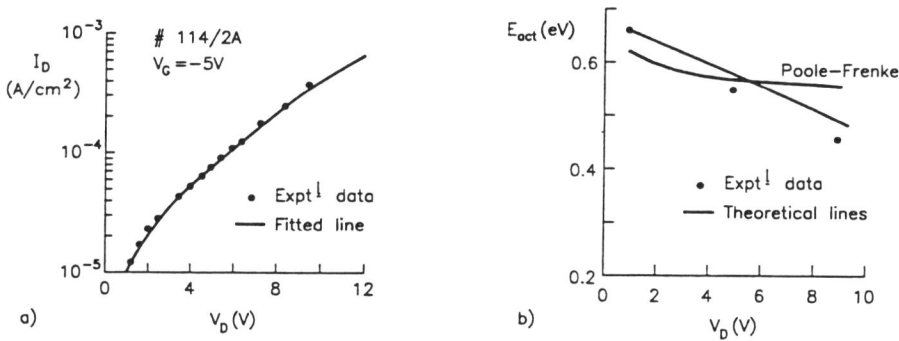

Fig. 12. a) experimental I_D-V_D off-state data and fitted phonon assisted tunnelling model, b) experimental variation of leakage current activation energy with drain bias and calculated curves for phonon assisted tunnelling and the Poole-Frenkel effect.

achieve the fields necessary to promote tunnelling. The drain space charge density due to the traps will be given by the trap density within $\sim 2kT$ of mid-gap. From the data in figure 10, this amounts to $\sim 1.5 \times 10^{16} cm^{-3}$ which would be insufficient to generate the necessary peak fields of $> 2 \times 10^5 V/cm$. However, numerical modelling of the TFT has shown that the required fields are produced by 2-D drain-gate coupling, which is particularly strong for the near abrupt junctions formed in these devices. This is illustrated by the results in figure 13 (for $V_D = 5V$) showing the dependence of the peak surface field on gate bias for the trap state density given in equation 4.2. It will be

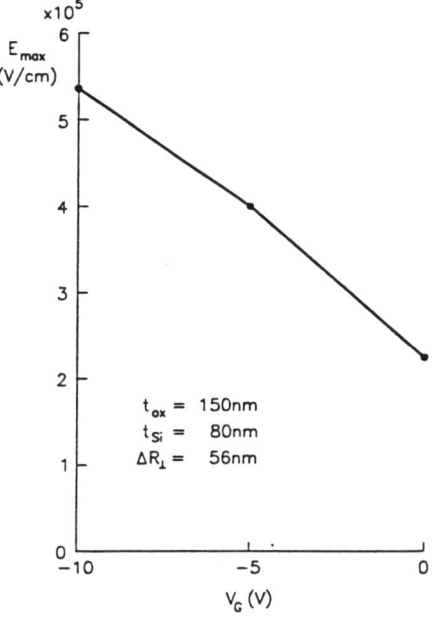

Fig. 13. Calculated variation of maximum drain field (V_D=5V) with gate bias for the trap state density given by equation 4.2. (A standard deviation of 56nm was used for the lateral straggle of the drain dopant beyond the defining window edge).

seen that, in the off-state, fields of the appropriate magnitude are produced, and, because of the dominace of 2-D gate-drain coupling effects, there was no change in the field for a trap state density 10 times lower. Clearly for sufficiently large mid-gap trap concentrations, the space charge itself would ultimately influence the field. (Although the traps may not be responsible for the magnitude of the field, they will of course constitute the generation centres). The 2-D coupling effect also explains why, when the trap state density is reduced, in for instance laser crystalled TFTs with low values of sub-threshold slope, the device still displays field enhanced leakage currents with an unchanged V_G and V_D dependence (as in figure 2).

6. INSTABILITY ISSUES

A number of instability mechanisms have been identified in poly-Si TFTs under different combinations of drain and/or gate bias stress including ionic drift and carrier trapping instabilities in the gate dielectric [35], water related instabilities in undensified gate oxide layers [45] and hot carrier induced instabilities [35,46]. The hot carrier instabilities arise from combined gate and drain bias stress and are qualitatively similar to instabilities reported in single crystal and SOI devices [47].

For on-state hot carrier stressing, the major effect has been an increase in leakage current and a reduction in on-current [35,46]. Both are associated with trapping state creation near the drain, and the dependence of the on-current reduction on gate and drain bias is shown in figure 14. The results in this figure were obtained from accelerated stress measurements in which the value of drain bias sufficient to cause a 30% reduction in on-current after 1 minutes stress is plotted against the gate bias stress. The shape of this curve indicates weak avalanche related effects at the drain and is comparable to the substrate current variation with gate bias reported in single crystal devices [47]. At gate bias values near the minimum of the I_D-V_G curve, few free carriers are available for multiplication, hence the device can sustain large

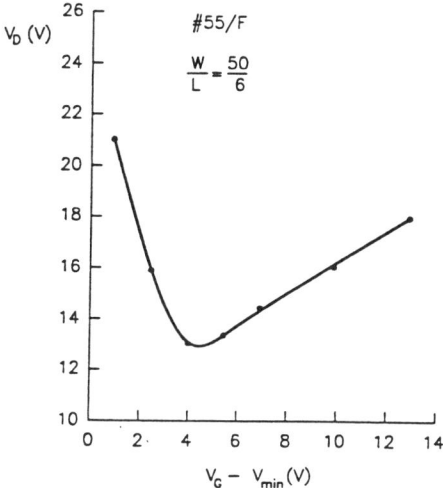

Fig. 14. Variation of drain bias, at which there is 30% hot carrier degradation of I_{ON} in 1 minute, with gate bias.

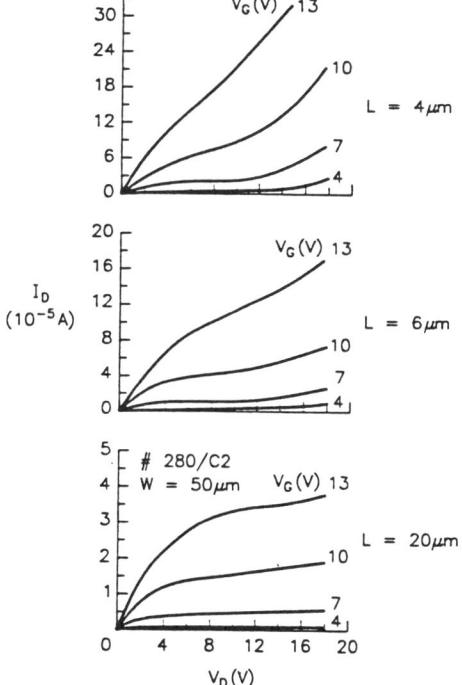

Fig. 15. Poly-Si TFT output characteristics showing channel length dependent effects.

values of drain bias and high fields. At large positive gate bias, the drain field, for a given value of drain bias, is reduced and although there is now a plentiful supply of free carriers, high drain bias can be sustained before the field is high enough to generate hot carriers. Between these two extremes, hot carrier degradation occurs more rapidly in the near-threshold regime because the gate bias produces only limited drain field relief, but is high enough to produce a significant density of free carriers. Given the exponential dependence of multiplication on drain bias, the stress induced ageing shows a similar dependence, in that the time required to achieve a given degree of degradation increases exponentially (at ~0.5-1 decade/volt) as drain bias is reduced [46].

For off-state stressing, at negative gate bias and positive drain bias hot hole injection [48] into the gate oxide near the drain has also been observed. This localised positive charge has had the effect of relieving the drain field in the off-state, and hence reducing the field enhanced leakage current.

One feature which distinguishes the poly-Si devices from single crystal devices is that the hot carrier ageing is channel length dependent [46], with short channel length devices degrading most rapidly. This is a reflection of a channel length dependent multiplication effect, as indicated by the output charactertistics shown in figure 15. The kick-up in the output characteristics at high drain bias has been referred to as the kink effect, but it does not have the charactertistic

S-shape associated with the classical floating body kink seen in SOI devices. In fact, the classical kink has been rarely reported in poly-Si devices, and, in our own structures, we have only observed it in the highest mobility devices, where $\mu \sim 200 \text{cm}^2/\text{Vs}$. The kick-up has been explained in terms of deep state trapping of avalanche induced holes at the drain [49], but the channel length dependence of this effect was not demonstrated. At the moment, improved understanding of the poly-Si TFT output charactertistic is required.

7. SOI AMLCD TECHNOLOGIES

The merit of the foregoing approach using poly-Si, either deposited or crystallised on transparent substrates, is that it readily lends itself to a large area, glass substrate technology. However, there have been some alternative approaches [3,4], particularly for small projection displays, in which a more conventional SOI approach has been adopted. In one case [4], direct bonding between an oxidised silicon wafer and a quartz substrate is used, together with a selective etch back procedure, to produce a thin single crystal silicon film on the quartz substrate. This layer is then subjected to an AMLCD process which is similar to the high temperature poly-Si process. The other technique [3], starts with an SOI wafer, formed by a ZMR type process, and the active elements of the display are fabricated in this layer which is then glued to a glass substrate and the silicon substrate removed by etching.

In both cases, superior device performance is achieved by the use of single crystal Si compared with the poly-Si alternative. However, the basic process steps are much the same for all the processes, except that the poly-Si deposition is replaced by the SOI layer formation together with its bonding to the transparent substrate. In the short term, the direct use of conventional SOI substrates [3] has the attraction of being readily implemented in a standard Si line, but both processes will be limited in substrate size and, ultimately, may only be justified if the application demands the superior device performance offered by crystalline silicon.

8. CONCLUSION

Poly-Si TFTs are emerging as an important technology for active matrix flat panel displays because they will enable the integration of various addressing circuits and, in the longer term, other circuit functions onto the display. This needs to be a large area, glass substrate compatible technology and in this paper we review a number of aspects of the technology and device behaviour of poly-Si TFTs. These have included the formation of poly-Si by solid phase and laser crystallisation; in both cases the film thickness was seen to affect the crystallite size and hence device properties. Field enhanced TFT leakage currents could be described by phonon assisted tunnelling and the fields necessary to produce this effect were due to 2-D gate-drain coupling enhanced by the abrupt nature of the doped regions. Finally, in common with single crystal and SOI devices the TFTs were shown to be susceptible to hot carrier damage at high drain bias.

9. Acknowledgements

We are grateful to Mrs A Gill and Mrs R Bunn for sample fabrication, Dr M J Trainor and Mr I D French for a-Si layers, Dr J P Gowers for electron microscopy and Mr J B Clegg for the SIMS measurements.

10. References

1. S D Brotherton, D J McCulloch, J B Clegg and J P Gowers. IEEE Trans ED-40, 2, 407, (1993).
2. T Sameshima. Mat Res Soc Symp Proc, 283, 679, (1993).
3. J P Salerno, D P Vu, B D Dingle, M W Batty, A C Ipri, R G Stewart, D L Jose and M L Tilton. SID '92 Digest, p63, (1992).
4. K R Sarma, C Rogers and C Chanley. SID '94 Digest, p419, (1994).
5. S D Brotherton, J R Ayres and N D Young, Solid State Electronics, 34, 7, 671 (1991).
6. K Sera, F Okumura, H Uchida, S Itoh, S Kaneko and K Hotta, IEEE Trans ED-36, 12, 2868 (1989).
7. S D Brotherton, D J McCulloch and M J Edwards. Solid State Phenomena, 37-38, 299, (1994).
8. P Mei, J B Boyce, M Hack, R A Lujan, R I Johnson, G B Anderson, D K Fork and S E Ready, Appl. Phys. Letts., 64, 1132, (1994).
9. P Joubert, B Loisel, Y Chouan and L Haji, Jnl. Electrochem Soc., 134, 10, 2541 (1987).
10. A T Voutsas and M K Hatalis, Jnl. Electrochem Soc., 139, 9, 2659, (1992).
11. N Kono, T Nagahara, K Fujimoto, Y Kashiwagi and H Kakinoki., Mat. Res. Soc. Symp. Proc. 283, 629, (1993).
12. A Mimura, N Konishi, K Ono, J-I Ohwada, Y Hosokawa, Y A Ono, T Suzuki, K Miyata and H Kawakami, IEEE Trans ED-36, 2, 351, (1989).
13. M K Hatalis and D W Greve, Jnl. Appl. Phys., 63, 7, 2260, (1988).
14. T Noguchi, H Hayashi and T Ohshima. Japan Jnl. Appl. Phys., 25, L121, (1986).
15. A Nakamura, F Emoto, E Fujii, Y Uemoto, A Yamamoto, K Senda and G Kano, Extended Abstracts of 20th Conference on Solid State Devices and Materials, Tokyo, 1988, p189.
16. S D Brotherton, IEE Colloquium Digest, 1993/067, p4/1, (1993).
17. T Sameshima, M Hara and S Usui, Jap. Jnl. Appl. Phys., 28, 1789, (1989).
18. Y Miyata, M Furuta, T Yoshioka and T Kawamura, Jap. Jnl. Appl. Phys., 31, part 1, 4559, (1992).
19. Y Nishihara, S Yamamoto, S Yamada, T Hikichi, I Asai and T Hamano, SID '92 Digest, p609 (1992).
20. Properties of Amorphous Silicon, EMIS data review 1, (IEE Inspec. Publication).
21. T Sameshima, Ph.D. Thesis, University of Shizuoka, 1991.
22. E Fogarassy, H Pattyn, M Elliq, A Slaoui, B Prevot, R Stuck, S de Unamuno and E L Mathe, Appl. Surf. Sci., 69, 231, (1993).

23. K Winer, G B Anderson, S E Ready, R Z Bachrach, R I Johnson, F A Ponce and J B Boyce, Appl. Phys. Letts, 57, 2222, (1990).
24. R Z Bachrach, K Winer, J B Boyce, S E Ready, R I Johnson and G B Anderson, Jnl. Electronic Materials, 19, 241, (1990).
25. D H Lowndes, G E Jellison, S J Pennycook, S P Withrow and D M Mashburn, Appl. Phys. Letts., 48, 20, 1389, (1986).
26. P Baeri and S U Campisano, Laser Annealing of Semiconductors, chapter 4 (Edited by J M Poate and J W Mayer, Academic Press, New York, 1982).
27. S de Unamuno and E Fogarassy, Appl. Surface Science, 36, 1, (1989).
28. E Fogarassy, B Prevot, S de Unamuno, M Elliq, H Pattyn, E L Mathe and A Naudon, Appl. Phys., A56, 365, (1993).
29. S Chen, J B Boyce, I-Wei Wu, A Chiang, R I Johnson, G B Anderson and S E Ready, Proc. 13th International Display Research Conf., P195 (31/8-3/9/93, Strasbourg, France).
30. J R Ayres, Jnl. Appl. Phys., 74, 1787, (1993).
31. G Fortunato and P Migliorato, Appl. Phys. Letts, 49, 16, 1025, (1986).
32. M Hack, J G Shaw, P LeComber and M Williams, Japanese Jnl. Appl. Phys., 29, 12, L2360, (1990).
33. M Shur and M Hack, Jnl. Appl. Phys., 55, 3831, (1984).
34. P Migliorato and M Quinn. Advanced Silicon and Semiconducting Silicon-Alloy Based Materials and Devices, p361, (Edited by J F A Nijs, published by Institute of Physics, Bristol, 1994).
35. J R Ayres, S D Brotherton and N D Young, Optoelectronics-Devices and Technologies, 7, 2, 301, (1992).
36. G A M Hurkx, D B M Klaasen and M P G Knuvers, IEEE Trans. ED-39, 331, (1992).
37. L J McDaid, S Hall, W Eccleston and J C Alderman, Micro-electronic Engineering, 13, 759, (1989).
38. J G Fossum, A Ortiz-Conde, H Shichijo and S K Banerjee, IEEE, Trans ED-32, 1878, (1985).
39. S K Madan and A D Antoniadis, IEEE Trans ED-33, 1518, (1986).
40. D W Greve, P A Potyraj and A M Guzman, Solid State Electronics, 28, 1255, (1985).
41. I-W Wu, A G Lewis, T-Y Huang, W B Jackson and A Chiang, IEDM-90, 867, (1990).
42. A Rodriguez, E G Moreno, H Pattyn, J F Nijs and R P Mertens, IEEE Trans. ED-40, 938, (1993).
43. S D Brotherton, INFOS '91, 117, (1991) (Edited by W Eccleston and M Uren. Published by Adam Hilger, Bristol, England, 1991).
44. G Vincent, A Chantre and D Bois, Jnl. Appl. Phys., 50, 5484, (1979).
45. N D Young and A Gill. Semicond. Sci. Technol., 7, 1103, (1992).
46. N D Young, A Gill, M J Edwards. Semicond. Sci. Technol., 7, 1183, (1992).
47. C Hu, S C Tam, F-C Hsu, P-K Ko, T-Y Chan and K W Terrill. Trans IEEE ED-32, 2,375, (1985).
48. G Tallarida, G Fortunato, L Mariucci, C Reita and P Migliorato. IEE Proc.-Circuits Devices Syst., 141, 1, 33, (1994).
49. M Hack and A G Lewis. IEEE EDL-12, 203, (1991).

HOT CARRIER RELIABILITY OF SOI STRUCTURES

D. E. IOANNOU
Electrical and Computer Engineering Department,
George Mason University, Fairfax, VA 22030

1. Introduction

Hot carrier effects have been a serious reliability concern in MOSFET's ever since the recognition in the mid-seventies, that they can significantly degrade the device characteristics during normal operation. Continuing reduction of device dimensions and increase in channel doping, to achieve higher chip density and speed, is making these reliability concerns only worse, because of the increasing electric fields. The problems have been (and continue to be) vigorously researched, and although some open questions remain, substantial progress has been made in: 1) establishing suitable device degradation monitors, 2) understanding the physical mechanisms involved and 3) developing technologies and designs to suppress these mechanisms [1].

Due to the rapidly growing interest in SOI technologies for high speed and low power applications, efforts to study the hot carrier effects in SOI MOSFET's have been intensified recently [2]. To date, experimental work has generated considerable interest on how fully depleted (FD) SOI, partially depleted (PD) SOI, and bulk MOSFET's compare with regard to their hot carrier degradation. It has been reported that [2]: 1) the front channel of FD transistors degrades less than that of PD ones [4], [5]; 2) the back channel of FD transistors degrades much more than the front one [6]; 3) for both PD and FD transistors, stressing one channel may also damage the opposite channel [7], [8]; and 4) SOI MOSFET's may degrade less [4], [5], or more [7], [8] than their bulk counterparts. These and other results may appear contradictory, partly because it is difficult to establish meaningful procedures for such comparisons.

This paper is a review of the state of the art of hot carrier induced degradation in SOI MOSFET's and addresses possible remaining concerns that require further study. First a brief description is given of some basic concepts relating to hot carriers in MOSFET structures and in particular in SOI MOSFET's. This is followed by a detailed discussion of experimental results on the degradation of typical partially depleted and then fully depleted devices. To understand these results it will be necessary to invoke the newly observed opposite channel based charge injection phenomenon, as well as the non-local nature of the impact ionization. Also, in the case of FD devices the importance of channel coupling both during stress and during measurement will become apparent. Some typical device lifetime results obtained recently for PD and FD transistors will be given, and the possibility of yet unknown degradation mechanisms for channel lengths in the 0.1μm region will be acknowledged. Finally the underlying mechanisms of degradation will be briefly discussed with special reference to the sequential stressing technique and its application in the study of interface states and oxide charges.

2. Hot Carriers in MOSFET's.

Carriers become hot when they absorb energy from large electric fields more quickly than they lose energy due to collisions with the lattice. The carrier energy increases until greater random motion leads to increased collision rates, which in turn lead to higher energy loss rates. A steady state is finally reached when the energy loss rate equals the energy absorption rate. At this steady state the electron temperature T_e, or random motion, exceeds that of the lattice, T_L. Fig.1 plots the lateral electric field along the channel for a 1 μm channel length MOSFET under typical hot carrier stress conditions. It is seen that the field peaks sharply near the drain end of the channel.

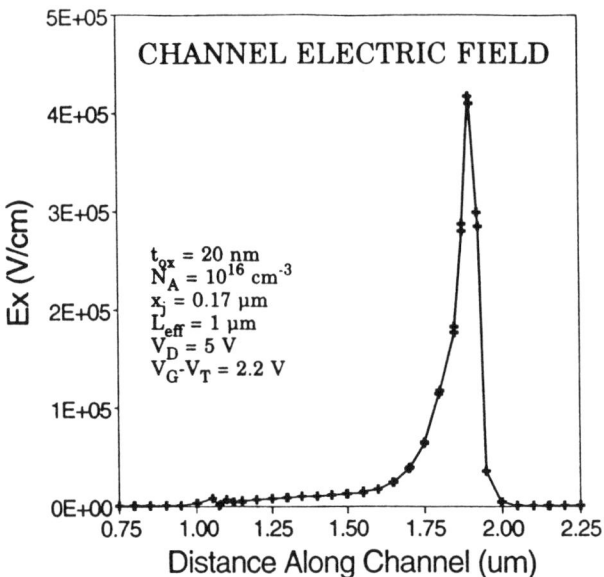

Figure 1. Lateral electric field profile in the channel of a typical MOSFET biased under hot electron stress

This type of field profile gives rise to several hot carrier processes, as follows: Process 1: electrons enter the gate oxide where they interact with and/or create oxide traps and interface states, and produce gate current I_G. Process 2: avalanche pair production near the drain. Process 3: holes produced by process 2 are collected by the substrate and give rise to substrate current I_{SUB}. Process 4: Process 3 forward-biases the source, which then injects electrons into the substrate. Process 5: some electrons from process 4 reach the drain, resulting in more impact ionization.

Fig.2 shows a typical SOI MOSFET structure [9]. It is clear that in order to conduct hot carrier studies in such structures, special consideration should be given to the behavior of the buried oxide, the effects of the floating body [10] and self-heating [11], the possibilities of partially depleted and fully depleted device operation [9], the interaction of the two (i.e. front and back) channels [12], [13] and lately, the trend towards thinner buried oxides [14].

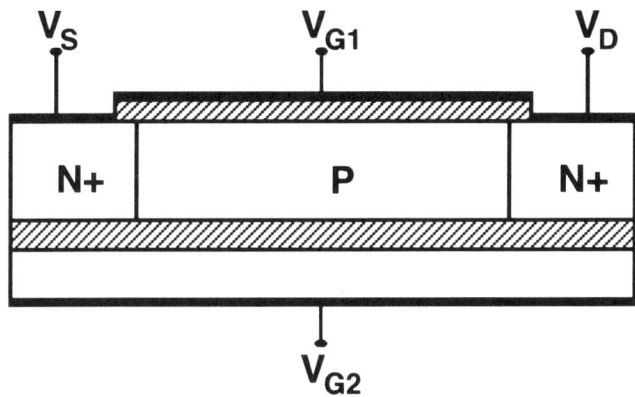

Figure 2. Typical SOI MOSFET structure showing front and back gates, buried oxide, and floating body.

Central to the understanding of hot carrier effects in SOI MOSFET's (as in bulk ones) is the impact ionization that generates e-h pairs near the drain of a device under stress, and the gate and substrate currents that result [15]. Either of these currents may be used as a monitor for nMOS devices, but the gate current is more appropriate for pMOS devices [16]. The substrate current is larger and thus easier to measure, but this requires a body (Si film) contact. Analytical expressions for these currents have been available for bulk devices for a long time and they have recently been adapted to SOI MOSFET's [17].

3. Partially-Depleted Devices

State of the art PD devices were subjected to hot carrier stress under realistic bias conditions, and the degradation of both channels was investigated [3]. The devices incorporated LDD's, their channel lengths were in the 1 μm range, and they had the following structural and doping parameter values: gate oxide thickness $t_{ox1}=20$ nm, buried oxide thickness $t_{ox2}=400$ nm, silicon film thickness $t_{Si}=300$ nm, and film doping $N_A=2 \times 10^{17}$ cm^{-3}. The front channel was stressed for 10^4s under the following conditions: $V_D=5.5$ V, $V_{G1}=2.5$ V, $V_{G2}=0$ V, and the front and the back channel parameter changes were measured. Then the back channel was stressed for various time periods (Fig.3) under $V_D=5.5$ V, $V_{G2}=55$ V, $V_{G1}=0$ V and both channels were examined. During the measurement $V_D=50$ mV, and the opposite gate was grounded. It was found that <u>following front channel stress</u> the front channel degrades very little ($\Delta V_{T1}=6$ mV, $\Delta g_{m1}=3 - 5$ %), and there is no observable change in the back channel. <u>Following back channel stress</u>, however, the back channel degrades very much ($\Delta V_{T2}=30$ V, and there is substantial transconductance g_m overshoot), but there is no front channel change.

With regard to the front channel, the question arises as to how the SOI devices compare to bulk ones. This is in fact not a straightforward question to answer: for instance in comparing FD and bulk devices, if the same level of channel doping is used,

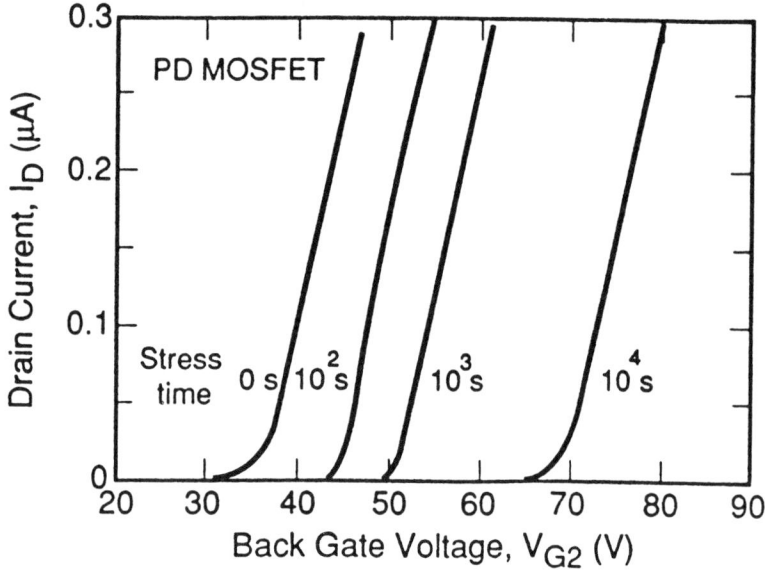

Figure 3. PD device transfer characteristics following various successive amounts of hot carrier stress

the threshold voltages, saturation currents and breakdown characteristics will be different. If the threshold voltages are similar, the doping profiles will be different etc etc. Most reasonable seems to be an approach taken recently [18] whereby for a given channel length each technology (device) is self-consistently optimized; the hot carrier degradation behavior of each one is then independently studied by using established (fundamental) degradation monitors; and finally comparisons are made. If such a comparison is made [18], it seems that presently SOI devices are somewhat inferior to their bulk counterparts, but if as expected in the near future, identical (low temperature) processes are used, SOI devices will outperform bulk ones, based on the nature of their structure and the resulting electric field profiles.

4. Fully-Depleted Devices

FD LDD devices were studied [3] with channel lengths down to 0.8 μm [3], and the following parameter values: t_{ox1}=15 nm, t_{ox2}=400 nm, t_{Si}=140 nm and N_A=4x10^{16} cm^{-3}. devices were subjected to various types of 10^4 s stresses (see Fig.4), and the degradation of both channels was studied. <u>Following front channel stress,</u> the front channel degrades very little (ΔV_{T1}=2 mV, Δg_{m1} less than 1%), and there is no change on back channel. <u>Following various back channel stresses</u>, the back channel degrades significantly (ΔV_{T2}=0.5 to 2 V), but much less than in the case of partially depleted devices, and there is no change on front channel.

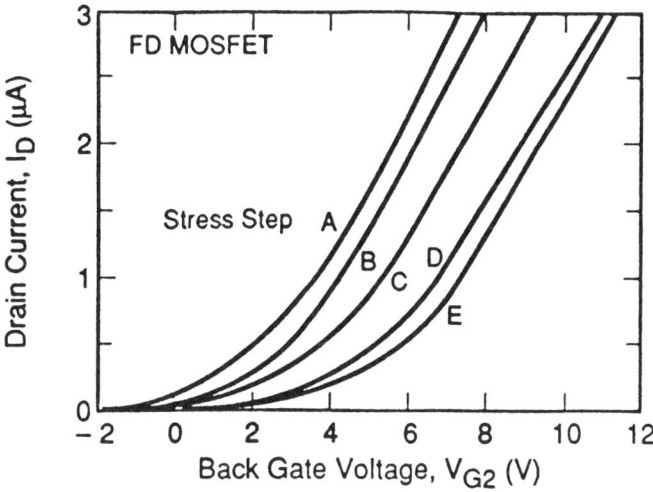

t_{ox1}=15 nm, t_{ox2}=400 nm; t_{Si}=140 nm, N_A=4x10^{16}cm^{-3} ; W/L = 3.6/1 μm, LDD

<u>Stress A:</u> V_D=5.5 V, V_{G2}=0 V, V_{G1}=1.5 V
<u>Stress B:</u> V_D=5.5 V, V_{G2}=27 V, V_{G1}=-0.5 V
<u>Stress C:</u> V_D=5.5 V, V_{G2}=55 V, V_{G1}=-1.5 V
<u>Stress D:</u> V_D=5.5 V, V_{G2}=55 V, V_{G1}=-5 V
<u>Stress E:</u> V_D=7.0 V, V_{G2}=55 V, V_{G1}=-5 V

Figure 4. FD device transfer characteristics following various successive amounts of hot carrier stress

A detailed study of FD devices was also conducted by Su et al [19], on non-LDD devices with the following characteristics: t_{Si}=70, 100, and 180 nm; N_A=5x10^{16}, 1x10^{17}, and 3x10^{17} cm^{-3}; t_{ox1}=10.8 nm, t_{ox2}=360 nm; and channel length L upwards from 0.25 µm. This group measured the gate current as a means to monitor the electric field. They observed that the worst case degradation occurs when stressing at V_G-V_T = 0.15 V and **not** at V_G = V_D/2, as is the case for bulk devices. They also found that the extend of the degradation depends on the state of the back channel (i.e., accumulation, depletion) during stress. Regarding the electric field measurement, significantly reduced gate current was observed for all SOI devices in comparison to bulk, with the thinnest film FD device exhibiting the least gate current.

More recently, Tsuchiya et al [20] conducted a study of ultra-thin, deep-submicron FD devices fabricated on thin buried layer SOI substrates. The devices studied were non-LDD devices with the following parameter values: t_{Si}=50 nm, t_{ox1}=5 nm, t_{ox2}=90 nm, and channel length L down to 0.1 µm. It was found that the damage mainly occurs in the channel under stress, and that under certain conditions, hole injection in the opposite channel was observed. The condition (i.e. depletion or accumulation) of the opposite channel during stress as well as during measurement was very important.

4.1 CHANNEL COUPLING

For the case of FD transistors the two channels may be electrostatically coupled [12], [13] and this has important consequences both during stress as well as during measurement [3]. It was found for instance [19] that the gate current (and thus the electric field) depends on the condition of the back interface, and decreases as the

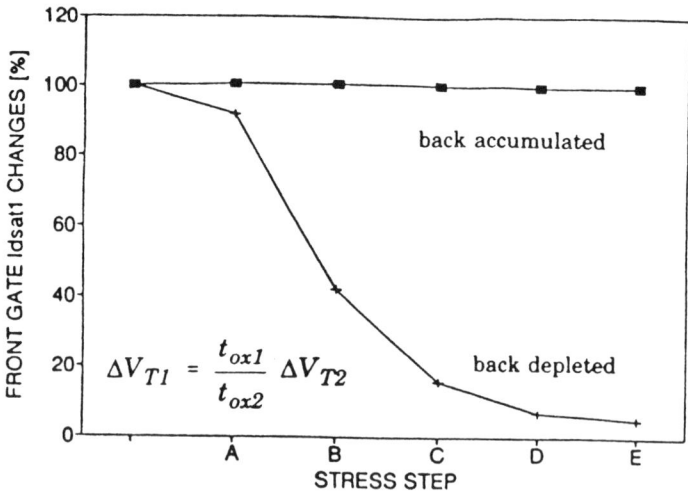

Figure 5. The effect of back channel degradation on the front channel parameters because of electrostatic coupling during measurement

interface moves from accumulation to depletion. Also, the front-gate linear current degradation as a function of back-gate voltage during front gate stressing was studied [19], and it was found that when the back-channel is accumulated during stress, increased degradation resulted, in agreement with the observation of increased gate current and electric field compared to stressing with grounded back gate. When the back channel is kept depleted during stress, however, even higher current degradation is observed initially, but with a weaker time dependence. This is clearly not in agreement with the observation of decreased gate currents and electric fields as the back channel is moved into depletion, and demonstrates the complexity of the channel interaction. It will be shown in section 5, for example, that opposite channel based charge injection can be very important.

It is also very important to recognize that correct results about the extend and the location of the damage can only be obtained if this channel coupling is switched off during measurement, by accumulating the opposite channel [3]. An example of this is given in Fig.5 where the degradation of the front channel current, following the various stress steps of Fig.4, is measured first with the back channel in accumulation and then in depletion: the results are dramatically different, and if not careful, damage sustained by the back channel could reflect itself on the front channel and influence its properties, thus lending itself to misinterpretations. For instance a back channel threshold voltage change ΔV_{T2} would induce a front channel threshold voltage change [12],[13] approximately equal to $\Delta V_{T1} = (t_{ox1}/t_{ox2})\Delta V_{T2}$.

4.2. NON-LOCAL FIELD EFFECTS ON SUBSTRATE CURRENT

Conventional numerical and analytical calculations of the substrate current rely on local electric field models of impact ionization. However, it has been shown recently that non-local effects are very important (even for non-submicron devices), which was also verified experimentally [19], [21], [22]. As it has already been mentioned above, a most interesting result is that the maximum device degradation occurs at low gate voltage values, about 0.2V above threshold, almost independently of the value of the drain voltage. Thus if SOI devices are stressed at gate voltages equal to about half the value of the drain voltage, as is common in bulk devices, the transistor lifetime will be overestimated. To conduct a correct analysis of the substrate current I_{SUB} and understand the above observations it is important to consider the non-local nature of the field dependence of impact ionization [23]. It has been shown that in small transistors the electron energy lags the electric field and that local field based analysis results in overestimating the impact ionization generation and substrate currents by several orders of magnitude. A simple post processing method to account for non-local effects in numerical simulation studies developed for bulk devices [24], has been recently applied to SOI MOSFET's by Omura et al [21]. In their study [21] FD devices with the following parameter values were used: t_{Si}=30 to 100 nm, N_A=8x10^{16} to 3x10^{17} cm^{-3}, t_{ox1}=7 nm, t_{ox2}=480 or 80 nm, and channel length L down to 0.1 μm. As to the drain design, SD, FGOLDD and SGOLDD structures were studied. Fig.6(a) shows typical experimental results demonstrating that maximum I_{SUB} is obtained for $V_G - V_T = 0.2$ V, which is in fact born out by the non-local effect based simulations shown in Fig.6(b).

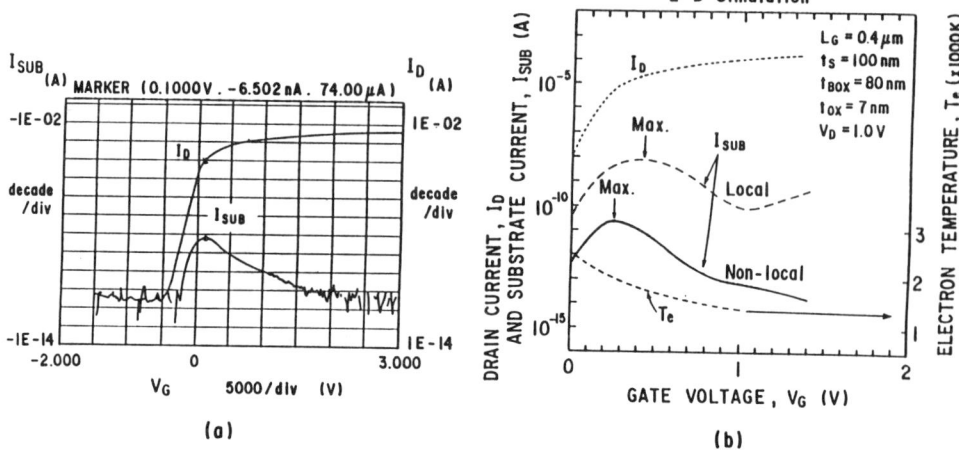

Figure 6. Measured (a) and simulated (b) substrate current variation with gate voltage [21]

It is also seen from Fig.6(b) that I_{SUB} is less than what is obtained under local-effects analysis. The authors also observed that for the SGOLD I_{SUB} goes down with t_{Si}, which hints to new challenges and opportunities for drain design engineering.

5. Opposite channel based charge injection

Fig.7 is a schematic illustration of the opposite channel based charge injection concept [25]. Here a typical SOI MOSFET is shown, biased under appropriate hot carrier stress conditions to set-up opposite channel based carrier injection. The biases in Fig.7(a) insure that the front channel conducts and operates in weak avalanche, whereas the back interface is kept accumulated. A result of the avalanche operation is the generation of electron-hole pairs near the drain. As shown, most of the electrons flow into the drain, but some are injected into the front gate oxide. The holes, however, are directed towards the back channel (kept accumulated), and aided by the favorable direction of the electric field they are finally injected into the buried oxide (Fig.7(a)). Similar phenomena occur when the roles of the two gates are interchanged, as shown in Fig.7(b).

6. Oxide Charge (Q_{ox}) or Interface States (D_{it})?

There are three acknowledged physical mechanisms which are responsible for the oxide damage and the corresponding degradation of the device characteristics: capture of injected carriers by preexisting (for instance process induced) oxide traps; generation, then filling of new (additional) oxide traps; and generation of Si/SiO$_2$ interface states. There is a heated debate as to which of the above mechanisms prevails, and opinion varies [26], [27]: Some state that the damage is due solely to interface state (D$_{it}$) generation (Ref 4-11 in [26]). Others maintain that oxide trapping (Q$_{ox}$) is responsible for device degradation (Ref. 12-17 in [26]). Still others suggest a combination of trapped charge and interface states (Ref. 18-21 in [26]).

The evidence however weighs on the view that all of them play some part, the importance of which depends on the particular bias conditions during stress as well as on the technology in hand. New light is expected to be shed into this question by the application of the newly developed sequential stressing technique, which exploits the opposite-channel-based carrier injection in SOI MOSFET's described above. For example, combining sequential stressing (Fig.7) with charge pumping measurements,

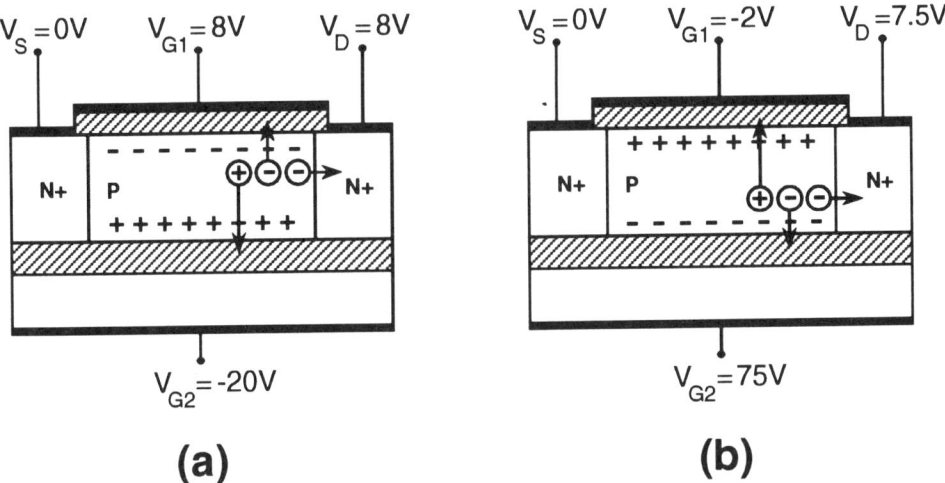

Figure 7. Opposite channel based charge injection

generation and annihilation of interface states was observed to occur under alternate hot electron/hole injection [28]. Combining it with basic transistor characteristics measurements on the other hand, revealed that hot carrier charging of the gate and buried oxides is due to a common origin [29].

7. Device Lifetime

Encouraging experimental results have been recently reported regarding device lifetimes. For example, Fig. 8 shows DC device lifetimes for state of the art FD, thin buried oxide SIMOX nMOS, pMOS and buried channel pMOS transistors [20]. The lifetime is here defined as the time required for the threshold voltage to shift by 10 mV. It is seen from this figure that the extrapolated upper limit of the supply voltage necessary to ensure a

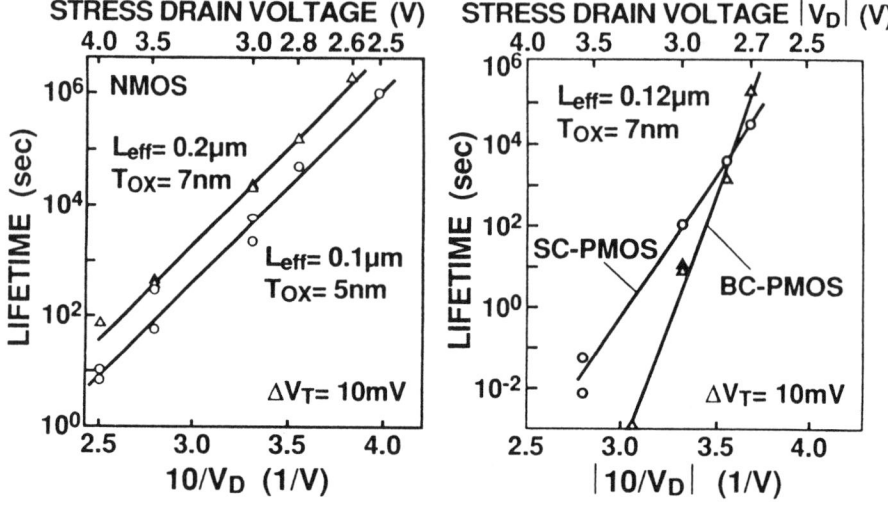

Figure 8. DC device lifetimes of state of the art FD SIMOX nMOS, pMOS, and buried channel pMOS transistors [20]

10-year lifetime is 2.1 V for the 0.1 μm nMOSFET, 2.3 V for the 0.12 μm, and 2.5 V for the 0.12 μm buried channel pMOSFET. Very long lifetimes have also been reported for PD devices [30].

8. Conclusions

The electric field decreases in moving from bulk, to partially depleted, to fully depleted devices. PD transistors seem to degrade more than FD ones, which in turn degrade more if the back channel is kept accumulated during stress. The worst case stress condition in FD devices occurs at gate voltages just above threshold. It is expected that thinner FD transistors with carefully engineered drains will exhibit improved resistance to hot carriers down to very short channel lengths. However, abnormally high degradation has been both predicted and observed when the channel length gets into the 0.1μ range. The condition of the opposite interface (i.e. depletion or inversion) is very important both during stress and during measurement for fully depleted devices. Non-local electric field effects are very important, as they lead to more optimistic expectations regarding degradation and open up new opportunities for innovative drain engineering. The opposite channel based carrier injection phenomenon is very important both as a new source of degradation as a probe for the study of the degradation mechanisms. Finally, good device lifetimes for both partially depleted and fully depleted devices, but there is a possibility of yet unknown degradation mechanisms for 0.1 μm, ultra thin devices.

9. Acknowledgements

Generous financial support from DNA, ONR and NRL over the years is gratefully acknowledged. Sincere thanks are due to my graduate students, past and present. Stimulating and continuous interactions with George Campisi and Hap Hughes of NRL have been crucial.

10. References

1. Wang, C.T. (1992) Hot Carrier Design Considerations for MOS Devices and Circuits, Van Nostrand Reinhold.
2. Ioannou, D.E. (1994) Current status of hot carrier effects in SOI MOSFET's, **1994** International IEEE SOI Conference Proceedings, 1-2.
3. Cristoloveanu, S., Gulwadi, S.M., Ioannou, D.E., Campisi, G.J., and Hughes, H.L. (1992) Hot-electron-induced degradation of front and back channels in partially and fully depleted SIMOX MOSFET's, IEEE Electron Device Lett. **13**, 603-605.
4. Colinge, J.-P. (1987) Hot-electron effects in silicon-on-insulator n-channel MOSFET's, IEEE Trans. Electron Dev. **34**, 2173-2177.
5. Fossum, J.G., Choi, J.-Y., and Sundaresan, R. (1990) SOI design for competitive CMOS VLSI, IEEE Trans. Electron Dev. **37**, 724-729.
6. Ouisse, T., Cristoloveanu, S., and Borel, G. (1991) Hot-carrier-induced degradation of the back interface in short-channel silicon-on-insulator MOSFET's, IEEE Electron Device Lett. **12**, 290-293.
7. Woerlee, P.H., van Ommen, A.H., Lifka, H., Juffermans, C.A.H., Plaja, L., and Klaassen, F.M. (1989) Half-micron CMOS on ultra-thin Silicon-on-Insulator IEEE IEDM **89**, 821-824.
8. Woerlee, P.H., Juffermans, C., Lifka, H., Manders, W., Oude Lansink, F.M., Paulzen, G.M., Sheridan, P., and Walker, A. (1990) A half-micron CMOS technology using ultra-thin silicon on insulator, IEEE IEDM **90**, 583-586.
9. Colinge, J.-P., (1991) Silicon-on-Insulator Technology: Materials to VLSI, Kluwer Academic Publishers, Dordrecht.
10. Choi, J.-Y. and Fossum, J.G. (1991) Analysis and control of floating-body bipolar effects in fully depleted submicrometer SOI MOSFET's, IEEE Trans. Electron Dev. **38**, 1384-1391.
11. Su, L.T., Goodson, K.E., Antoniadis, D.A., Flick, M.I., and Chung, J.E. (1992) Measurement and modeling of self-heating effects in SOI NMOSFET's, IEEE IEDM **92**, 357-360.
12. Lim, H.-K. and Fossum, J.G. (1983) Threshold voltage of thin-film silicon-on-insulator (SOI) MOSFET's, IEEE Trans. Electron Dev. **30**, 1244-1251.
13. Mazhari, B., Cristoloveanu, S., Ioannou, D.E., and Caviglia, A.L. (1991) Properties of ultra-thin wafer-bonded silicon-on-insulator MOSFET's, IEEE Trans. Electron Dev. **38**, 1289-1295.
14. Auberton-Herve, A., and Lamure, J. (1994) Low dose SIMOX for ULSI applications, this book.
15. Hu, C., Tam, C.S., Hsu, F.-C., Ko, P.-K., Chan, T.-Y., and Terril, K.W. (1985) Hot-

electron-induced MOSFET degradation-- model, monitor, and improvement, IEEE Trans. Electron Dev. **32**, 375-385.
16. Ong, T.-C., Ko, P.-K., and Hu, C. (1990) Hot-carrier current modeling and device degradation in p-MOSFET's, IEEE Trans. Electron Dev. **37**, 1658-1666.
17. Ma, Z.J., Wann, M.C., King, J.C., Cheng, Y.C., Ko, P.K., and Hu, C. (1994) Hot-carrier effects in thin FD SOI MOSFET's, IEEE Electron Device Lett. **15**, 218-220.
18. Reinbold, G. and Auberton-Herve, A.-J. (1993) Aging analysis of nMOS of a 1.3µm partially depleted SIMOX SOI technology comparison with a 1.3µm bulk technology, IEEE Trans. Electron Dev. **40**, 364-370.
19. Su, T.L., Fang, H., Chung, J.E., and Antoniadis, D.A., (1992) Hot-carrier effects in fully-depleted SOI MOSFET's, IEEE IEDM **92**, 349-352.
20. Tsuchiya, T., Ohno, T., Kado, Y., and Kai, J. (1994) Hot-carrier-induced degradation in ultra-thin, fully-depleted, deep-submicron NMOS and PMOS SOI transistors, **1994** IEEE International Reliability Physics Symposium Proceedings, 57-64.
21. Omura, Y. and Izumi, K. (1994) Physical background of substrate current characteristics and hot-carrier immunity in short-channel ultrathin-film MOSFET's/SOI, IEEE Trans. Electron Dev. **41**, 352-358.
22. Krishnan, S. and Fossum, J.G. (1993) Non-local modeling of impact ionization for optimal device/circuit design in fully depleted SOI CMOS technology, **1993** International IEEE SOI Conference Proceedings, 114-115.
23. Agostinelli, V.M., Hansat, K., Bordelon, T.J., Lemersal Jr, D.B., Tasch, A.F., and Maziar, C.M. (1994) Sensitivity issues in modeling the substrate current for submicron n- and p-channel MOSFET's, Solid-State Electronics **37**, 1627-1632.
24. Slotboom, J.W., Streutker, G., von Dort, M.J., Woerlee, P.H., Pruijmboom, A., and Gravesteijn, D.J. (1991) Non-local impact ionization in silicon devices, IEEE IEDM **91**, 127-130.
25. Zaleski A., Ioannou, D.E., Campisi, G.J., and Hughes, H.L. (1993) Successive charging/discharging of gate oxides in SOI MOSFET's by sequential stressing of front/back channel, IEEE Electron Device Lett. **14**, 435-437.
26. Doyle, B., Bourcerie, M., Marchetaux, J.-C., and Boudou, A. (1990), Interface state creation and charge trapping in the medium-to-high gate voltage range during hot-carrier stressing of n-MOS transistors, IEEE Trans. Electron Dev. **37**, 744-754.
27. Heremans, P., Bellens, R., Groeseneken, G., and Maes, H.E. (1992) Comment on "The generation and characterization of electron and hole traps created by hole injection during low gate voltage hot-carrier stressing of n-MOS transistors, IEEE Trans. Electron Dev. **39**, 458-464.
28. Sinha, S.P., Zaleski, A., Ioannou, D.E., Campisi, G.J., and Hughes H.L., (1994) Generation and annihilation of interface states under alternate hot electron/hole injection in SOI MOSFET's, **1994** Intern. IEEE SOI Conf. Proceedings, 123-124
29. Zaleski, A., Sinha, S.P., Ioannou, D.E., Campisi, G.J., Jenkins, W.C., and Hughes H.L. (1994) Common origin of hot carrier charging and discharging of gate and buried oxide in SOI MOSFET's, **1994** Intern. IEEE SOI Conf. Proceedings, 126-125.
30. Jenkins, W.C. and Liu, S.T. (1994) Hot electron lifetime of 0.8 mm CMOS transistors fabricated in SIMOX, Proceedings of the 6th Intern. Symp. on SOI Technology and Devices, Electrochemical Society Proceedings vol. **94-11**, 333-399.

NOVEL TESC BIPOLAR TRANSISTOR APPROACH FOR A THIN-FILM SILICON-ON-INSULATOR SUBSTRATE

C. J. PATEL, N.D. JANKOVIC† AND J-P COLINGE‡
Microelectronics Centre, Middlesex University, Bounds Green Road, London Nl1 2NQ, United Kingdom.
†*Microelectronics Department, Faculty of Electronic Engineering, University of Nis, Beogradska 14, 18000 Nis, Serbia.*
‡*Microelectronics Laboratory, Universite Catholique de Louvain, Maxwell-DICE, B-1348 Louvain-la-Neuve, Belgium.*

Abstract. Silicon-on-Insulator (SoI) technology offers an advanced electronic material structure suitable for realisation of a high performance bipolar transistor (BT). In this work, we demonstrate a route to a high performance thin-film BT on SoI fabricated using SoI-CMOS process for future thin-film SoI-BiCMOS circuits. The proposed novel approach to a thin-film BT has a device structure with a highly efficient Top poly-silicon Emitter and a low resistance N+ Side-Collector (**TESC**). An npn TESC-BT was fabricated on an 85nm thinned silicon overlayer of SIMOX material. Good common-emitter output characteristics of the npn TESC-BT were obtained, demonstrating the viable underlying concept of a TESC approach for a bipolar transistor fabricated on thin-film SoI substrate. The evaluated lateral pinched base resistance of 4kΩ for a 20µm long TESC device is reflected in the collector current role-off at a very high current density (6000Acm^{-2}). A reduction in the base Gummel number, as a consequence of back-gate biasing the TESC device dramatically enhances the current gain and induces drift current favourable for achieving higher f_t. Under such baising conditions or with a suitable doping profile, a 2-dimensional bipolar operation can occur. The TESC device offers the potential for realising vertical bipolar operation in a very thin-film silicon overlayer by inverting the back-interface, which acts as an extended side collector, thus improving the collecting efficiency. The TESC approach to bipolar transistor in a thin-film SoI was found to be a versatile device which can sustain, both the lateral and vertical bipolar operation.

The advantages of SoI substrates are well established, and have been demonstrated in recent years through high performance SoI-CMOS circuits. High speed, low power logic is one of the primary motivations for developing sub-quarter-micron SoI-CMOS technologies. Future BiCMOS circuits on SoI substrates will require thin-film bipolar transistors with a high current driving capability and high transconductance which will allow circuit operation with small voltage swing, thus offering a low power-delay product.

The conventional thin-film SoI lateral bipolar transistor (LBT) with N+-P-N+ lateral homojunctions achieve lower collector resistance compared to vertical bipolar transistor. Unfortunately, the high base resistance in a thin film (several kilo-ohms per square) leads to a severe current crowding effects since the base contact is placed outside the active base region [1]. Figure 1. illustrates the physical and structural problem associated with conventional SoI-LBT. The I-R voltage drop renders a major part of the conventional SoI-LBT ineffective at large emitter currents.

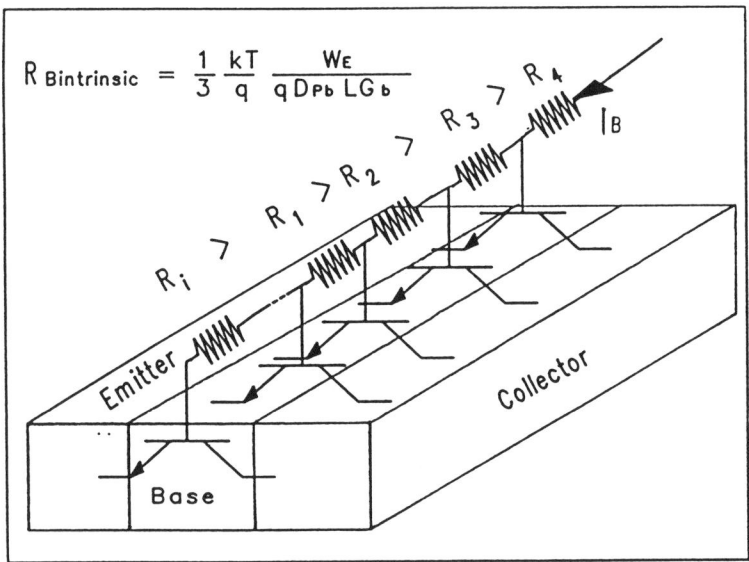

Figure 1. Illustration of the lateral I-R voltage drop in SoI-LBT leading to severe current crowding. (Instrinsic base resistance, $R_{Bint.}$ for single-sided contact, where L_e is the basewidth, other terms have their usual meaning).

Various approaches to SoI bipolar transistors (SoI-BTs) have been proposed for a possible SoI-BiCMOS integration, differing in transistor technology and design [2]. To overcome the problem of spreading base resistance mentioned above, various techniques of self-aligned base contacts have been proposed [3-6]. These alternatives provide a metal, or highly doped polysilicon, layer contact to the entire base region, but this complicates the fabrication process. One of these solutions, attempt to eliminate high collector and base resistance by prolonged silicon etching after the contact opening step [3]. Another interesting attempt, known as a pedestal-like base contact [4], is depicted in Figure 2.

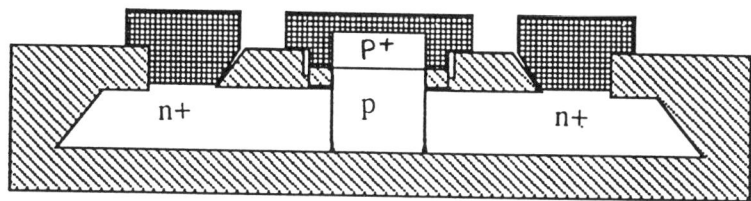

Figure 2. Cross-section of pedestal-like base SoI-LBT [4].

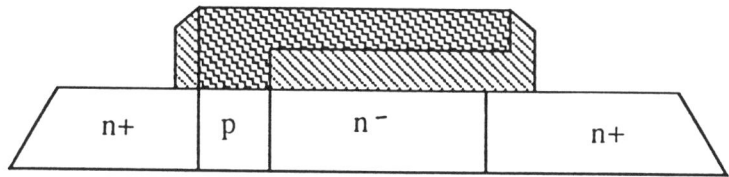

Figure 3. Cross-section of the double diffused collector SoI-LBT [5]

In recent years a new and promising deep-submicron base region, using a double diffused collector has been introduced to push the collector-emitter breakdown voltage, BV_{ceo} to higher values (Figure 3) using various self-aligned techniques on SoI [5-9] substrates. The N- drain extension permits the optimisation of maximum operation frequency, f_t versus BV_{ceo} for each SoI-BT.

With the advent of 'deep' sub-micrometre photolithography, however the proposed TESC approach overcomes inherent current crowding effects in the conventional SoI-LBTs by placing the base contact alongside the entire active region. This introduces a new concept for SoI-LBT design.

The TESC device structure with a highly efficient Top poly-silicon Emitter and a low resistance N+ Side-Collector is fabricated on 85nm thinned silicon overlayer of SIMOX material. The SoI layer received a boron implant to yield $1.5e17$ cm^{-3} active base doping. Figure 4., shows the layout of the TESC-BT with 1 x 16 cm^2 emitter opening through a gate oxide followed by polysilicon deposition, arsenic implantation and out-diffusion to form a shallow ($\approx$100Å) mono-crystalline silicon emitter.

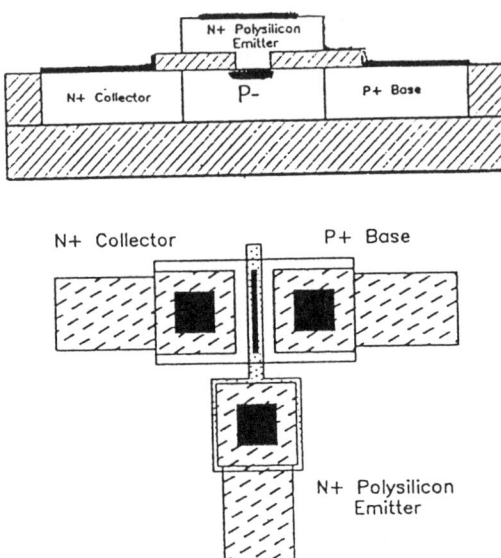

Figure 4a. Schematic cross-section of the TESC bipolar transistor.
Figure 4b. Top view of the TESC bipolar transistor layout.

In comparison with SoI-CMOS process the TESC-BT fabrication requires only one additional mask. An effective lateral base width of 0.8μm is obtained finally after the lateral diffusion of the collector dopants. However, due to the unique approach to the TESC structure, a vertical base width defined by the SoI thickness exist.

A good common-emitter characteristics of the npn TESC-BT is measured with grounded substrate, shown in Figure 5. The Early voltage, V_A, of -7V is obtained due to a low doped base and a high doped collector. A respectable collector breakdown voltage, BV_{ceo}, of 4.5V is obtained. The TESC structure offers the possibility of a further improvement in the BV_{ceo} with the use of well known LDD technique for MOSFETs. Figure. 6, shows a Gummel characteristics of the TESC device with grounded substrate and zero bias on the collector. The evaluated lateral pinched base resistance of 4kΩ for a 20μm long TESC device is the lowest value so far obtained for CMOS-BT fabricated on a 85nm thin film. Thus the collector current role-off occurs at very high current density (6000Acm^{-2}).

The base current ideality factor of 1.6 down to small emitter biases is indicative of a highly efficient poly-emitter contact for reducing the base injection currents [10]. In addition the TESC structure offer a considerable improvement in current gain with positive back-biasing as shown in Fig. 6. An increase in current gain, by six times (h_{fe}=300) is measured with 6V back-biasing. The induced current gain is obtained as a consequence of collector extension, reduction in the base Gummel number and due to the adding of the electric field,

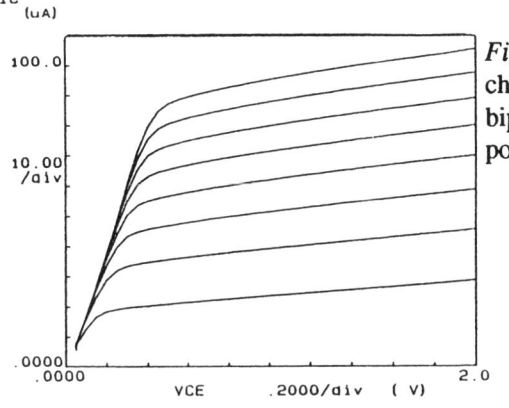

Figure 5. Common-emitter output characteristics of the TESC bipolar transistor with 16x1μm² poly-emitter area.

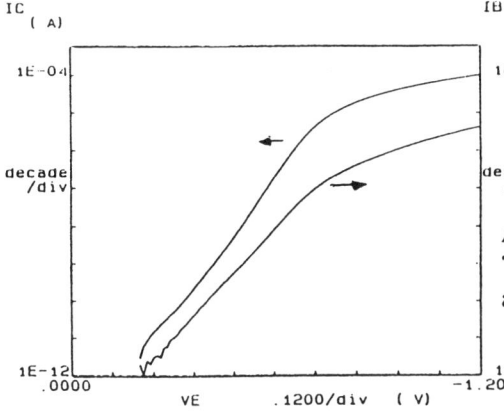

Figure 6. The Gummel plot of the TESC bipolar transistor ($V_{BG}=3V$, and $V_C=0V$).

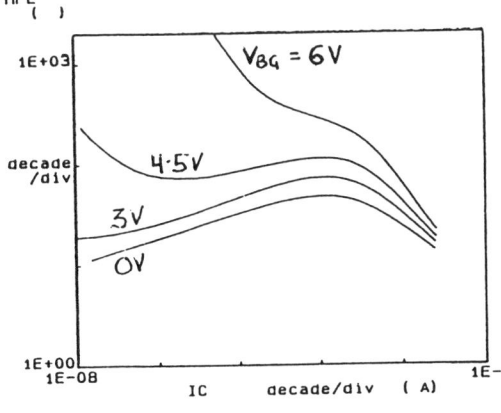

Figure 7. Measured current gain h_{fe}, cersus collector current of the TESC bipolar transistor with back-gate bias as a parameter ($V_{BG}=0, 3, 4.5$ and 6 volts).

providing drift current component as a result of the back-interface inversion layer. Thus improving the collecting efficiency and demonstrating the TESC device capability of sustaining a vertical bipolar transistor in a very thin SoI layers. However, the increase in the h_{fe} at low collector currents and high back-gate voltages, V_{bg}, is primarily due to the emitter-collector leakage at the base surface, under the pedestal poly-emitter. This leakage arises due to full depletion of the thin silicon ba.se layer at the side-collector, leading a shortening of the lateral base width. The predicted leakage has been verified by TITAN6.4, a 2D process and device simulation [11]. This artifact can be corrected by optimising base doping.

In summary, a novel TESC approach to a bipolar transistor in a thin silicon film on insulator has been demonstrated with a full functionality and a tremendous potential for both the future SoI-BiCMOS and SoI-bipolar circuit exist. The advantage of an N+ side-collector to give enhanced current gain and improved drive capability is verfied with TESC approach to thin-film bipolar transistor in SoI. The new mode of operation of the TESC device further enhances the advantages of reduced junction capacitances associated with the multi-layered SoI material for SoI-BiCMOS technology. The TESC bipolar transistor approach offers a route to a high performance vertical bipolar operation in a very thin silicon-on-insulator substrate.

References

[1] J-P Colinge, "Half-Micrometre-Base Lateral Bipolar Transistors made in Thin Silicon-on-Insulator", Electron. Lett., Vol. 22, pp. 886-887, 1986.

[2] N.D. Jankovic and C.J. Patel, "Bipolar Transistors in SoI Technologies", Proceedings of MIOPEL'93, pp. 151-157, (Nis, Serbia) 1993.

[3] T.H. Ning and D.D. Tang. "High Performance Lateral Bipolar Transistor on Insulating Substrate", IBM Tech. Disclos. Bull, Vol. 26, pp. 5858-5862, 1984.

[4] J. C. Strum et. al, "A lateral Silicon-On-Insulator Bipolar Transistor with a Self-Aligned Base Contact", IEEE Vol., EDL-8, pp. 104-107, 1987.

[5] U. Magnusson et. al, "A Lateral Bipolar transistor Concept on SoI using Self-Aligned Base defination technique", Microelectron. Eng., No.15, pp. 341-344, 1991.

[6] B. Edholm et. al, "A Self-Aligned Lateral Bipolar Transistor Realized on SIMOX-material", IEEE Trans. Electron. Dev., Vol. 40, No. 12, pp.2359-2360, 1993.

[7] G.G. Shahidi et. al, "A Novel High-Performance Lateral Bipolar on SoI", Proceedings IEDM-91, pp. 663-666, 1991.

[8] N. Higaki et. al, "A New SoI-Lateral Bipolar Transistor for High Speed Operation", Japanese J. Appl. Phys., Vol. 30, pp. L2080-L2082, 1991.

[9] S.A. Parke et. al, "A High-Performance Lateral Bipolar on SIMOX", IEEE Vol., EDL-14, pp. 33-35, 1993.

[10] S.A. Ajuria et. al, "Quantitative Correlations Between the Performance of Polysilicon Emitter Transistors and the Evolution of Polysilicon/Silicon Interfacial Oxides upon Annealing", IEEE Trans. Elec. Dev., Vol. 39, No. 6, pp. 1427, 1992.

[11] A. Poncet, "Software Tools for Silicon Device Optimization", Proceedings of ISSSE'92, pp.600-604, (Paris), 1992.

PROBLEMS OF RADIATION HARDNESS OF SOI STRUCTURES AND DEVICES

A.N.NAZAROV
Institute of Semiconductor Physics, National Academy of Sciences of Ukraine
252650, Kiev-28, Prospect Nauki, 45, Ukraine

This paper reviews the problems arising in SOI systems and SOI-based CMOS devices under the action of the total dose of ionizing irradiation. The effects of ionizing irradiation upon the gate, lateral, and buried insulating layers of SOI CMOS transistors are analyzed, and methods for improving the radiation hardness are reported. Particular attention is paid to structure, defect composition, and optimization methods improving the stability of buried insulating layers produced by various technologies. The most promising technologies of SOI fabrication are compared from the standpoint of stability and hardness against ionizing irradiation.

1. Introduction

Intensive development of SOI technology and SOI-based devices in the last 15 years is associated mainly with the unique properties which can be achieved for structures with full internal insulation. Their most essential advantages are:
1. Improved immunity of CMOS circuits under the conditions of pulsed radiation interference [1].
2. Extended temperature range of operation of SOI devices and circuits - up to 300°C [2].
3. Significantly increased operating frequency of SOI CMOS IC's, with simplified circuit design and higher packaging density of components on the chip [3].
4. Possibility of three-dimensional integration of microelectronic devices and circuits [4].

Most research and development efforts have been directed towards the creation of SOI devices and circuits designed to operate in a hostile environment. Nevertheless, the impact of full accumulated dose of ionizing radiation or various bias-temperature stress cycles can be significant.

This is mainly due to the fact that in SOI MOS transistors the active region of the device is surrounded from three sides by insulating layers (Fig.1a): (1) the gate insulator, (2) the protective insulating layer at the sidewall of the silicon island, and (3) the underlying buried insulating layer. Since insulating layers and insulator-semiconductor interfaces are most sensitive to various treatments, it is natural to expect that such devices should be less stable (in comparison with MOS elements based on the bulk silicon) to accumulated irradiation dose, bias-temperature stress, and injection treatments.

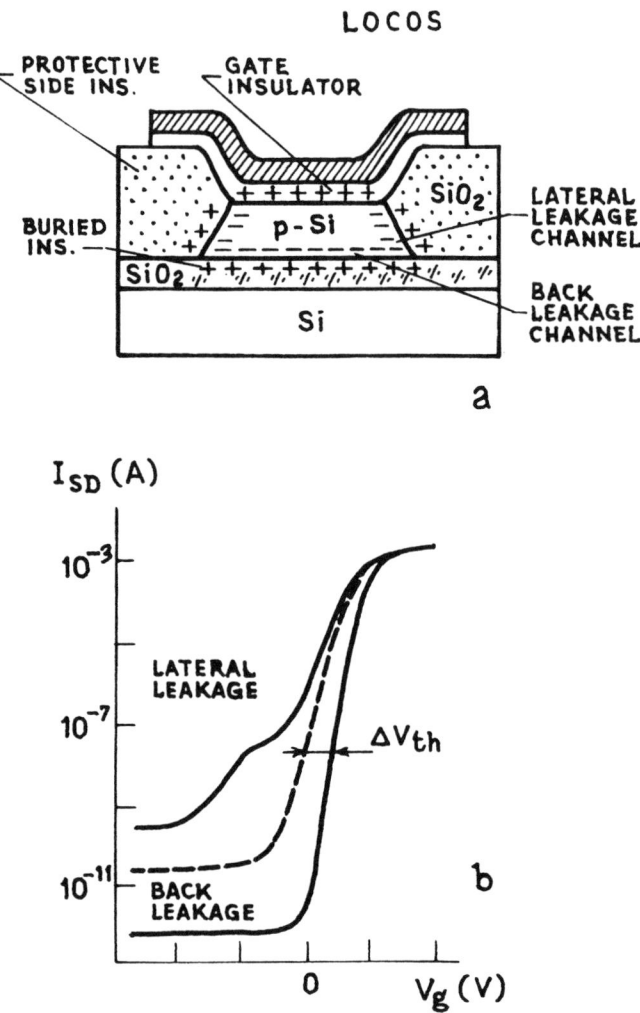

Figure 1. Formation of leakage channels in SOI transistor after effect of ionizing irradiation (a) and influence of charge accumulation in different parts of SOI structure on current-voltage characteristic of SOI MOS transistor (b).

2. Effect of the total irradiation dose on the operation of an SOI MOS transistors

Let us consider briefly the effect of ionizing irradiation on the electrical characteristics of an SOI MOS transistor (assuming the insulator to be silicon dioxide). At doses below 10^5 rad (Si), the exposure to ionizing radiation results into the introduction in SiO_2 of a positive charge localized predominantly (for the gate insulator) near the SiO_2-Si interface [5]. Accumulation of the positive charge in the gate insulator leads to a negative shift of the threshold voltage of the transistor (Fig.1b).

At the doses above 10^5 rad (Si), electronic interface states are being induced at the SiO_2-Si interfaces [6]. In the case of a p-channel transistor under normal operating conditions the interface states capture holes, so that the charges in the insulator and on the interface states (and the corresponding shifts of the threshold voltage) add up. In the case of an n-channel transistor, the interface states capture electrons, such that the shifts of the threshold voltage caused by the two charges are in opposite directions. So, p-channel MOS transistors should generally have higher threshold voltage shift induced by irradiation than n-channel ones. However, n-channel transistors operate under a positive gate bias, which enhances generation of SiO_2-Si interface states. Therefore, n-channel transistors can exhibit a noticeable degradation of the drain-gate transconductance under the action of irradiation (Fig. 1b).

Accumulation of positive charges in the side insulator of n-channel SOI MOS transistors induces a lateral leakage channel which shows up as a significant flattening of the drain-gate I-V curve in the sub-threshold region (Fig. 1b).

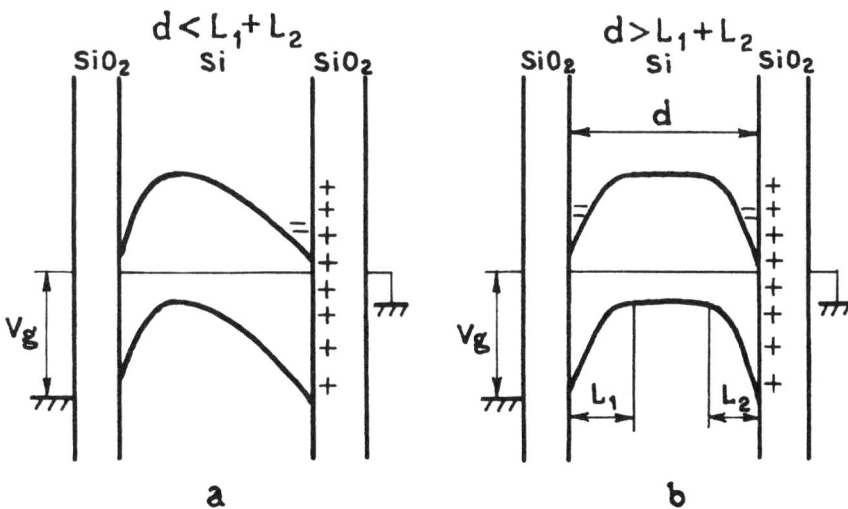

Figure 2. Influence of charge in the buried dielectric on the potential distribution through silicon for the cases of "thin" (a) and "thick" (b) films.

Accumulation of positive charges in the buried oxide layer can, depending on the thickness of the semiconductor film, result in the following effects: in case of an "electrically thin" film (such that its thickness (d) is smaller than the total thickness of the depletion layers of the upper (L_1) and lower (L_2) space charge regions (Fig.2a)), any charge variation in the buried insulator can significantly change the threshold voltage of the operating transistor through a charge coupling effect between the top and the bottom interface [7]. In case of a "thick" film (Fig.2b), for an n-channel transistor accumulation of a positive charge induces a leakage channel at the lower interface (Fig. 1b). In case of a p-channel SOI MOSFET, accumulation of the positive charge in the side and buried insulators does not usually result in leakage channels.

3. Improvement of radiation stability of the gate and side insulators

Let us consider several possible techniques to reduce the effect of the total dose of ionizing radiation on the electrical parameters of CMOS SOI transistors.

The gate insulator is located just above the operating region of the transistor, and thus it should satisfy the most strict requirements: (1) minimum charge accumulation; (2) minimum interface state generation; (3) minimum charge injection from the semiconductor or metal into the insulator; (4) possibility of operation under the highest electric field intensities.

Nearly three decades of studies in this field have demonstrated that the best insulator for radiation-hard technology is silicon dioxide [8]. To reduce charge accumulation in SiO_2 and to improve the stability against radiation and high temperatures, one needs an insulator with a low content of contaminants, an abrupt insulator-semiconductor interface, and the oxide must be as thin as possible. As a result ion implantation into or through the insulator should be avoided [8]. For SiO_2, the most favorable thickness for this purpose is 25 to 40 nm, which alows, on one hand, for the charge to move out of the insulator during the irradiation, while, on the other hand, it allows the film to withstand sufficiently high gate voltages. A thorough optimization of the technological process has shown that the best for these purposes is SiO_2 either thermally grown in dry oxygen at 1000°C [9] or produced by wet pyrogenic oxidation at T=850°C [10]. Usually, SOI CMOS technology uses one of these two insulators.

In recent years, much research has been done on the applications of thin (about 10 nm) nitrided and subsequently oxidized oxides (ONO) [11, 12] for submicron devices. These feature an improved mechanical strength and low charge trapping characteristics. However, this technology is far from being optimized.

To create radiation-hard sidewall insulation (field isolation), technological and device design methods are used. The main technological methods are as follows. In case of LOCOS technology, two-layer insulators are usually employed [13], such as SiO_2 thermally grown in wet oxygen and phosphosilicate glass, SiO_2 thermally grown in dry oxygen and Si_3N_4, or SiO_2 thermally grown in dry oxygen and CVD SiO_2. The latter combination of insulators exhibits the best stability with respect to total-dose irradiation [13, 14].

Improvement of the radiation stability was reported for thick insulators in case of additional implantation of B^+ [15], F^+ [16], or Cr^+ [17] ions into wet oxide, thereby creating extra electron traps in the insulator.

An effective design technique for suppressing the influence of charge accumulation in the side insulator onto the channel leakage is creation of the dividing layer of doped polysilicon, connected to ground [18] (Fig. 3a). Futhermore widely used techniques of leakage suppression in the side channel of CMOS SOI devices are creation of side p-n junctions [1] (Fig.3b) and fabrication of edgeless transistors [19] (Fig. 3c); however, the latter technique significantly reduces the packing density of elements on the chip.

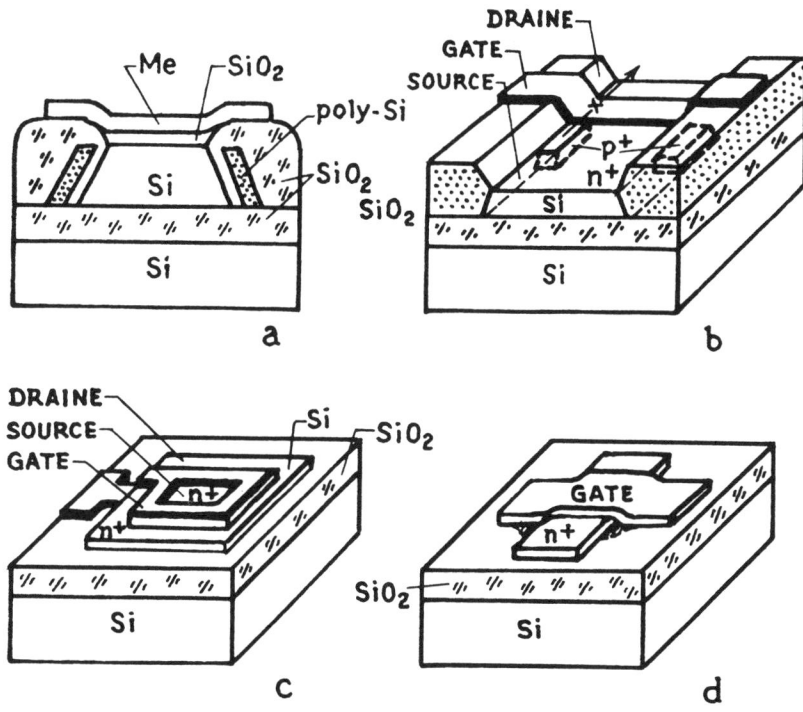

Figure 3. Different types of protection of SOI MOS transistors against the lateral leakage by: creation of polysilicon dividing layers (a), side p-n junctions (b), fabrication of ring-type transistor (c) and "gate-all-around" transistor (d).

A new design of edgeless SOI MOS transistor has been proposed in recent years (Fig.3d), which allows for higher packing density of elements in a circuit. This is a transistor with the gate completely surrounding the operating channel ("gate-all-around" (GAA) transistor [20]). It combines the advantages of the double-gated and edgeless SOI transistors, exhibiting improved electrical parameters, and decreased generation of carriers into the insulating layers [21].

4. Effect of ionizing radiation onto the buried insulating layer of SOI structures

4.1. COMPARATIVE ANALYSIS OF TECHNOLOGIES

The buried insulating layer of SOI devices is the most sensitive to various treatments, because it has been subjected to the largest number of high-temperature processing operations.

Now, the leading technologies for preparation of SOI structures are Separation by IMplantation of OXygen (SIMOX) and Bonded and Etchback Silicon-On-Insulator (BESOI) [22,23]. The former technology is most promising for creation of CMOS VLSI IC's with elements having deep-submicron channels, because it allows for the fabrication of high-quality silicon layers down to 80 nm in thickness, with excellent thickness uniformity across the wafer. The latter technology is most promising for bipolar and power devices due to the high quality of the material produced. However, the problem of fabricating thin silicon layers using the BESOI technique is not fully solved. The third technology, which has been actively developed in Japan, is Zone Melting Recrystallization (ZMR). Now this technology is aimed at the creation of 3-D IC's, but the quality of the films presently produced allows creation of inexpensive MSI CMOS IC's with improved general and radiation stability [24]. In this paper, we will focus on the properties of buried insulating and active silicon layers produced by these three technologies (Table I). The main defects in the silicon films of SOI structures fabricated by the ZMR technique are small-angle subgrain boundaries, i.e. dislocation networks [24]. The most common defects in the silicon film of SIMOX-produced structures are dislocations and stacking faults of various sizes [25], whereas in the BESOI structures the dominant defects are voids at the bonded surfaces [23].

To improve the quality of ZMR silicon layers, the substrate is usually heated to about 1300 C, and the recrystallization process takes several minutes [24]. In case of SIMOX technology, in order to reduce the density of dislocations and oxygen inclusions and to obtain an abrupt Si-SiO_2 interface, implantation is performed at a temperature of about 600°C, with a subsequent annealing for several hours at 1320°C. To decrease the size of the voids and to achieve high-quality wafer bonding, the BESOI process is performed at about 900°C, with a subsequent annealing (see Table 1).

Therefore, the strict temperature limitations, as well as the technique of SIMOX production of insulating layers by itself, testify that the properties of buried insulators produced by most SOI technologies can be substantially different from those established for gate or lareral isolation dielectrics, such that special studies of their general and radiation stability should be performed.

TABLE 1. Comparison of SOI technologies.

Parameters	SIMOX	BESOI	ZMR
Thickness of Si film	< 0.6 μm upto 0.08 μm	usially > 2 μm Plasma thinned film [25] > 0.37 μm	from 0.1 to 2 μm
Variation of Si film	0.01 μm	usially 0.2 μm Plasma thinned film 0.02 μm	0.02 μm
Thickness of buried oxide	< 0.8 μm	> 0.01 μm	from 0.3 to 3.0 μm
Temperature of production	~ 600 °C	~ 900 °C	from 800 to 1300 °C
Annealing temperature	1300 °C	~ 1000 °C	
Time of high temperature annealing	to 10 h	from 20 min to 2 h	10 min
Main defects in Si	Dislocations [25] $N \sim 1\text{-}5\times10^5$ cm^{-2} Stalking faults $N \sim 6\times10^4$ cm^{-2}	Dislocations [23] $N < 10^2$ cm^{-2}	Subgrain boundaries, dislocation $N < 5\times10^4$ cm^{-2}, twins [24]
Main defects in buried dielectric	Silicon agregates [25]	Voids on bonded interfaces [2], electron states on bonded interfaces	Defects of SiO$_2$ network
Impurities in Si	Cr, Mn, Fe, Co, Ni, Cu, Mo, W, Au, Re [27], O$_2$ [25]		O$_2$ (~ 10^{17} cm^{-3}) [2]

Charge accumulation and relaxation after irradiation in various buried insulating layers are compared in Table 2. It was found that buried oxides synthesized by single ion implantation have a very high density of positive charge traps uniformly distributed across the film [28,29]. Electron capture was also observed, leading to a significant compensation of the positive charge [28]. Such behavior is essentially different from that of the thermally grown oxide produced by radiation-stable technology, which mainly exhibits capture of holes near the Si-SiO$_2$ interfaces [29], with an insignificant compensation of electron traps in the adjacent region of SiO$_2$ [30].

The behavior of ZMR SOI systems with respect to accumulation and diffusion of the radiation-induced charge is similar to that of a soft thermal oxide [28], with the only difference, namely the existence of insignificant rapid re-capture of electrons into shallow traps, which has also been reported for SIMOX and BESOI systems [31].

TABLE 2. Radiation response characteristics of different SOI structures.

	SUPOX SIMOX [33]	SIMOX [28]	BESOI [32]	Thermal oxide [29]	ZMR [28]
Hole traps	Little trapping in bulk, variable trapping near Si/SiO$_2$ interface	Strong (100%) trapping in bulk	Little trapping in bulk, variable trapping near Si/SiO$_2$ interface	Variable (1 - 100%) trapping near Si/SiO$_2$ interface	Variable trapping near Si/SiO$_2$ interface
Electron traps	Not significant trapping in bulk and near Si/SiO$_2$ interface	Strong transient trapping / detrapping in bulk	Strong trapping on internal WB interface	Usually not significant near Si/SiO$_2$ interface [30]	Usually not significant
Time-dependent recovery	Hole transport and electron thermal detrapping	Fast electron detrapping. Slight thermal detrapping of holes	Slow hole transport. Thermal annealing of trapped holes. Not significant fast electron detrapping [30].	Fast hole transport. Tunnel and thermal anneal of trapped holes at interface polySi-SiO$_2$-Si (T_{an}=1320 °C) [31]. Not significant fast electron detrapping	Not significant fast electron detrapping [31].

Wafer-bonded systems reveal only a small radiation-induced capture of holes in the bulk, similarly to conventional thermally grown oxide. However, capture of electrons at the wafer-bonding interface has been observed [32].

Generation of high densities of positive charge traps and accumulation of a high concentration of positive charge in the ion-beam synthesized oxides after irradiation were demonsrated by different researchers [34-36]. These phenomena were ascribed to a substantial concentration of strained Si-O-Si bonds and to an oxygen deficiency in the SiO$_2$ layer [37], resulting in a high density of oxygen vacancies in SiO$_2$ (at E'-centers) detected by ESR after irradiation or additional hole injection [38-40]. It has been demonstrated [39,41] that generation of E'-centers in SIMOX buried oxides is tens of times more intensive than for the thermally-grown oxide. The presence of electron traps in SIMOX oxides, which compensate the positive charge in the dielectric [42], may be associated with a substantial concentration of Si-H and Si-OH bonds [41,43] which are able to capture electrons [44] and the amphoteric nature of E'-centers [45].

TABLE 3. Defect parameters obtained by ESR technique for different SOI structures.

Thermal oxide (polySi-SiO$_2$-Si)	SIMOX single	SIMOX multiple	BESOI	Behavior	Proposed model
g=2.0005 [39,46]	g=2.0005 [39,46] (VUV illumination) N~1.0x10^{18} cm^{-3}	g=2.0005 [39] N~0.35x10^{18} cm^{-3}	g=2.0005 [46] (VUV illumination) N~4x10^{12} cm^{-2}	Hole trap	E'$_\gamma$ - center [49] $O_3 \equiv Si^\bullet \ ^+Si \equiv O_3$
	74G doublet (after treatment in forming gas 10% H$_2$, 90% N$_2$) [43]				$O_2 = Si$ [43] $\mid$ H
	10G doublet [43] (after treatment in forming gas)				$O_2 = Si$ [43] $\mid$ OH
			g=2.0025 [46] 23.1G doublet N~6x10^{11} cm^{-2}	Hole trap	EH-center [46]
g=1.99994 [46] (after annealing T=1320 °C t=6h)	g$_{[001]}$=1.99971 [38] γ-irradiation 10 Mrad N=1.1x10^{12} cm^{-2}	g=1.99971 [38] γ-irradiation 10Mrad N=2x10^{12} cm^{-2}	g$_\parallel$=1.99994, g$_\perp$=1.99958 [46]		UL1 [38,48] O-related donor in Si
g=2.0021 [46]	g=2.0021 [40,46]		g=2.0021 [47] (Ar glow discharge at cathodic potential)	Hole trap	E'$_\delta$-center [40] Si $O_3 \equiv Si \ Si^+ Si \equiv O_3$ Si
	g=2.0058 [38] g=2.0061 [46] N~7x10^{13} cm^{-2}	N=0 [38]		Hole trap	D$_0$ -center [38,46] Si-Si$^\bullet$ $^+$Si$\equiv$O$_3$

Parameters of ESR-detected centers and their identification for various buried oxides are presented in Table 3. It can be seen that most defects in SIMOX systems are associated with the excess of silicon. In BESOI systems, in addition to the inclusions of amorphous silicon (D_0-centers), a research group of Sandia National Laboratory has detected a hydrogen-bonded paramagnetic EH-center localized at the wafer-bonding interface [46].

In polySi-SiO_2-Si structures subjected to an additional six-hour high-temperature annealing in Ar ambient at 1320°C, paramagnetic E'_δ-centers were also detected, distributed over the entire insulator [46] similar to the distribution of the centers in SIMOX buried oxides [40]. It has been noted in many papers [25,31,46] that such high-temperature annealing, in the case where the oxide layer is not in contact with the external oxygen ambient, leads to the outdiffusion of oxygen from the silicon dioxide, oxygen deficiency in the SiO_2 layer (formation of E'_δ-centers [46,47] or shallow electron traps [31]) and saturation of the silicon film with oxygen (formation of oxygen donors [38,48]). This has been observed not only in SIMOX systems, but also in BESOI, ZMR, and in thermally grown oxides covered with polysilicon. A comparison of thermal annealing of shallow electron traps [31] and E'_δ-centers [46] has shown that E'_δ-centers are more stable and, most probably, are not directly associated with shallow electron traps.

TABLE 4. SOI trap parameters obtained by thermally activated methods after irradiation.

	Thermal oxide	SIMOX single impl.	SIMOX double impl.	ZMR	BESOI	Nature
Ea_{max1} (eV)		0.57 [53]				Deep hole trap
Ea_{max2} (eV)	1.00 [52] (without field correction)	1.15 [53] 1.30 [50]	1.10 [53]	1.05 [51]	?	Deep hole trap
Ea_{max3} (eV)	1.40 [52]	1.35 [50]	?	1.45 [51]	?	Deep hole trap
Ea_{max4} (eV)	Can appear after thermal annealing of polySi-SiO_2-Si structure at 1320 °C	0.41 [31] $N=1.5 \times 10^{11}$ cm^{-2}	?	$N=4.2 \times 10^{10}$ cm^{-2} [31]	0.42 $N=2 \times 10^{11}$ cm^{-2} [31]	Shallow electron trap

Table 4 shows the parameters of traps in SiO_2 determined by thermally activated techniques for SOI systems produced by various technologies. It can be seen that deep hole traps of the same type with an energy maximum at 1.05 and 1.45 eV were detected in SIMOX [50] and ZMR [51] systems and in the conventional thermally grown oxide [52]. Moreover, shallow electron traps with an activation energy of 0.42 eV were detected in standard SIMOX systems, BESOI, ZMR, and polySi-SiO_2-Si structures subjected to high-temperature annealing. The shallow traps concentration in ZMR systems is much lower than in SIMOX and BESOI systems [31].

Deep hole traps observed after the irradiation (E_{a1}=1.0 eV, E_{a2}=1.45 eV) are probably associated with E'_γ-centers (oxygen vacancy in the SiO_2 network, $O_3 \equiv Si'$ $+Si \equiv O_3$) detected by ESR and efficiently capturing holes in case of optical injection ($\sigma \sim 10^{-12}$-$10^{-14} cm^2$) [37].

4.2. OPTIMIZATION OF TECHNOLOGIES

As seen from the above consideration, the most promising SOI material for CMOS IC applications is SIMOX, but the quality of the buried insulator in such structures is far from being perfect. So, in the last six years it was attempted to optimize or modify this technology. It has been shown [54,55] that reduction of the oxygen implantation dose

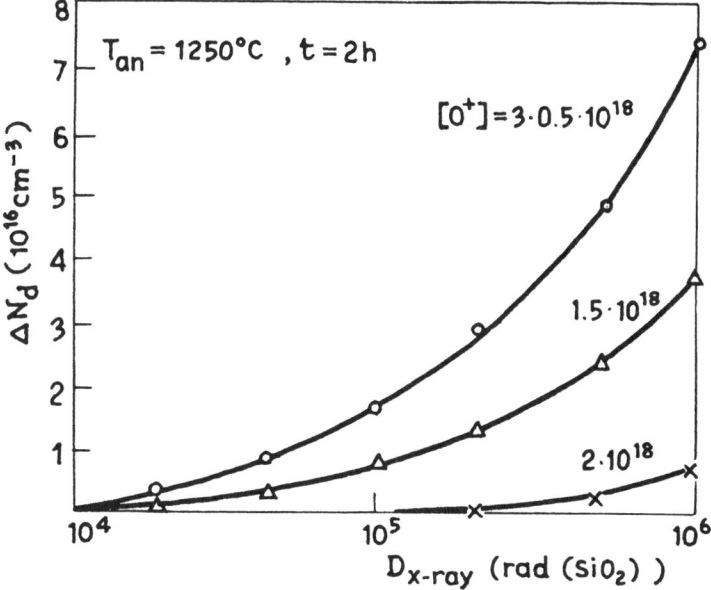

Figure 4. Density of donor levels obtaining after x-ray irradiation of SIMOX SOI structure in silicon layer as a function of x-ray dose [57].

and energy allows for a significant improvement of the quality of the silicon film and fabrication of uniform thin SiO_2 layers with a low content of silicon inclusions. The use of multiple oxygen implantation with a reduced dose decreases the density of silicon dangling bonds at the amorphous inclusions in SiO_2 [38] and the current flowing through the insulator [56]. However, for SOI systems produced by this technique and irradiated with increased doses, formation of donor centers in the silicon film has been reported (Fig. 4) [57], which may result in degradation of charge carrier mobility in the film and irradiation-induced shift of the transistor threshold voltage.

It was shown [31,58] that, in order to decrease the content of silicon inclusions in the buried oxide, an additional annealing in oxygen ambient is needed. However, such a treatment is effective only in case of the absence of the top silicon layer. The most promising technique for improving the quality of the buried oxide turned out to be additional small-dose ($\sim 1 \times 10^{17} cm^{-2}$) implantation of oxygen ions into the insulator with a subsequent annealing at 1000°C [33,59]. Such an implantation reduces non-stoichiometry of the oxide due to additional oxidation of silicon islands in the insulator. In this case, rapid recapture of electrons is significantly suppressed [33], shallow electron traps are annealed, and concentration of deep hole traps is also substantially reduced [59] (Table 2).

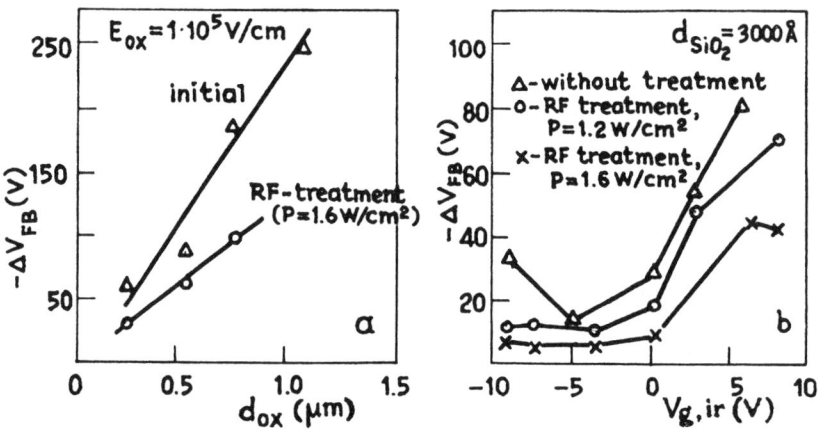

Figure 5. Shift of flat-band voltage of polySi-SiO_2-Si structures after γ-irradiation gate (b) before (Δ) and after (o, x) RF treatment.

A technique capable of decreasing the density of strained silicon bonds in silicon dioxide is RF plasma annealing [60]. It can, at the final stage of device fabrication, improve the quality of the SiO_2-Si interface [61] and eliminate radiation-induced defects in thin layers of Si and SiO_2 [62,63]. It has been demonstrated [64] that such treatment of polySi-SiO_2-Si systems essentially improves radiation stability of thick insulating layers. Fig. 5a,b shows the results of gamma-irradiation of structures with different thickness of the buried insulator before and after the plasma treatment and dependence of charge accumulation in such structures on the gate voltage during the irradiation. It is seen that the plasma treatment substantially reduces accumulation of the positive charge.

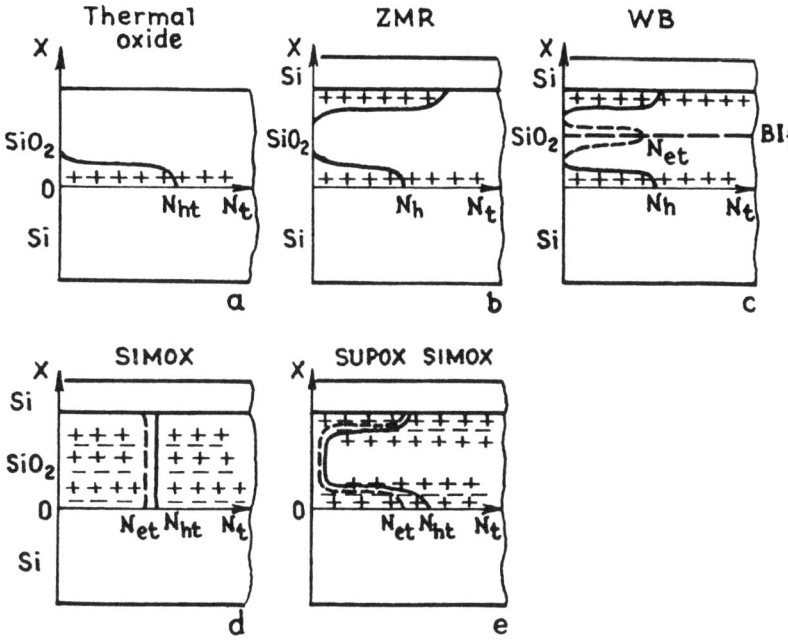

Figure 6. Illustration of charge trap distribution in the silicon oxides obtained by different SOI techniques.

Fig. 6 illustrates the main features of distribution of the captured charge in the insulators of various SOI systems. It shows that the from the viewpoint of charge capture parameters the most similar to the thermally grown oxide (Fig.6a) are oxides of ZMR (Fig.6b) and BESOI systems (Fig.6c). In the barrier oxides of ZMR structures, the positive charge is captured near the upper interface more efficiently than at the lower one, and charge redistribution is easily controlled by the external electric field [64], like in the conventional thermally grown oxide. Therefore, the most thorough optimization of the SOI SIMOX technology (Fig.6e) can produce buried insulators with the parameters close to those of radiation soft thermally grown oxide.

4.3. METHODS OF IMPROVING RADIATION STABILITY AND SUPPRESSING LEAKAGE IN THE LOWER CHANNEL OF SOI MOS TRANSISTORS

The simplest technique for reducing charge accumulation in the buried insulator is decreasing its thickness. Since a reduction of the implantation dose and energy results in a higher quality of the SIMOX silicon and insulating films, this method can be regarded as justified. However, this also leads to some decrease of the device operation speed and maximum breakdown voltage. Due to the latter effect, the circuits should be designed for lower voltage operation, exclusively.

An alternative technique for improving the radiation stability of insulating films is fabrication of two-layer insulating films like SiO_2-Si_3N_4 [66] or oxinitride films [67]. Nitride or oxinitride layers have close concentrations of electron and hole traps, which can significantly reduce the radiation-induced charge in the insulator.

To improve radiation stability of the buried oxide of SIMOX systems, an additional small-dose nitrogen implantation can be carried out during the oxygen implantation [68]. This significantly reduces the radiation-induced shift of the threshold voltage for the lower channel of the transistor (Table 5). However, the introduction of nitrogen resultes into a decrease of the threshold voltage of the n-channel transistors and increase of the threshold voltage for the p-channel transistors. Research work aiming at the creation of buried nitride or oxinitride layers by ion beam synthesis are still performed by a number of groups [69-71]. However, no significant success has been achieved so far in fabrication of high-quality silicon films by this technique. Ion-beam synthesis of multilayered insulators is an extremely difficult problem [72].

TABLE 5. Influence of additional nitrogen implantation during the ion-synthesis of buried dielectric on radiation shift of threshold voltage (ΔV_{th}) of bottom n-channel transistors (γ, Co^{60}, $D=10^6$ rad (Si)) [68].

Implanted dose (cm^{-3})	ΔV_{th} (V)	
	$V_{sub}=0$ V	$V_{sub}=-3$V
$[O^+]=2.25 \times 10^{18}$	-25.0	-8.5
$[O^+]=2.0 \times 10^{18}$ $[N^+]=5 \times 10^{16}$	-14.9	-3.7
$[O^+]=1.8 \times 10^{18}$	-17.2	
$[O^+]=1.7 \times 10^{18}$ $[N^+]=1.0 \times 10^{17}$	-13.1	-5.1

Similarly, it was proposed to fabricate nitride- and oxinitride-based multilayered buried insulators in ZMR SOI structures [73,74] and BESOI structures [75]. Fig. 7 shows the bar chart of the flatband voltage shift for the top insulator- semiconductor interface after gamma-irradiation of a capacitor n-Si(film)-insulator-n-Si(substrate) with a negative bias applied to the top electrode, providing conditions for maximizing effect of the induced charge on the potential of the silicon film-insulator interface. It is seen that an additional oxinitride layer reduces the flatband voltage shift twofold, and a nitride layer - by 3.6 times in comparison with SiO_2.

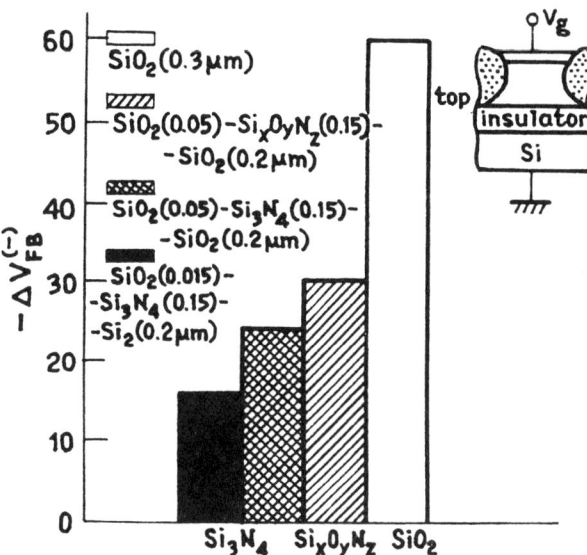

Figure 7. Diagram of flat-band shift for top insulator-semiconductor interface after γ-irradiation of ZMR SOI structures (n-Si(film) - insulator - n-Si(substrate)) with different types of buried dielectrics [74].

Table 6 shows similar results for BESOI structures with various types of buried insulators [75]. Irradiation was carried out with an X-ray source (10 keV) up to the dose of 5×10^3 rad (SiO_2) under the external electric field of $\pm 1 \times 10^6$ V/cm. In this case, the structures with the nitride buried layer also display the smallest shift of potential at the top insulator-semiconductor interface. For comparison, the table shows also the results for the lower insulator-semiconductor interface of a SIMOX structure, illustrating its lower radiation hardness even with respect to the reference BESOI system (with a pure thermally grown oxide). Therefore, the proposed techniques can produce sufficiently thick radiation-stable buried insulating layers.

To suppress the effect of the positive charge accumulation in the buried insulator onto the threshold voltage of the operating transistor and to reduce the leakage currents, additional measures can be taken. The devices can be operated with a negative voltage applied to the silicon substrate [76], or the threshold voltage of the back-channel transistor can be increased using ion implantation. However the first variant can bring about some detrimental effects to p-channel transistors, such as threshold voltage shifts or, even the formation of a back channel. Thus an optimization of the applied voltage to the substrate is required for each specific design of devices. The latter problem issolved by several techniques. The first and the most widely employed technique involves additional doping of the bottom SiO_2-Si interface [19]. The second technique, used in SIMOX technology, involves creation of a narrow defect-rich layer at the bottom SiO_2-Si interface, providing efficient screening of charge variation in the buried insulator [77]. However, this technique is very difficult to implement in industry and does not allow fabrication of such promising components as double-gate transistors.

TABLE 6. Measured voltage shifts of X-ray irradiated samples ($5 \cdot 10^3$ rad(SiO_2)) [75]

№	Sample	ΔV^+_{mgtop}(V)	ΔV^-_{mgtop}(V)
1.	BESOI (0,05μm SiO_2+0,36μm SiO_2)	+1,25	-10,0
2.	BESOI (0,05μm nitrided SiO_2+0,36μm SiO_2)	-5,5	-15,7
3.	BESOI (0,05μm nitrided SiO_2+0,36μm nitrided SiO_2)	-4,5	-10,0
4.	BESOI (0,05μm SiO_2-0,16μm Si_3N_4 +0,299μm SiO_2)	+1,5	-1,0
5.	SIMOX (Ibis)	+2,5 *)	-12,75 *)

*) Data for bottom interface

5. Comparison of parameters of radiation-stable CMOS SOI transistors

By now, a number of radiation-stable SOI components and medium-scale integration circuits have been produced using both ZMR and SIMOX technologies. However, comparison of parameters of components produced by various technologies before and after irradiation yields rather controversial results. It was reported [78] that after irradiation the performance of ZMR SOI MOS transistors is better than that of their SIMOX counterparts. Figure.8 (a and b) shows the irradiation dose dependence of the threshold voltage shift in ZMR and SIMOX n- and p-channel SOI MOS transistors. It is seen that ZMR SOI transistors have a lower threshold voltage shift both for both n-channels and p-channels. The transistors shown in Fig. 8 were fabricated using an additional doping of the lower Si-SiO_2 interface (n-channel devices). In this case, no effect of the buried layer charge on the threshold voltage shift is observed.

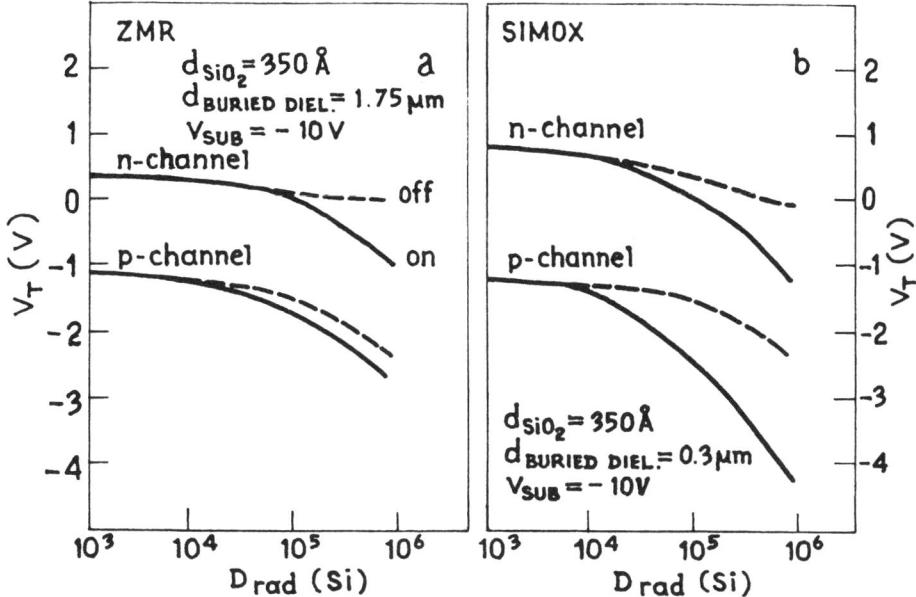

Figure 8. Threshold voltage as a function of γ-irradiation dose for SOI MOS transistors obtained by ZMR (a) and SIMOX (b) techniques [78].

Fig. 9 shows the results of studies of lower-channel leakage currents in edgeless transistors with an additional doping of the inner Si-SiO$_2$ interface, fabricated on ZMR and SIMOX SOI substrates. It is seen that leakage currents of these structures after gamma-irradiation with a dose of 2×10^5 rad (Si) are essentially identical [19].

The effect of gamma-irradiation on the main parameters of ZMR SOI MOS transistors is also illustrated by Table 7. Small shifts of the threshold voltage and small changes of the transconductance of n- and p-channel are associated with the use of a thin (~10 nm) radiation-stable gate insulator, whereas the small leakage currents can be assiociated with the additional doping of the lower Si-SiO$_2$ interface [78]. Therefore, a possibility was demonstrated of creation of ZMR SOI MOS transistors stable against ionizing irradiation with the total dose up to 10^7 rad (Si) [78].

The use of a radiation protection of the edge channels of a transistor using doped polysilicon layers and the suppression of the lower Si-SiO$_2$ interface leakage by using of a defect-rich layer has allowed fabrication of SIMOX CMOS SOI transistors stable against the total dose of ionizing irradiation up to 2×10^6 rad (Si) [18] (see Table 7).

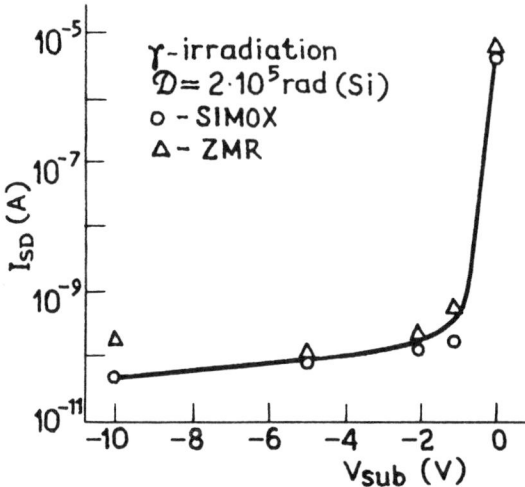

Figure 9. Back leakage channel current of SOI MOS transistors (n-channel) as a function of irradiated voltage on the silicon substrate [19].

TABLE 7. Effect of γ-irradiation on the properties of the ZMR and SIMOX SOI MOSFET's.

	ZMR		SIMOX	
References	[78]		[18]	[79]
$d_{gate\ SiO_2}$ (Å)	100		170	230
$d_{buried\ SiO_2}$ (Å)	5000		1600	3800
W/L (μm)	46/3.5		100.8/5	10/1.2
D_γ, rad (Si)	10^6	10^7	2×10^6	10^7
ΔV_{th} n - channel (V) p - channel	-0.04 -0.11	-0.18 -0.46	-0.12 -0.20	+0.5 -0.8
$\Delta g_m/g_m$ n - channel p - channel		5% 5%	14% 3%	
$I_{leak.}$ (A/μm)		$>2 \times 10^{-13}$	$\sim 1 \times 10^{-11}$	$\sim 5 \times 10^{-10}$

Recent results of development of a radiation-stable technology of SIMOX-based CMOS SOI transistors have shown that such elements can withstand ionizing irradiation doses of the order of 10^7 rad (Si) [79]. A microprocessor containing 11,000 transistors produced by such technology had no failures under pulse irradiation with a flow up to 7×10^{10} rad/s and was still operating after exposure to a total dose of 10^8 rad has been demonstrated.

6. Conclusion

The analysis of structural and electrical properties of SOI systems and SOI-based IC components shows that it is possible to produce IC components and MSI circuits resistant against the total dose up to 10^7 rad (Si) based on either SIMOX or ZMR structures. However, properties of the devices are strongly dependent on processing parameters and the equipment used for their fabrication.

The stability and hardness of the buried insulating layer is the crucial problem for creation of radiation-stable IC components.

The most promising trend for development of stable and radiation-resistant SOI CMOS IC's is fabrication of circuits with gate-all-around transistors and radiation-resistant gate and buried insulators. For this purpose, specially optimized multilayered insulators may be suitable. From this point of view, ZMR technology shows certain promises, because it allows fabrication of inexpensive and stable MSI circuits.

7. References

1. Auberton-Herve, A.J. (1990) SIMOX-SOI technologies for high-speed and radiation-hard technologies: status and trends in VLSI and ULSI application, in Silicon-on-Insulator Technology and Devices, Montreal, pp.437-454.
2. Colinge, J.-P. (1991) Silicon-on-Insulator Technology: Materials to VLSI, Kluwer Academic Publishers, Dordrecht.
3. Stanley, T.D. (1988) The state-of-art in SOI technology, IEEE Trans. Nucl. Sci. 35, 1346-1349.
4. Kawamura, S., Sasaki, N., and Iwai, T. (1983) Three-dimensional CMOS IC's fabricated by using beam recrystallization, IEEE Electr. Dev. Lett. 4, 366-368.
5. Nichols, D.K. (1980) A review of dose rate dependent effects of total ionizing dose irradiations, IEEE Trans. Nucl. Sci. 27, 1016-1024.
6. Winokur, P.S., and Boesch, H.E. (1980) Interface-state generation in radiation-hard oxides, IEEE Trans. Nucl. Sci. 27, 1647-1650.
7. Lim, H.-K., and Fossum, J.G. (1983) Threshold voltage of thin-film silicon-on-insulator (SOI) MOSFET's, IEEE Trans. Electr. Dev. 30, 1244-1251.
8. Borkan, H. (1977) Radiation hardening of CMOS technologies - an overview, IEEE Trans. Nucl. Sci. 24, 2043-2046.
9. Aubuchon, K.G. (1971) Radiation hardening of p-MOS devices by optimization of the thermal SiO_2 gate insulator, IEEE Trans.Nucl.Sci. 18, 117-125.
10. Naruka, K., Yoshida, M., Maeguchi, K., and Tango, H. (1983) Radiation-induced interface states of poly-Si gate MOS capacitors using low-temperature gate oxidation, IEEE Trans. Nucl. Sci. 30, 4054- 4058.
11. Dunn, G.J., and Wyatt, P.W. (1989) Reoxidized nitrided oxide for radiation-hardened MOS devices, IEEE Trans. Nucl. Sci. 36, 2161-2168.
12. Yang, W., Jayaraman, R., and Sodini, C.G. (1988) Optimization of low-pressure nitridation/ reoxidation of SiO_2 for scaled MOS devices, IEEE Trans. Electr. Dev. 5, 935-944.
13. Watanabe, K., Kato, M., Okabe, T., and Nagata, N. (1986) Radiation effects of double layer dielectric films, IEEE Trans. Nucl. Sci. 33, 1216-1222.
14. Watanabe, K., Kato, M., Okabe, T., Nagata, N., (1985) Radiation hardened silicon devices using a novel thick oxide, IEEE Trans. Nucl. Sci. 32, 3971-3974.

15. Yoshii, I., Hama, K., and Maeguchi, K. (1989) Radiation effects on p+-poly gate MOS structures with thin oxides, IEEE Trans. Nucl. Sci. 36, 2124-2130.
16. Kato, M., Watanabe, K., and Okabe, T. (1989) Radiation effects on ion implanted silicon dioxide films, IEEE Trans. Nucl. Sci. 36, 2199-2204.
17. Wang, S.T., Royce, B.S.H., and Pussel, T.J. (1975) The effect of ion implantation on oxide charge storage in MOS devices, IEEE Trans. Nucl. Sci. 22, 2168-2173.
18. Ohno, T., Izumi, K., Shimaga, M., and Shiono, N. (1987) Total-dose effects of gamma-ray irradiation on CMOS/SIMOX devices, IEEE Circ. and Dev. Mag. 3, 21-26.
19. Tsao, S.S., Fleetwood, D.H., Weaver, H.T., Pfeiffer, L., and Celler, G.H. (1987) Radiation-tolerant, side wall - hardened SOI/ MOS transistor, IEEE Trans. Nucl. Sci. 34, 1686-1691.
20. Colinge, J.-P., and Terao, A. (1993) Effects of total-dose irradiation on gate-all-around (GAA) devices, IEEE Trans. Electr. Dev. 40, 78- 82.
21. Flandre, D., Francis, P., Colinge, J.-P., and Cristoloveanu, S. (1993) Comparison of hot-carrier effects in thin-film SOI gate-all-around accumulation-mode p-MOSFET's, Proceedings of IEEE Int. SOI Conference, pp.160-161.
22. Colinge, J.-P. (1993) SOI technology for deep-submicron CMOS applications, in G.K.Celler, E.Middlesworth and K.Hoh (eds.), ULSI Science and Technology, 93-13, Electrochem. Soc. Inc., NJ., pp.39-55.
23. Hosack, H.H. (1992) Silicon-on-insulator materials, in W.E.Bailey (ed.), Silicon-on-Insulator Technology and Devices 92-13, Electrochem. Soc. Inc., NJ., pp.5-18.
24. Limanov, A.B., and Givargizov, E.I., (1991) Laser zone recrystallization of thin silicon films: method, structure, mechanisms of crystallization, Microelectronica. 20, 3-11 (in Russian).
25. Stoemenos, J., Aspar, B., and Margail, J. (1994) Mechanisms of SIMOX synthesis and related microstructural properties, in S.Cristoloveanu (ed.), Silicon-on-Insulator Technology and Devices. 94-11, Electrochem. Soc. Inc., NJ., pp.16-27.
26. Clapis, P.J., Ledger, A.M., and Daniell, K.E. (1993) Analysis of color variation in bonded SOI wafers, Proceedings of IEEE Int.SOI Conference, pp.66-67.
27. Sundaresan, R., Mao, B.-Y., Matloubian, M., Chen, C.-E. and Pollack, G.P. (1989) Characterization of leakage current in buried-oxide SOI transistors, IEEE Trans. Electron. Dev. 36, 1740-1745.
28. Boesch, H.E.Jr., Taylor, T.L., Hite, L.R., and Beiley, W.E. (1990) Time-dependence hole and electron trapping effects in SIMOX buried oxide, IEEE Trans. Nucl. Sci. 37, 1982-1989.
29. Pennise, C.A., and Boesch, H.E.Jr. (1990) Determination of charge trapping characteristics of buried oxides using a 10-keV X-ray source, IEEE Trans. Nucl. Sci. 37, 1990-1994.
30. Fleetwood, D.M., Reber, R.A., and Winikur, P.S. (1991) Effect of bias on thermally stimulated current (TSC) in irradiated MOS devices, IEEE Trans. Nucl. Sci. 38, 1066-1077.
31. Stahlebash, R.E., Campisi, G.J., McKitterick, J.B., Marzara, W.R., Roitman, P., and Brown, G.A. (1992) Electron and hole trapping irradiated SIMOX, ZMR and BESOI buried oxides, IEEE Trans. Nucl. Sci. 39, 2086- 2097.
32. Boesch, H.E., and Taylor, T.L. (1992) Time-dependent radiation-induced effects in wafer-bonded SOI buried oxides, IEEE Trans. Nucl. Sci. 39, 2103-2113.
33. Boesch, H.E., Taylor, T.L., and Krull, W.A. (1993) Charge trapping and transport properties of SIMOX buried oxide with a supplemental oxygen implant, IEEE Trans. Nucl. Sci. 40, 1748-1754.
34. Brady, F.T., Krull, W.A., and Li, S.S. (1989) Total dose radiation effects for implanted buried oxides, IEEE Trans. Nucl. Sci. 36, 2187-2191.
35. Lawrence, R.K., Hughes, H.L., and Stahlbash, R.E. (1990) Radiation sensitivity of buried oxides, J. Electron. Materials. 19, 665-670.
36. Boesh, H.E., Taylor, T.L., and Brown, G.A. (1991) Charge buildup at high dose and low fields in SIMOX buried oxides, IEEE Trans. Nucl. Sci. 38, 1234-1239.

37. Revesz, A.G., Brown, G.A., and Hughes, H.L. (1993) Properties of buried SiO_2 films in SIMOX structures, in Mat. Research Soc. Symp. Proceedings. 264, pp.255-265.
38. Stesman, A., Derine, R., Revesz, A.G., and Hughes, H. (1990) Irradiation- induced ESR active defects in SIMOX structures, IEEE Trans. Nucl. Sci. 37, 2008-2012.
39. Conley, J.F., Lenahan, P.M., and Roitman, P. (1991) Electron spin resonance study of E' trapping centers in SIMOX buried oxides, IEEE Trans. Nucl. Sci. 38, 1247-1252.
40. Vanheusden, K., and Stesmans, A. (1993) Characterization and depth profiling of E' defects in buried SiO_2, J. Appl. Phys. 74, 275-283.
41. Zvanut, M.E., Stanbush, R.E., Carlos, W.E., Hughes, H.L., Lawrence, R.K., and Brown, G.A. (1991) SIMOX with epitaxial silicon: point defects and positive charge, IEEE Trans. Nucl. Sci. 38, 1253-1258.
42. Conley, J.F., Lenahan, P.M., and Roitman, P. (1992) Evidence for deep electron trap and charge compensation in separation by implanted oxygen oxides, IEEE Trans. Nucl. Sci 39, 2114-2120.
43. Conley, J.F. and Lenahan, P.M. (1992) Room temperature reactions involving silicon dangling bond centers and molecular hydrogen in amorphous SiO_2 thin films on silicon, IEEE Trans. Nucl. Sci. 39, 2186-2191.
44. Sugano, T. (1989) Carrier trapping in silicon MOS devices, Acta Politechnica Scandinavica Electrical Engineer. Ser. N64, 220-241.
45. O'Reilly, E.P. and Robertson, J. (1983) Theory of defects in vitreous silicon dioxide, Phys. Rev. B 27, 3780-3786.
46. Warren, W.L., Shanneyfeld, M.R., Sohwank, J.R., Fleetwood, D.M., Winokur, P.S., Devine, R.A.B., Maszara, W.P., and McKitterick, J.B. (1993) Paramagnetic defect centers in BESOI and SIMOX buried oxides, IEEE Trans. Nucl. Sci. 40, 1755-1764.
47. Vanheusden, K., and Stesman, A. (1994) Similarities between separation by implanted oxygen and bonded and etchback silicon-on-insulator materials as revealed by electron spin resonance, in S.Cristoloveanu (ed.), Silicon-on- Insulator Technology and Devices 94-11, Electrochemical Society Publishers, NJ, pp.197-202.
48. Vauheusden, K., and Stesma, A. (1993) Electron spin resonance characterization and localization of a thermally generated donor inherent to the separation by implantation of oxygen process, J. Appl. Phys. 73, 876-889.
49. Feigl, F.J., Fowler, W.B., and Yip, K.L. (1974) Oxygen vacancy model for E' center in SiO_2, Solid State Commun. 14, 225-229.
50. Flament, O., Herve, D., Massean, O., Bonnel, Ph., Raffaelli, M., leray, J.L., Margail, J., Giffard, B., and Auberton-Herve, A.J. (1993) Field dependent charge trapping effects in SIMOX buried oxides at very high dose, IEEE Trans. Nucl. Sci. 40, 1765-1771.
51. Nazarov, A.N., Lysenko, V.S., Gusev, V.A., and Kilchitskaya, V.I. (1994) C-V and thermally activated investigation of ZMR SOI meza structures, in S.Cristoloveanu (ed.), Silicon-on-Insulator Technology and Devices. 94-11, Electrochemical Society Publishers, NJ, pp.245-252.
52. Fleetwood, D.H., Miller, S.L., Reber, R.A., McWhorter, P.J., Winokur, P.S., Shaneyfelt, M.R., and Schwank, J.R. (1992) New insights into radiation-induced oxide-trap charge through thermally-stimulated-current measurement and analysis, IEEE Trans. Nucl. Sci. 39, 2192-2203.
53. Pennise, Ch.A., and Boesch, H.E. (1991) Thermal annealing of trapped holes in SIMOX buried oxides, IEEE Trans. Nucl. Sci. 38, 1240-1246.
54. Mao, B.Y., Sundaresan, R., Chen, C.-E.D., Matloubian, M., and Pollack, G.P. (1988) The characterisrics of CMOS devices in oxygen implanted silicon-on-insulator structures, IEEE Trans. Electron. Dev. 35, 629-633.
55. Aspar, B., Pudda, C., Papon, A.M., Auberton-Herve, A.J., and Lamure, J.M. (1994) Ultra thin buried oxide layers formed by low dose SIMOX processes, in S.Cristoloveanu (ed.), Silicon-on-Insulator Technology and Devices, 94-11, The Electrochem. Soc. Publishers, NJ, pp.62-69.

56. Annamalai, N.K., Bockman, J.F., McGraer, N.E., and Chapski, J. (1990) A comparision of buried oxide characteristics of single and multiple implant SIMOX and bond and etch back wafers, IEEE Trans. Nucl. Sci. 37, 2001-2006.
57. Brady, F.T., Li, S.S., and Krull, W.A. (1990) Study of the effects of processing on the response of implanted buried oxides to total dose irradiation, IEEE Trans. Nucl. Sci. 37, 1995-2000.
58. Revesz, A.G., Brown, G.A., and Hughes, H.L. (1993) Bulk electrical conduction in buried oxide of SIMOX structures, J. Electrochem. Soc. 140, 3222-3229.
59. Stahlbush, R.E., Hughes, H.L., and Krull, W.A. (1993) Reduction of charge trapping and electron tunneling in SIMOX by suplemental implantation of oxygen, IEEE Trans. Nucl. Sci. 40, 1740-1754.
60. Ma, T.P., and Ma, W.H.-L. (1978) Effects of RF annealing on excess charge centers MIS dielectrics, IEEE J. Sol. St. Circ. 13, 445-454.
61. Lysenko, V.S., Sytenko, T.N., Snitko, O.V., Zimenko, V.I., Nazarov, A.N., Osiyuk, I.N., Rudenko, T.E., and Tyagulskii, I.P. (1986) Interrelation between surface states and transition layer defects in Si-SiO_2 structures, Solid State Commun. 57, 171-174.
62. Lysenko, V.S., Lokshin, M.M., Nazarov, A.N., and Rudenko, T.E. (1985) RF plasma annealing of implanted MIS structures, Phys. Stat. Sol. (a) 88, 705-712.
63. Lysenko, V.S., Nazarov, A.N., Valiev, S.A., Zaritskii, I.M., Rudenko, T.E., and Tkachenko, A.S. (1989) EPR and TSCR investigations of implanted Al-SiO_2-Si systems treated with RF plasma discharge, Phys. Stat. Sol. (a) 113, 653-664.
64. Nazarov, A.N., Lysenko, V.S., Mikhailov, S.N., Tkachenko, A.S., Pavlyuk, M.I., Molostov, A.N., and Kilchitskaya, V.I. (1994) Effect of RF plasma treatment on moving and accumulation of charge in SiO_2 layers of Al-polySi-SiO_2-Si structures, Mikroelectronika 232, N3, 39-46 (in Russian).
65. Nazarov, A.N., Mikhailov, S.N., Lysenko, V.S., Givargizov, E.I. and Limanov, A.B. (1992) Investigation of processes of the charge moving and accumulation in dioxode buried layers in ZMR SOI structures, Mikroelectronika 21, N3, 3-13 (in Russian).
66. Saks, N.S. (1978) Response of NMOS capacitors to ionizing radiation at 80K, IEEE Trans. Nucl. Sci. 25, 1226-1232.
67. Panchly, R.K., and Erdman, F.M. (1983) Radiation effects on oxynitride gate dielectrics, IEEE Trans. Nucl. Sci. 30, 4141-4145.
68. Mao, B.Y., Chen, G.E., Pollack, G., Hyghes, H.L., and Davis, G.E. (1987) Total-dose hardening of buried insulator in implanted siilicon-on-insulator structures, IEEE Trans. Nucl. Sci. 34, 1692-1697.
69. Schork, R., and Ryssel, H. (1992) Formation of ultra thin SOI layer by implantation of nitrogen, Proceedings of IEEE Int. SOI Conference, pp.16-17.
70. Polchlopek, S.W., Bernstein, G.H., and Kwor, R.Y. (1993) Properties of nitrogen implanted SOI structures, IEEE Trans. Electron. Dev. 40, 385-391.
71. Danilin, A.B. (1994) Peculiarities of behavior of nitrogen and oxigen atoms sequentially implanted into silicon, in J.-P.Colinge, V.S.Lysenko, A.N.Nazarov (eds.), Physical and Technical Problems of SOI Structures and Devices, Crimea, Ukraine, pp.21-23.
72. Skorupa, W. (1994) Ion beam processing for Silicon-on-Insulator, this volume.
73. Lysenko, V.S., Nazarov, A.N., Rudenko, T.E., Rudenko, A.N., Kilchitskaya, V.I., Givargizov, E.I., and Limanov, A.B. (1994) Properties of SOI MOSFET fabricated on multilayer buried dielectrics, in S.Cristoloveanu (ed.), Silicon-on-Insulator Technology and Devices 94-11, Electrochem. Soc. Publishers, NJ, pp.333-339.
74. Djurenko, S.V., Kilchitskaya, V.I., Lysenko, V.S., Nazarov, A.N., Rudenko, A.N., Rudenko, T.E., Yurchenko, A.P. (1994) Physical properties of ZMR SOI materials with multilayered buried dielectrics, in J.-P.Colinge, V.S.Lysenko, A.N.Nazarov (eds.), Physical and Technical Problems of SOI Structures and Devices, Crimea, Ukraine, pp.67-69.

75. Pennise, C.A., Boesch, H.E., Goetz, G. and McKitterick, J.B. (1993) Radiation-induced charge effects in buried oxides with different processing treatments, IEEE Trans. Nucl. Sci. 40, 1765-1773.
76. Mao, B.Y., Chen, C.-E., Matloubian, M., Hite, L.R., Pollack, G.P., Hughes, H.L., and Maley, K. (1986) Total dose characterization of CMOS devices in oxygen implanted silicon-on-insulator, IEEE Trans. Nucl. Sci. 33, 1702-1705.
77. Nakashima, S., Maeda, Y., and Akiya, M. (1986) High-voltage CMOS SIMOX Technology and its application to a BSH-LSI, IEEE Trans.Electron.Dev. 33, 126-132.
78. Tsaur, B.Y., Sterino, V.J., Choi, H.K., Chen, C.K., Montanain, R.W., Shot, J.T., Shedd, W.M., La Pierre, D.C., and Blundchard, R. (1986) Radiation hardened JFET devices and CMOS circuits fabricated in SOI films, IEEE Trans.Nucl.Sci. 33, 1372-1380.
79. Leray, J.L., Dupont-Nivet, E., Peri, J.F., Coic, Y.M., Rafaelli, M., Auberton-Herve, A.J., Bruel, M., Giffard, B., and Margail, J. (1990) CMOS/SOI hardening at 100 Mrad (SiO_2), IEEE Trans. Nucl. Sci. 37, 2013-2020.

FABRICATION OF SIMOX STRUCTURES AND IC'S TEST ELEMENTS

G.G.VORONIN, L.V.DEGTYARENKO, I.G.LUKITSA,
V.G.MALININ, V.V.STARKOV, Y.V.FEDOROVITCH,
L.N.FROLOV

Russian scientific–research institute "Electronstandart"
St.Petersburg, Russia

This work investigates SIMOX technology and mainly explores tow aspects of it:
the development of high-current ion implanter dedicated for use in laboratory environment;
the optimization of SIMOX formation process
A schematic diagram of the equipment is shown in fig.1. The microwave ion source of the implanter uses the idea of electron cyclotron resonance (ECR source). Several test-runs of this source have demonstrated a number of its advantages for operation with such chemically active gas as oxygen compared with others sources like duoplasmatron, high frequency source and duopigatron . The absense of hot cathode, the high density of plasma in the area of ion extraction and the high percentage of monatomic ions are among these advantages. The use of such source allows one to increase significantly the ion beam current in process chamber (more than 50 mA) and to enhance the performance of the implanter providing reliable operation during 6-8 hours without interruption and rising the lifetime up to several hundred hours.

The oxygen ions of the beam from the ECR source are accelerated up to 190 keV, subjected to the electromagnetic separation and, in the end, are brought into focus on the specimen in the process chamber by a quadrupole triplet electromagnetic system. The required uniformity of dose distribution on the surface of the specimen is achieved through electromagnetic scanning of the beam. The extraction system provides effective selection of the ions from the plasma, monitoring the current of the beam and its initial focusing.

For manufacturing of SIMOX wafer 76 or 100 mm in diameter silicon wafers were placed in the mounts of the process chamber. The process may be carried out both for individual wafers and for batch of wafers (up to 8).

The implantation was performed by the beam of monatomic oxygen ions. The scanning of the beam was arranged along two directions: sufficiently slow along the direction which is perpendicular to the wafer flat, and with the frequency of 50 Hz along the direction wich is parallel to the wafer flat.

The temperature on the wafer before and during implantation was maintained at the level of $600°$ C . Built-in measuring system of beam current provided dose value control and uniformity of dose across the wafer. Experimental implantation of single wafers with current of 1 mA revealed that such a current was high enough to cause cracks on the surface of the wafer.The value of this current was optimized with accepting as figure of merit the achieving the steady-state temperature of the wafer. Obtained value of current corresponded the current density in the range of $12\text{-}15\,\mu A/cm^2$.

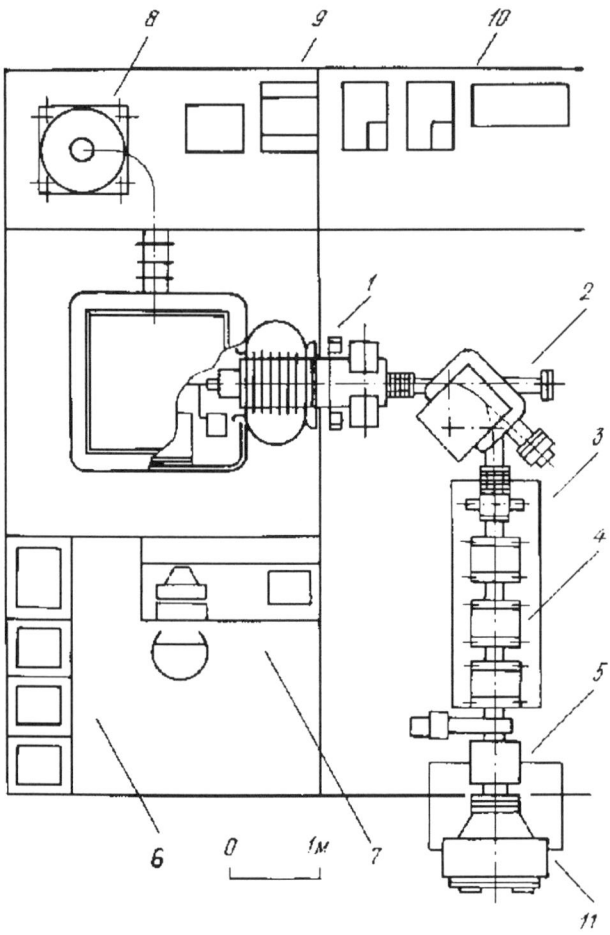

Figure. 1. Schematic diagram of implanter "Quartz-15"

1 - oxygen ion accelerator; 2 - analysing magnet; 3 - slotted aperture; 4 - quadrupole lens; 5 - scanning unit; 6,7 - power and control systems; 8 - high-voltage installation; 9,10 - power supplly; 11 - process chamber

The experimental work for optimization of the implantation process was carried out on n-type silicon with [100] orientation, 76 mm wafer diameter and a specific resistivity of 7.5 Ohm·cm. Starting parameters for investigation were: conductivity, amount of oxygen and wafer thickness as well as uniformity of these parameters along the wafer. Process conditions were : energy $E=(200\pm5)$ keV, current density $j = 12-15$ $\mu A/cm^2$, temperature during implantation $T_s = (600\pm50)^o$ C. Time for the accumulation of dose $2\cdot10^{18} cm^{-2}$ was equal to 5 hours. In order to decrease the probability for contamination during implantation, the wafers were preoxidized to obtain an oxide cap with 20 nm thickness.

Oxygen dose and its space distribution were monitored through RBS method (Rutherford Backscattering). The RBS spectrum of a SIMOX structure after annealing is shown in fig.2.

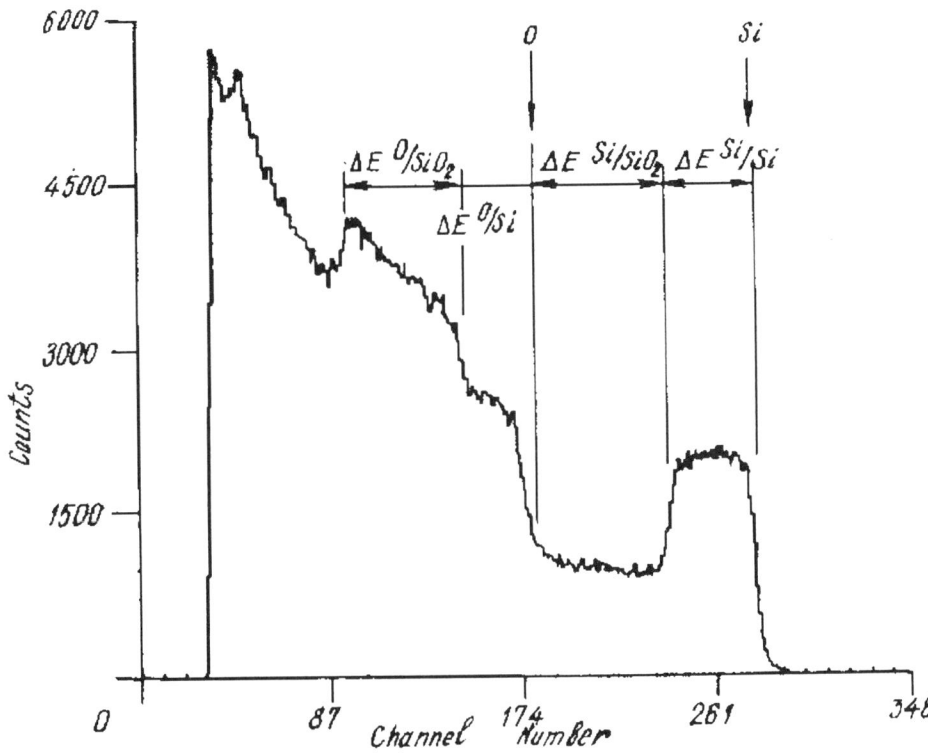

Figure.2. RBS unchanneled spectrum of SIMOX structure after annealing

The uniformity of embedded oxygen in the case of scanning along the wafer flat was high enough (approximately 2%). As for the direction, which is perpendicular wafer flat, the uniformity was in the range of 5—20 % and probably was caused not by dose variation but by inhomogeneous heating of the wafer. Such heating was due to slow scanning of beam along this direction.

The investigations have revealed, that the chosen conditions of implantation have formed the oxid buried layer with a thickness of 0.35—0.5 μm at a depth of 0.35 μm. Post-implantation annealing was carried out at temperature (1320 ±0.5)°C in an ambient of dry nitrogen for 5 hours.

The comparison of channelled and unchannelled RBS spectra for annealed wafers has evaluated the crystallinity degree of silicon overlayer to be X=3.6%. This value corresponds to values presented by early publications and does not differ significantly from the value of bulk substrates (3.1%). Therefore, the chosen annealing condition provides the recovery of the silicon overlayer crystal structure. Concentration profiles of SIMOX samples are revealed by second ion mass spectrometry (SIMS) and presented on fig. 3,4.

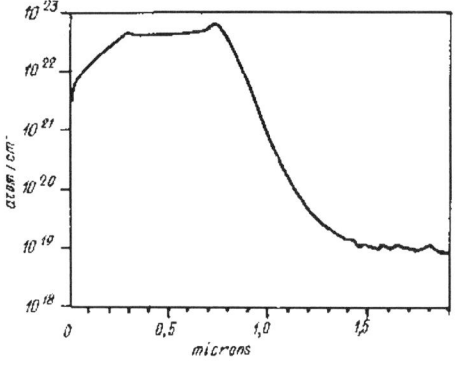
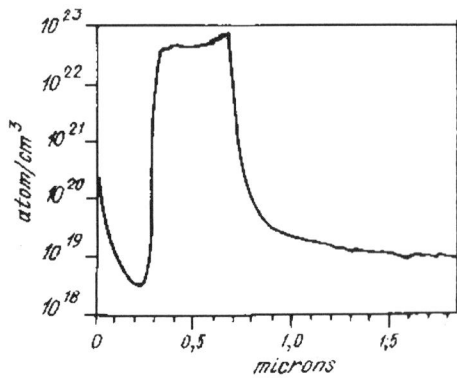

Figure.3. SIMS profile of as implanted oxygen

Figure.4. SIMS profile of oxygen after annealing

The shown profiles give evidence that essential oxygen segregation took place after annealing. In addition, the oxygen concentration profile clearly shows that the oxygen doping level is enough for stoichiometric conformity of buried oxide to SiO_2. Therefore, this analysis confirms the adequacy of chosen implantation dose and its diagnostic methods.

As shown with the help of cross sectional analysis, the uniformity of thickness distribution in SIMOX structures is enhanced after annealing. The specimen structure is radicaly improved, yielding the thickness of SiO_2 buried layer in the range of 300—350 nm with sharp and flat interfaces. The SiO_2 layer still has the inclusion of silicon in the state of hexagonal islands with dimensions from 10 to 100 nm, whose crystallographic orientation is similar to the orientation of the substrate. These inclusions are arranged predominantly at the distance of 50—150 nm from top as well as from bottom interfaces of oxide layer and practically do not interfere with its dielectric properties. Further improvement of SIMOX structures is possible with the implementation of multiple sequential implantation and annealing.

Test n- and p- channel transistors used in this work were fabricated using a slightly modified conventional CMOS technology. The drain junction perimeters of n- and p- channel transistors were respectively 260 and 1080 μm, the channel length was 8 μm.

Table 1 shows the basic characteristics of test transistors.

TABLE 1. Electrical characteristics of insulation layer and CMOS transistors.

Parameters	Well to well insulation	
Breakdown voltage, V	200—300	
Insulation leakage current at V = 100V, A	$<10^{-9}$	
Parameters	P — channel transistor	n — channel transistor
Threshold voltage, V	(-1,6)—(-2,2)	1,6— 2,9
Breakdown voltage between the drain and the transistor body	25—27	20—22
Drain leakage current at V=5V,A		
at room temperature	10^{-12}	10^{-13}
at 120°C	10^{-9}	10^{-10}

From the table 1 we notice that the breakdown voltage of the well to well insulator which features the dielectric properties of the buried oxide is in the range of 200-300 V. This testifies that the buried oxide is sufficiently uniform and completely adequate for most IC design specifications.

The characteristics of the n- and p- channel transistors differ little from those fabricated in monosilicon. This is revealed by a number of calculated characteristics presented in table 2.

TABLE 2. Physical and electrical characteristics of SIMOX test structures.

Parameters	Values	
Field intensity of well to well breakdown, V/cm	$6 \cdot 10^6$	
Fixed charge in gate oxide, cm^{-2}	$(3—4) \cdot 10^{11}$	
Parameters	P — channel transistor	n — channel transistor
Mobility of channel carriers, cm^2/Vs	240	650
Drain leakage current density at room temperature, A/cm^2	$3 \cdot 10^{-7}$	10^{-7}
Generation centers concentration in ovelayer substrate, cm^{-3}	$3 \cdot 10^{14}$	10^{14}
Estimated carrier lifetime, μs	0,3	0,9

Fixed charge and channel carrier mobility values demonstrate, that the interface between the gate oxide and the silicon overlayer has the similar properties with conventional interface $Si-SiO_2$.

The comparison between similar characteristics of the bulk and investigated SOI devices suggests that the amount of generation centers in the latter is from two to three times greater than in bulk devices. However, future development of implantation and annealing processes will decrease with certainty the dislocation density and improve the quality of the top silicon film.

The investigations of drain leakage currents in SOI MOSFETs as a function of temperature show the considerable extension of temperature range for reliable operation. This is attributable to the decreasing of p-n junction areas by more than two orders of magnitude.

The diagram on the fig.5 demonstrates the similarity of energy characteristics for n- and p- channel SIMOX transistors in wider temperature range with leakage currents from two to three orders of magnitude lower than for bulk devices.

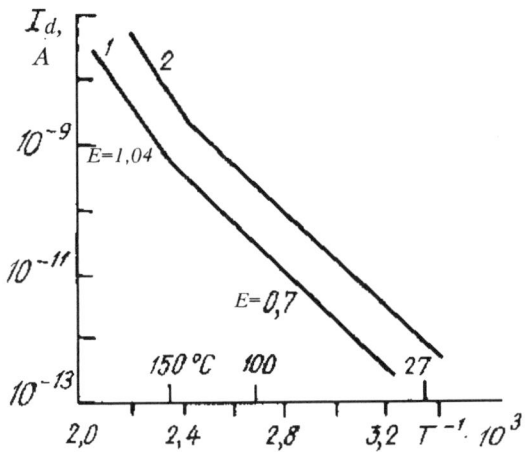

Figure.5. Drain leakage currents of n- and p- channel transistors as a function of temperature.
curve 1 — n-channel transistor
curve 2 — p-channel transistor
E — activation energy, eV

The feasibility of high temperature performance of SIMOX circuits was estimated by the output characteristics of p-channel transistor at temperature 300°C. The transistor is entirely fit to work at such temperature. On the contrary, the leakage current of bulk transistor in this situation reaches the value of several milliampers thus preventing the operation of the transistor completely.

However, it should be emphasized that further improvement of SIMOX structure quality and optimization of IC process are needed for the manufacturing of high temperature SOI devices.

LOW-FREQUENCY NOISE CHARACTERISATION OF SILICON-ON-INSULATOR DEPLETION-MODE p-MOSFETS

N.B. Lukyanchikova, M.V. Petrichuk and N.P. Garbar
Institute of Semiconductor Physics
252650, Kiev, Ukraine
E. Simoen and C. Claeys
IMEC
Kapeldreef 75, B-3001 Leuven, Belgium

This paper discusses the different noise sources in thick- and thin-film depletion mode SOI p-MOSFETs. As is shown, for front- or back-channel operation, the low-frequency noise spectrum is predominantly 1/f-like. This noise is usually attributed to carrier interactions with the near-interface oxide traps. For a buried-channel mode of operation, additional generation-recombination (GR) noise is observed. Detailed analysis reveals that the GR noise amplitude and time constant are a strong function of either the front- or the back-gate bias. A model will be proposed which enables in principle the extraction of the underlying density of interface traps.

1. Introduction

Depletion-mode, or buried-channel silicon MOSFETs are unique devices in this sense that they combine compatibility with CMOS technology, with superior, JFET-like low-frequency (LF) noise performance for specific operation conditions [1]-[3]. This renders such type of devices quite attractive for analog applications. When fabricated in an SOI technology, the role of the substrate is replaced by the back-gate electrode. From a LF noise viewpoint, this can introduce an additional source of fluctuations [4]-[6]. Therefore, it is the aim of this work to study in detail the different front- and back-gate related noise sources in depletion-mode p^+pp^+ SOI MOSFETs, which are fabricated in different SOI CMOS technologies. The emphasis is mainly on the excess generation-recombination (GR) noise which is often found in linear operation [4]-[6] and which is most likely related to either the front- or the back-interface traps.

2. Experimental

The devices studied have been processed in a 1 and a 0.5 micron SOI CMOS technology on SIMOX substrates. The gate-oxide thickness t_{oxf} and film thickness t_f are, respectively, 20 nm and 180 nm (1 micron), and 15 nm and 100 nm (0.5 micron). The body doping density is such that the 1 μm CMOS transistors are partially depleted (thick film) buried-channel (BC) type [7], while the 0.5 μm structures are thin film fully depleted - and therefore accumulation-mode (AM) devices [8]. The device width W is 20 μm and transistors with various lengths L have been studied. The LF noise is measured in the 1 Hz - 10 kHz range in linear operation, for a fixed drain bias V_D and for different combinations of the front- and of the back-gate bias (V_{Gf} and V_{Gb}, respectively).

3. Results

A typical input characteristic in linear operation for a L=1 µm BC p-MOSFET is shown in Fig. 1, with the back-gate in depletion. The front threshold voltage (V_{Tf}) is designed to be around -0.8 V, while the back V_T is ≈-5 V, for zero front-gate bias [7]. The corresponding current noise spectral density S_I is represented in Fig. 2, for three measurement frequencies f. From the figure, it is clear that different noise regimes, characterised by a different type of spectrum can be distinguished, similar as for bulk BC MOSFETs [1]-[3]. First, a 1/f-like regime is observed for $V_{Gf} \ll V_{Tf}$. In this case, a surface channel exists and the noise is determined by the interactions between the holes and the near-interface oxide traps, with density D_{ot}. From the input-referred noise spectral density S_{VGf}, D_{ot} can be derived, for instance from [9]:

$$D_{ot} = \frac{8 S_{VGf} f (W \times L) C_{oxf}^2}{q^2 kT} \quad (eV^{-1} cm^{-2}) \tag{1}$$

yielding typical values in the range 10^9 eV^{-1}cm^{-2} for the BC devices [6]-[7] and 10^{10} eV^{-1}cm^{-2} for the AM p-MOSFETs. Hereby is q the electron charge, kT the thermal energy and C_{oxf} the front-oxide capacitance per unit of area. A similar analysis can in principle be performed for the back-channel noise.

When V_{Gf} approaches V_T (for constant V_{Gb}), the nature of the noise spectrum changes and becomes Lorentzian-like. At the same time, conduction occurs now in a buried channel, while the surface is depleted. A detailed investigation of the underlying GR noise is reported elsewhere [6],[10]. The main results can be summarised as follows. The parameters of the GR noise component, which is described by:

$$S_I(f) = \frac{S_I(0)}{[1+(2\pi f\tau)^2]} \tag{2}$$

are strong functions of one of the gate biases. Figs. 3 and 4 refer to the case where both the amplitude $S_I(0)$ and the GR time constant τ are exponential functions of V_{Gf}. In the case of Fig. 3, this occurs for front-channel, in the second case for back-channel conduction. Hereby is τ given by the relationship $\tau=1/2\pi f_0$, with f_0 the corner frequency of the Lorentzian spectrum, for which $S_I(f_0)=S_I(0)/2$.

The dependence of the characteristic time constant τ and of the normalised amplitude $S_I(0)/I_D^2$ on V_{Gf}, corresponding with the spectra of Fig. 3, are shown in Figs 5 and 6. To a first approximation, it is found that τ and $S_I(0)$ follow a law according to:

$$\tau \approx \tau_0 \exp(\alpha V_{Gf}) \tag{3a}$$
$$S_I(0) \approx S_{I0} \exp(\beta V_{Gf}) \tag{3b}$$

respectively.

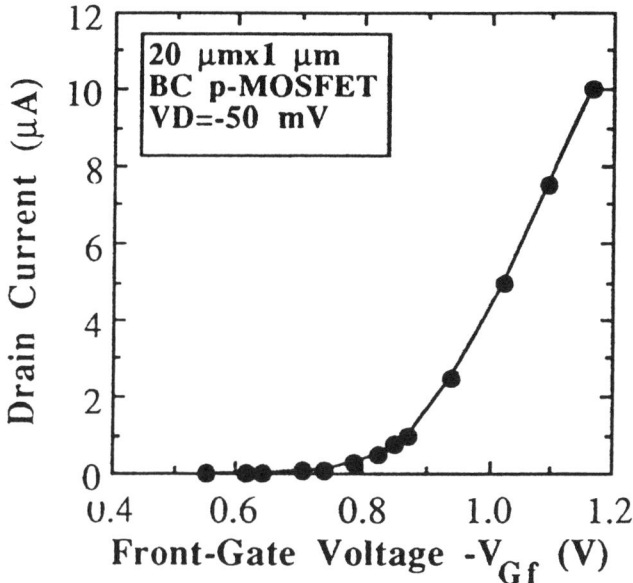

Figure 1. Input characteristic of a L=1 µm BC p-MOSFET. V_D=-50 mV, V_{Gb}=0 V.

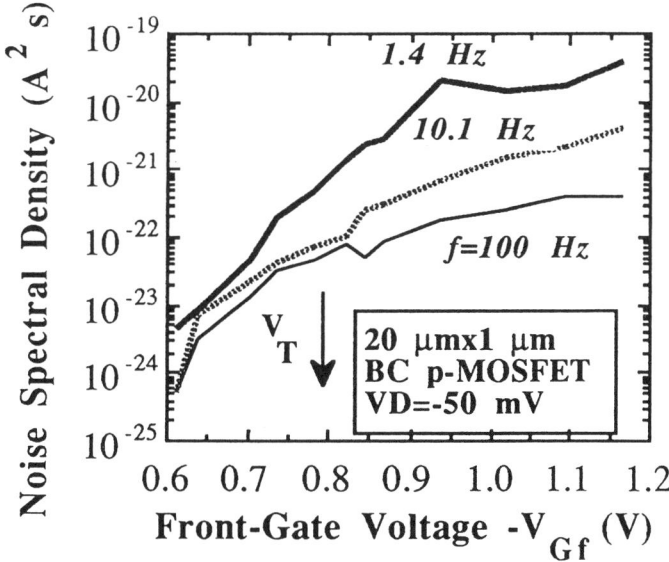

Figure 2. Current noise spectral density in linear operation for a L=1 µm BC p-MOSFET. V_{Gb}=0 V; V_D=-50 mV.

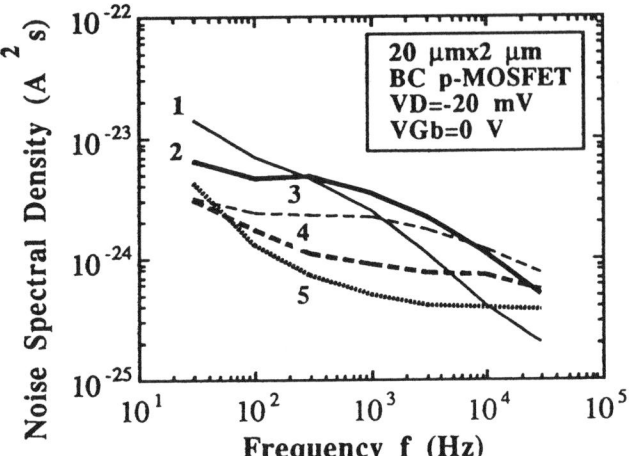

Figure 3. Current noise spectral density in linear operation for a L=2 μm BC p-MOSFET. V_{Gb}=0 V and V_{Gf}=-0.65 V (1); -0.7 V (2); -0.75 V (3); -0.8 V (4) and -0.85 V (5).

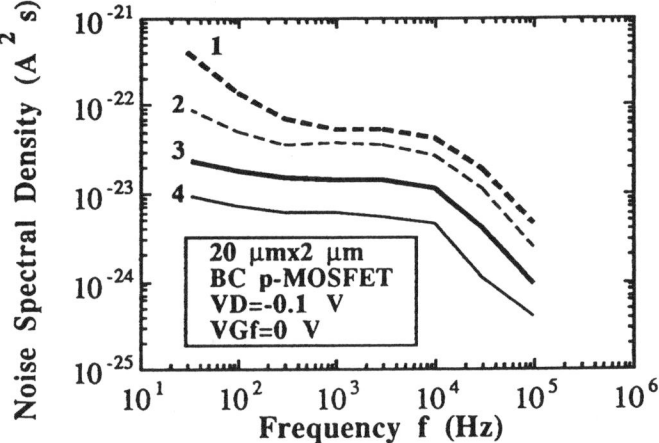

Figure 4. Current noise spectral density in linear operation for a L=2 μm BC p-MOSFET. V_D=-0.1 V. V_{Gf}=0 V and V_{Gb}=-8.85 V (1); -6.54 V (2); -5.38 V (3) and -4.47 V (4).

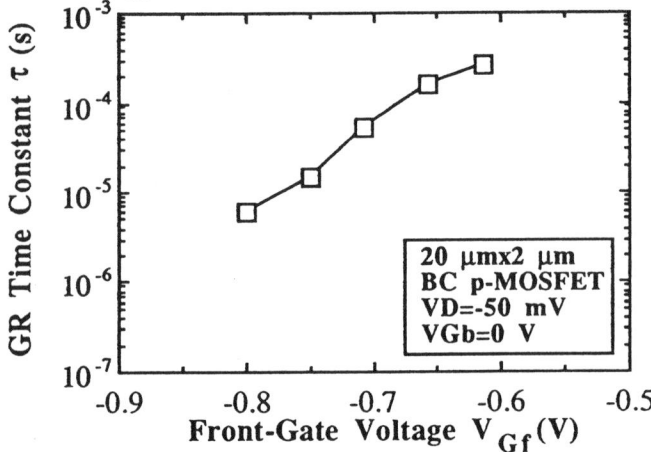

Figure 5. GR time constant as a function of the front-gate voltage for a L=2 μm BC p-MOSFET. V_D=-50 mV.

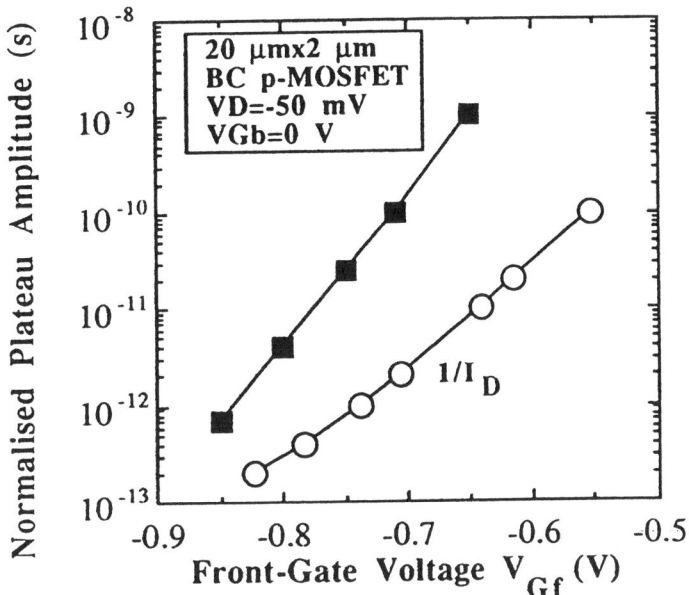

Figure 6. Dependence of the normalised noise amplitude on the front-gate voltage for a L=2 μm BC p-MOSFET. $V_{Gb}=0$ V; $V_D=-50$ mV.

4. Discussion

The empirical relationships of eq. (3) can be understood physically as follows. The exponential dependence of the GR time constant on the front- or the back-gate bias stems from the exponential V_{Gf} (V_{Gb}-) dependence of the free carrier density, when the device is in a buried-channel operation mode and if it is assumed that the slowest, dominant process in the GR of carriers is the capture step. This condition yields a τ proportional to 1/n (n free carrier density). To evaluate the capture cross section, an accurate value of the relevant n is required, which may be found if the (surface) Fermi level is known, as a function of V_{Gf}. Furthermore, the experimental α coefficients are very close to the front-subthreshold swing S which is in the range 22-25 V^{-1} [6],[10], for the front-gate related GR noise. This relationship is potentially very interesting but not at all straightforward and needs further elaboration.

A typical I_D^{-3} dependence is observed for the plateau amplitude, which can be explained if it is assumed that the relevant density of traps shows a dependence on V_{Gf} (or V_{Gb}) [6]. In analogy with the case for bulk BC p-MOSFETs, and based on the observed empirical trends, traps located at the front- (or back-) interface are highly suspected as the origin of the GR noise. Following the model of Jones and Taylor [3], a density of front interface traps D_{itf} can be calculated from:

$$D_{itf} = (\frac{C_{oxf}}{q})^2 \frac{LW}{kT} \frac{S_{VGf}(0)}{\tau} \qquad (4)$$

yielding densities in the range $2\text{-}3\times10^{11}$ $eV^{-1}cm^{-2}$ for the front- and a few 10^{10} $eV^{-1}cm^{-2}$ for the underlying back-interface traps, respectively. Hereby is $S_{VGf}(0)$ the input-referred plateau amplitude. It should be remarked finally that similar trap densities were obtained in depletion-mode SOI n-MOSFETs [4]. The large difference between D_{ot} and D_{it} derived from the 1/f noise and the GR noise, respectively, indicates that different types of trap centres are monitored in both cases.

5. Conclusion

In summary, it can be concluded that for SOI depletion-mode p-MOSFETs strong GR noise can be observed, which is influenced either by the front- or the back-gate electrode. Studying this GR noise enables to estimate the front- and back-interface quality, which can be compared with the results of other, more standard techniques.

6. References

1. Watanabe, T. (1985) Low-Noise Operation in Buried-Channel MOSFET's, *IEEE Electron Device Letters* **EDL-6**, 317-319
2. Carruthers, C. and Mavor, J. (1992) Noise Characteristics of n-Channel Deep-Depletion Mode MOS Transistors, *IEE Proc. G* **139**, 377-383.
3. Jones, B.K. and Taylor, G.P. (1992) Spectroscopy of Surface States Using the Excess Noise in a Buried-Channel MOS Transistor, *Solid-State Electron.* **35**, 1285-1289.
4. Elewa, T., Boukriss, B., Haddara, H.S., Chovet, A. and Cristoloveanu, S. (1991) Low-Frequency Noise in Depletion-Mode SIMOX MOS Transistors, *IEEE Trans. Electron Devices* **ED-38**, 323-327.
5. Matloubian, M., Scholz, F., and Lum, L. (1994) Low-Frequency Noise in Fully-Depleted SOI PMOSFETS, *IEEE Trans. Electron Devices* **ED-41**, to be published in the November issue.
6. Lukyanchikova, N.B., Petrichuk, M.V., Garbar, N. P., Simoen, E. and Claeys, C. Investigation of Generation-Recombination Noise in Buried-Channel p-MOSFETs, Paper submitted to Solid-State Electron.
7. Simoen, E., Magnusson, U., Van den bosch, G., Smeys, P., Colinge, J.P., and Claeys, C. (1993) Adaptation of a Standard 1 μm, Double Layer Metal CMOS SOI Technology to Space Applications, Proc. of the 2nd ESA Electronic Components Confe rence ESA WPP-063, 295-300.
8. Colinge, J.P. (1990) Conduction Mechanisms in Thin-Film Accumulation-Mode SOI p-Channel MOSFET's, IEEE Trans. Electron Devices ED-37, 718-723.
9. Christensson, S., Lundstrom, I. and Svensson, C. (1968) Low-Frequency Noise in MOS Transistors - I. Theory, *Solid-State Electron.* **11**, 797-805.
10. Lukyanchikova, N.B., Petrichuk, M.V., Garbar, N.P., Simoen, E. and Claeys, C., Strong Low-Frequency Noise in Buried-Channel pMOSFETs Under Inversion Conditions, Paper to be published in AIP Proc. of the 6th van der Ziel Symposium on Low-frequency Noise, St. Louis (Mo), 27-28 May, 1994.

Section 4:

SOI Circuits

SOI DEVICES AND CIRCUITS: AN OVERVIEW OF POTENTIALS AND PROBLEMS

Jean-Pierre Colinge

Microelectronics Laboratory (DICE)
Université Catholique de Louvain (UCL)
Bâtiment Maxwell, Place du Levant 3
B-1348 Louvain-la-Neuve, Belgium

Silicon-on-Insulator (SOI) technology has evolved from a mere laboratory curiosity in the early '80s to a technology in which large circuits such as 1 Mbit SRAMs can be made. The flexibility provided by full dielectric isolation and the quasi-ideal properties of the SOI MOSFET (sharp subthreshold slope, low body-effect coefficient,...) have given rise to new fields of applications for SOI devices. Beside high-temperature and radiation hard niche applications, SOI technology is now increasingly used for the fabrication of low-voltage, low-power CMOS circuits, high-frequency (microwave) devices, and high-temperature circuits. Some problems related to the physics of SOI and to SOI circuits still continue to raise questions about the future of SOI technology. These issues will be addressed in this paper.

TABLE 1: Some milestones of SIMOX CMOS technology

Year	Company	Milestone	Ref.
1978	NTT	SIMOX acronym, First SIMOX circuits	1
1987	Hewlett-Packard	Thin-film (fully depleted) CMOS circuits	2
1988	Hewlett-Packard	2 GHz thin-film CMOS circuits	3
1989	TI / Harris	64kb SRAM	4
1989	LETI	Thin-film 16k SRAM	5
1989	AT&T	6.2 GHz thin-film CMOS circuits	6
1989	NTT	21 ps CMOS ring oscillator	7
1990	TI	Commercial 64 SRAM	
1991	TI	256kb SRAM	8
1991	IBM	256kb fully-depleted SRAM	9
1991	Westinghouse	14 GHz f_T microwave SOI MOSFETs	10
1993	Westinghouse	23 GHz f_T microwave SOI MOSFETs	11
1993	TI	1Mb SRAM, 20 ns access time @ 5V	12
1993	IBM	512 kb SRAM, 3.5 ns access time @ 1V	13
1993	Mitsubishi	1 M gate array	14

1. SOI devices: Potentials

SOI (and in particular SIMOX) CMOS technology has greatly evolved over the last ten years. Some of the important milestones of this progress are listed in Table 1.

Figure 1 presents equivalent capacitor networks based on the physics of bulk and fully depleted (FD) SOI devices, which illustrate how well the surface potential is controlled by the gate.[15,16] In a bulk device, part of the gate bias serves to build up a depletion layer, which is not (or hardly) the case in a FD SOI device. The presence of a buried oxide in FD devices also allows for the good control of the surface potential by the gate voltage.

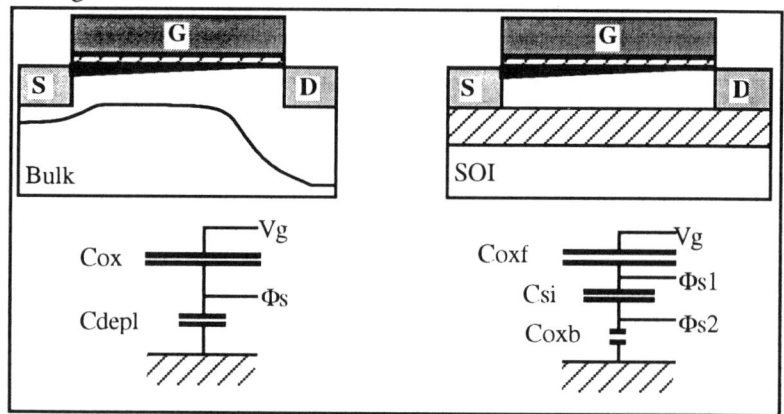

FIGURE 1: Capacitor networks illustrating the different body-effect coefficients in bulk and fully depleted (inversion-mode) SOI devices

1.1. DRAIN CURRENT

The influence of the body-effect coefficient on the current drive of the device can best be understood by using the simple device model presented in Figure 1. The saturation drain current is given by the following expression:

$$I_{Dsat} = \frac{1}{2\lambda} \mu C_{ox} \frac{W}{L} (V_G - V_{TH})^2 \qquad (1)$$

where λ is the linearized body-effect coefficient. In a bulk transistor, λ is given by

$$\lambda = 1 + \frac{\varepsilon_{si}}{C_{ox} X_{dmax}} \cong 1.15 \text{ to } 1.5, \text{ typically} \qquad (2)$$

In a SIMOX thin-film FD transistor, on the other hand, the body-effect coefficient is given by:

$$\lambda = 1 + \frac{\frac{\varepsilon_{si}}{t_{si}} C_{oxb}}{C_{oxf} \left[\frac{\varepsilon_{si}}{t_{si}} + C_{oxb}\right]} \cong 1.05, \text{ typically} \qquad (3)$$

where C_{oxf}, C_{oxb} and t_{si} are the gate oxide capacitance, the buried oxide capacitance and the silicon film thickness, respectively. From the above equations, it is evident that the saturation drain current is 20 to 30% higher in an SOI FD device than in a bulk device with similar parameters (Figure 2).[17,18] It is worth mentioning that the body-effect coefficient depends on the doping level in bulk devices, while it is independent of the doping in FD SOI devices. As the dimensions of bulk devices are scaled down, the substrate doping level is increased, and the body-effect coefficient increases as well. In FD SOI devices, the body-effect coefficient does not increase significantly as the devices are scaled down.

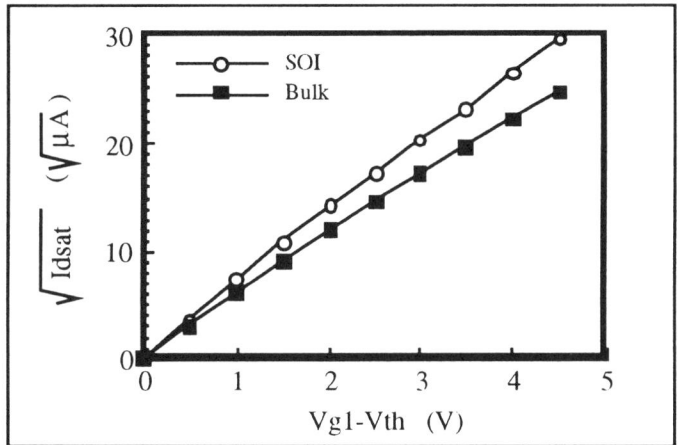

FIGURE 2: $\sqrt{I_{Dsat}}$ as a function of V_{G1}-V_{th} in a bulk and a thin-film, fully depleted SOI device having similar technological parameters.

The superior current drive capability of FD SOI devices gets somewhat degraded when short gate lengths are considered, because of velocity saturation effects. It can be seen, however, from Figure 3, that FD SOI devices still present a 25% current drive improvement over bulk devices for a gate length of 0.2 μm.[19]

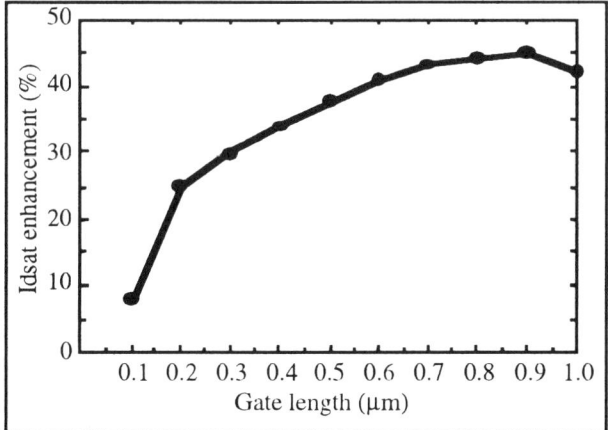

FIGURE 3: Drain saturation current enhancement (FD SOI vs. bulk) as a function of gate length.

1.2. SUBTHRESHOLD SLOPE

The subthreshold swing (inverse subthreshold slope, or, in short, the subthreshold slope) of a MOSFET is also affected by the body effect. Indeed, the subthreshold swing is given by the following expression:[20]

$$S \text{ (mV/dec)} = \lambda \frac{kT}{q} \ln(10) \qquad (4)$$

if the influence of the interface traps is neglected. The low value of λ in FD SOI devices yields an improvement of the subthreshold slope over bulk devices (Figure 4). The subthreshold slope is degraded in all devices (bulk and SOI) as the length of the channel is reduced, but this degradation is less severe in FD and especially in volume-inversion (VI) SOI devices than in bulk transistors.[21,22]

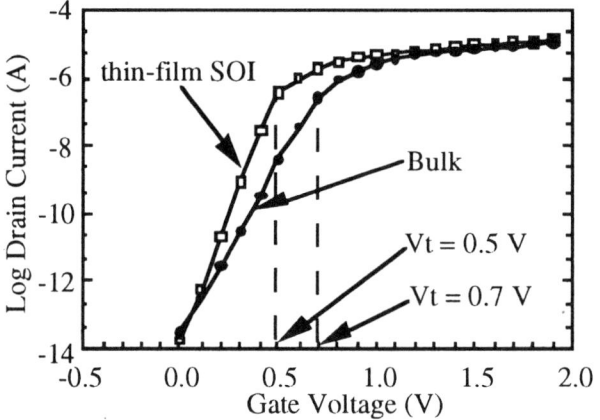

FIGURE 4: Subthreshold characteristics of bulk and FD SOI n-channel MOSFETs.

1.3. G_M/I_D

The gain of a transistor in an analog circuit is given by $\frac{g_m}{I_D} V_A$, where g_m is the transconductance and V_A is the Early voltage of the device. Since FD SOI devices are virtually free of kink effect, their Early voltage is similar to that of bulk transistors. Below threshold (in the weak inversion regime) the value of g_m/I_D can be rewritten:

$$\frac{g_m}{I_D} = \frac{dI_D}{I_D \, dV_G} = \frac{\ln(10)}{S} = \frac{q}{\lambda kT} \qquad (5)$$

Again, we see that the reduced body-effect coefficient of FD SOI transistors provides us with increased performance over bulk. Figure 5 presents the variation of g_m/I_D as a function of drain current in bulk and FD SOI n-channel transistors.

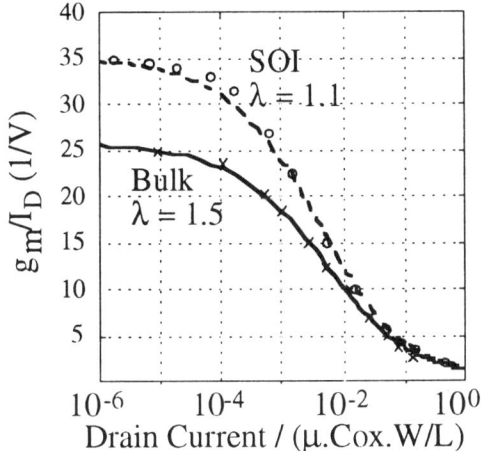

FIGURE 5: g_m/I_D ratios in saturation ($V_D = 2.5$ V) for bulk ($\lambda=1.5$) and SOI ($\lambda=1.1$) MOSFETs.

2. SOI devices: Problems

Some problems are associated with the full dielectric isolation of SOI devices. Among these problems, we will discuss the reduced drain breakdown voltage caused by parasitic lateral bipolar transistor action and the self-heating of the transistors. Hot-carrier degradation problems will be discussed as well.

2.1. REDUCED DRAIN BREAKDOWN VOLTAGE

The presence of a parasitic lateral bipolar transistor in an SOI MOSFET without body tie reduces the drain breakdown voltage. In an n-channel device the hole base current is supplied by impact ionisation near the drain. The base current is given by $I_{body} = (M-1) I_{Dsat} \cong (M-1) I_{ch}$ where I_{ch} is the channel current, and the resulting increase of drain current is given by $\Delta I_D = b \, I_{body} = ß (M-1) I_{ch}$ (Figure 6).

From bipolar transistor theory, the collector (drain) breakdown voltage with open base, BV_{CE0}, is smaller than when the base is grounded (BV_{CB0}). The breakdown voltages are related as follows:

$$BV_{CE0} = \frac{BV_{CB0}}{\sqrt[n]{ß}} \qquad (6)$$

where ß is the gain of the bipolar device, and n ranges typically between 3 and 6. The above relationship is quite a simplification of the breakdown mechanisms occurring in SOI MOSFETs, since both ß and M-1 (the multiplication factor) depend on the drain voltage in a highly nonlinear fashion. Assuming an emitter efficiency close to unity one can also write:

$$ß \cong 2 (L_n/L_B)^2 - 1 \qquad (7)$$

where L_B is the base width, which can be assumed, to a first approximation, to be equal to the effective channel length, L, and L_n is the electron diffusion length: $L_n^2 = D_n \tau_n$, where D_n and τ_n are the diffusion coefficient and the lifetime of the minority carriers in the base, respectively. Taking BV_{CEO} as the drain breakdown voltage with floating body, equal to BV_{DS}, one obtains:

$$BV_{DS} = \frac{BV_{CB0}}{\sqrt[n]{\frac{2D_n \tau_n}{L^2} - 1}} \qquad (8)$$

where L is the gate length.

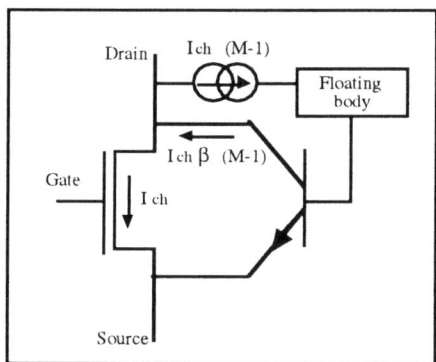

FIGURE 6: Parasitic bipolar transistor of the SOI MOSFET. I_{ch} is the channel current.

One major concern with SOI devices is their low drain breakdown voltage. This low BV_{DS} is caused by the presence of a parasitic lateral bipolar structure with floating base in the SOI MOSFET. Any hole current generated near the drain by impact ionization gives rise to a drain (collector) current through the bipolar effect. This collector current itself gives rise to more impact ionization, and this positive feedback loop causes the transistor to latch. In some cases, the device can no longer be turned off by the gate (single device latch-up).[23,24,25] Although techniques such as LDD, LDS (lightly doped source) and emitter efficiency reduction by using silicides can slightly improve the drain breakdown voltage,[26] the BV_{DS} of SOI devices remains substantially lower than that of bulk MOSFETs, rendering even 1-μm devices unsuitable for 5-volt operation. As the channel length is decreased, however, punchthrough starts to lower the BV_{DS} of bulk devices more rapidly than the parasitic bipolar effect lowers the BV_{DS} of SOI devices. As a result, there is a gate length (around 0.3 μm) below which the drain breakdown voltage of SOI devices is actually larger than that of the corresponding bulk transistors.[27,28] This is illustrated in Figure 7 where the drain breakdown voltage of SOI and bulk transistors (either measured or simulated) is plotted as a function of gate length.[29] It can be seen that punchthrough reduces the BV_{DS} of bulk devices more than the parasitic bipolar effect reduces the BV_{DS} of SOI devices for gate lengths below 0.35

µm. Both types of devices can, of course, be optimized to improve the BV_{DS}. For example, Figure 7 presents data points from IBM and Matsushita (bulk devices) with $BV_{DS} > 3.5$ volts for L=0.15 µm (HS-GOLD structure)[30] and $BV_{DS} = 4.5$ volts for L=0.25 µm (LATID structure)[31] have been reported. Similarly, a BV_{DS} of 4.2 volts for L=0.21 µm has been obtained in SOI devices using a gate-overlapped LDD structure, source and drain silicidation or low-lifetime SIMOX material (HP).[32] Figure 7 indicates a trend suggesting that SOI is advantageous for short-channel, low-voltage applications.

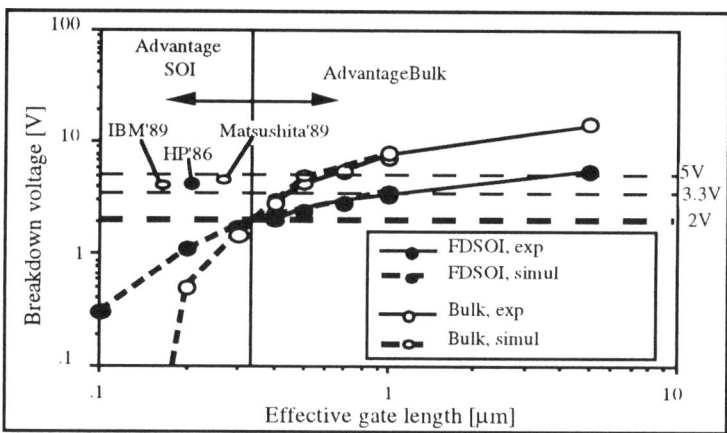

FIGURE 7: BV_{DS} vs. gate length in bulk and FD SOI n-channel MOSFETs

2.2. HOT-CARRIER DEGRADATION

It seems that there is no consensus on the hot-carrier degradation issues in SOI devices. Several papers have been published on that topic, leading to contradictory conclusions. Several publications support the idea that SOI devices are less prone to hot-carrier degradation than bulk devices, while some others draw the opposite conclusion. These discrepancies are partly due to the fact that it is almost impossible to make one-to-one comparisons between fully depleted SOI and bulk devices. Indeed, if the same channel doping concentration is used in both types of devices, the threshold voltages and saturation currents will be different. If the threshold voltages are similar, the doping profiles are different, etc... In addition, the drain breakdown characteristics are different as well. People seem to agree, however, on the following point: the peak drain electric field is lower in fully depleted SOI devices than in bulk devices, leading to less hot-carrier generation.[33,34,35,36]

However, because the degradation mechanism does not depend only on the hot-carrier generation rate, but also on many other different parameters (the quality of the buried oxide, the direction and magnitude of the vertical electric field, ...) the hot carriers degrade SOI devices in a different way than they do in bulk devices [37,38,39], mainly due to carrier injection in the buried oxide.

Figure 8 presents the threshold voltage shift and the transconductance degradation in SOI thin-film (80 nm) and bulk transistors having a gate length of 0.5 µm. A much larger initial transconductance degradation, a more pronounced initial threshold voltage shift, and a weaker time dependence are observed in the SOI devices.

The parameter which is degraded the most in SOI is the threshold voltage. The degradation depends on both front- and back-gate bias.[40]

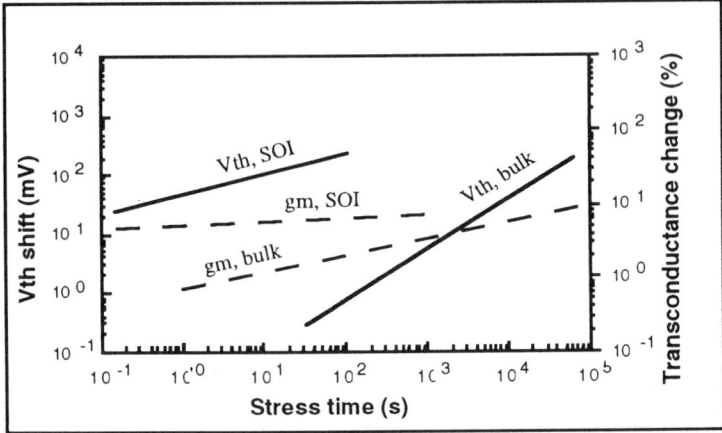

FIGURE 8: Threshold voltage shift and transconductance degradation in 0.5 μm thin-film SOI and bulk n-channel transistors as a function of stress time. V_{GS}=2 V, V_{DS}=6 V.[41]

2.3. SELF-HEATING EFFECTS

SOI transistors are thermally insulated from the substrate by the buried insulator. As a result, removal of excess heat generated by the Joule effect within the device is less efficient than in bulk, which yields to substantial elevation of device temperature. The conduction paths for excess heat are several: heat diffuses vertically through the buried oxide and laterally through the silicon island into the contacts and the metallization.[42]

The negative resistance[43] which can be seen in the output characteristics of SOI MOSFETs is due to a mobility reduction effect caused by device heating[44]. This effect is clearly visible on the output curves once sufficient power is dissipated in the device. Because of the relatively low thermal conductivity of the buried oxide, the devices heat up by 50 to 150°C and a mobility reduction is observed[45]. This effect is quite dramatic and it looks as if it really could jeopardize SOI technology.

One should not forget, however, that this effect takes place as power is dissipated into the device. This is the case when the device is measured in a quasi-dc mode with a curve tracer or an HP4145, but not in an operating CMOS circuit. Indeed, in an operating CMOS circuit, there is virtually no current flowing through the devices in the standby mode, and power is dissipated in the devices during switching for only brief periods of time only (< 1 nanosecond).

It has been shown by a pulsed measurement technique[46] that the time constants involved in the self-heating of SOI transistors are on the order of several tens of nanoseconds, and that no negative resistance effect is observed when the devices are measured in the pulse mode, because the time during which power is dissipated is much shorter than the thermal inertia of the devices. A similar experiment reports that the self-heating does not influence the output characteristics of transistors if the measurement is carried out at a slew rate higher than 20 V/μs.[47]

It seems thus that the negative resistance is not a problem for digital circuits (with the possible exception of output buffers). There might, however, be an influence of the duty cycle (the frequency at which the devices are switched) on the overall local temperature, which could modify the mobility. As far as analog circuits are concerned, transistors are usually operated at relatively low gate voltage overdrives (=V_G-V_T) or in weak inversion. This limits the power dissipated in each device and reduces the heating effects.

3. SOI Circuits

SOI devices can be used to produce integrated circuits for some types of niche applications, such as high-temperature electronics and radiation-hardened electronics.[48,49] We will not consider these types of applications here and we will focus on a field which has a chance to become mainstream in future years: low-voltage, low-power electronics.

3.1. LOW-VOLTAGE, LOW-POWER DIGITAL CIRCUITS

As we have seen earlier, owing to better coupling between the gate voltage and the surface potential than in bulk devices, the body-effect coefficient of SOI transistors, λ, is significantly lower than that of bulk devices ($\lambda \cong 1.05$ in a SIMOX device and $\lambda=1.15$ to 1.5 (depending on substrate doping) in a bulk MOSFET). This has a significant impact on both the subthreshold swing, S, and the drain saturation current.

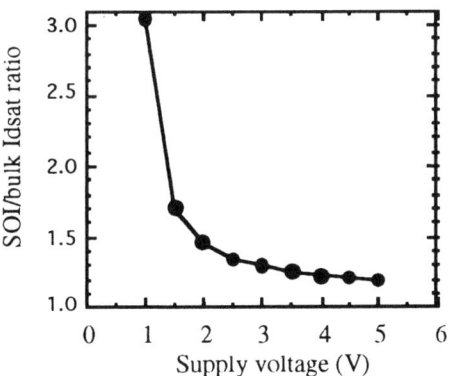

FIGURE 9: SOI/bulk I_{dsat} ratio vs. supply voltage. $\lambda_{SOI}=1.05$ and $\lambda_{bulk}=1.15$

The good subthreshold swing allows one to reduce the threshold voltage to, say, 500 mV compared to 700 mV in bulk without increasing the off current, compared to a bulk device (see Figure 4). This reduction of threshold voltage has a positive, but modest effect on the speed of the circuits (through an increase of I_{Dsat}, Eqn. (1)) when $V_{DD}=5V$. This effect, however, becomes more pronounced as the supply voltage is reduced. Figure 9 presents the ratio of saturation currents between an SOI MOSFET with $\lambda=1.05$ and $V_T=0.5V$ and a bulk transistor with $\lambda=1.15$ and $V_T=0.7V$. The I_{Dsat} enhancement due to the V_T reduction is quite remarkable at low supply voltage.

In parallel to the reduction of V_{DD}, a reduction of power consumption is desirable. The dissipated power of a circuit is roughly proportional to $P=f \cdot C \cdot V^2$ where f is the frequency, V is the supply voltage, and C is the sum of all the capacitances in the circuit. Since all capacitances but the gate oxide capacitance are smaller in SOI than in bulk,[50] less power is dissipated in SOI circuits. Figure 10 presents the power consumption of circuits functioning at different supply voltages, and for different values of total capacitance. These values are normalized such that the power consumed by a bulk circuit (capacitance = C) operating under 5 volts corresponds to P=100%. SOI circuits present capacitance values corresponding to a third (C/3) or even a quarter (C/4) of the bulk value. One can see that, by reducing the supply voltage to, say, 1 volt, and using SOI devices with C/3, the power consumption is about 1% that of the bulk circuit with $V_{DD}=5$ V.

This estimation is supported by experimental evidence. It has been reported, for instance, that a frequency divider implemented on SOI is twice as fast and consumes half the power of the equivalent bulk circuit.[51] Table 2 presents the performances of some recent low-voltage, low-power (LVLP), high-speed SOI CMOS circuits. The choice between fully depleted (FD) and partially depleted (PD) devices is not fully resolved yet. Some favour the better performances of the FD devices, and others prefer the better uniformity control of PD devices.

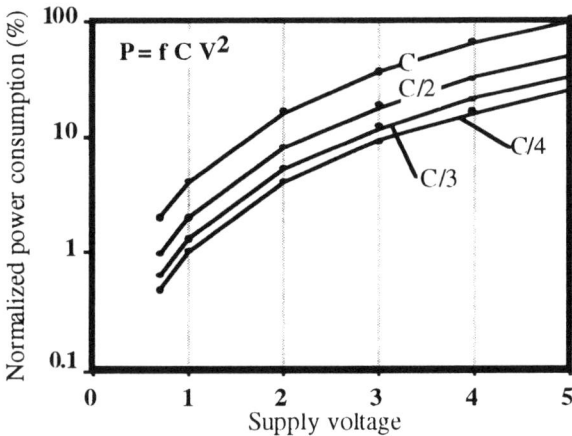

FIGURE 10: Power consumption versus supply voltage and circuit capacitance.

TABLE 2: Low-voltage, low-power, and high-speed SOI CMOS circuits

Circuit	L(μm)	Vdd	Frequency	Power	Company	Ref.
Frequ. Divider	0.15	1V	1.2 GHz	50 μW	NTT	52
Frequ. Divider	0.15	2V	2.5 GHz	130 μW	NTT	idem
Frequ. Divider	0.1	1V	1.2 GHz	60 μW	NTT	idem
Frequ. Divider	0.1	2V	2.6 GHz	350 μW	NTT	idem
Prescaler	0.4	1V	1 GHz	0.9 mW	NTT	53
Prescaler	0.4	2V	2 GHz	7.2 mW	NTT	idem
PLL	0.4	1.2 V	1 GHz	1.4 mW	NTT	idem
PLL	0.4	2V	2 GHz	8.4 mW	NTT	idem
PLL	0.24	1.5V	2.2 GHz	4.5 mW	NTT	54
512k SRAM	0.2	1V	3.5 ns access time		IBM	55
DRAM	0.6	2V			Mitsubishi	56

3.2. MICROPOWER ANALOG CIRCUITS

As far as analog micropower circuits are concerned, it is known that the maximum voltage gain of MOSFETs is obtained when the value of g_m/I_D is largest. This condition appears in the weak inversion regime for MOS transistors.[57]

The value of $\frac{g_m}{I_D} = \frac{q}{\lambda kT}$ is given by Eqn (5).

The low body-effect coefficient of SOI devices should thus enable devices obtaining near-optimal micropower designs. (g_m/I_D values of 35 V^{-1} are obtained, while g_m/I_D reaches typical values of 25 V^{-1} in bulk MOSFETs), see Figure 11.

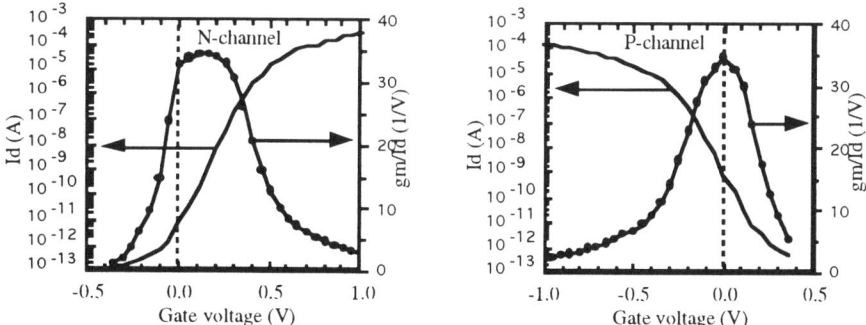

FIGURE 11: g_m/I_D and I_D in 600μm/20μm SOI transistors. V_{back}= 0V (fully depleted n-channel) and V_{back}=-1.5V (accumulation-mode p-channel). Silicon and gate oxide thicknesses are 80 and 30 nm, respectively. g_m/I_D reaches values of 35 V^{-1} in both devices just below threshold.

The CMOS analog switch (pass-gate) provides a rough estimation of the lowest acceptable analog supply voltage, V_{dd}, that can be used in a circuit (Figure 12). In order to transmit a signal without alteration through a switch with the gate of the nMOS transistor held at V_{dd} and that of the pMOS transistor at 0 V, V_{dd} must be larger than a minimum value which can be expressed as a function of the threshold voltage V_T and body factor λ of the n- and p-MOSFET. The following relationship must, therefore, be satisfied:[58]

$$V_{dd} \geq \frac{V_{Tp}.\lambda_n + V_{Tn}.\lambda_p}{\lambda_p + \lambda_n - \lambda_p.\lambda_n}$$

In bulk CMOS with V_T being close to 0.7 V and λ close to 1.5, V_{dd} is classically limited to a minimum supply voltage of 3 V. In fully depleted SOI CMOS with λ equal to 1.1 and the possibility to lower V_T to 0.5 V without degrading the leakage current performance, V_{dd} can be as low as 1.2 V. This confirms the potential of FD SOI CMOS for low-voltage analog applications, for example when using switched-capacitor circuits.

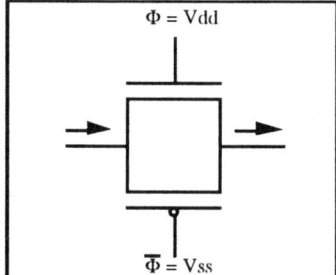

FIGURE 12: CMOS analog switch

CMOS operational amplifiers (opamps) can be designed[59] on the basis of the knowledge of g_m/I_D and the Early voltage, V_A, as a function of the scaled drain current, $I_D/(W/L)$. Figure 13 shows the results of such a design, carried out under the assumptions that λ is close to 1.1 and 1.5 for fully depleted SOI and bulk MOSFETs, respectively, and that V_A is similar for both device types. The design was first used to synthesize a simple 1-stage opamp with specified active device size (W/L) and output capacitance (C_L). Figure 13 clearly shows that the A_{v0} of the SOI amplifier is 25 to 35 % larger than in the bulk counterpart, while the static bias current I_{dd} is divided in the same proportion.

Taking into account the reduction of parasitic source/drain and intrinsic gate capacitances[60] found in SOI in addition to the lower λ value, it can be further demonstrated that a 2-stage Miller opamp can be synthesized to achieve a A_{v0} 15 dB larger and a I_{dd} 3 times smaller in SOI than in bulk for similar area and identical bandwidth considerations (Figure 14). Conversely, total area and I_{dd} can be divided by a factor up to 2-3 when all other specifications are kept constant. These theoretical predictions have been confirmed by the realization of a 100 dB-A_{v0}, 3 µA-I_{dd} Miller opamp using a fully depleted SOI CMOS process (Figure 15). These results establish the potential of FD SOI CMOS to boost the speed (f_T), precision (A_{v0}) and power (I_{dd}) performances of opamps well over bulk implementations.

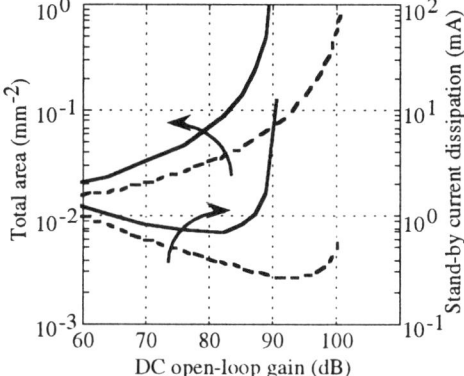

FIGURE 13: Synthesis of bulk (—) and SOI (- -) Miller amplifiers for C_L = 10 pF and f_T = 10 MHz.

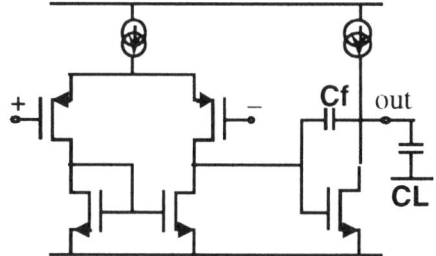

FIGURE 14: Layout of a simple micropower Miller operational amplifier. V_{DD}=3V and I_{DD}=3µA (1µA in each branch of the circuit).

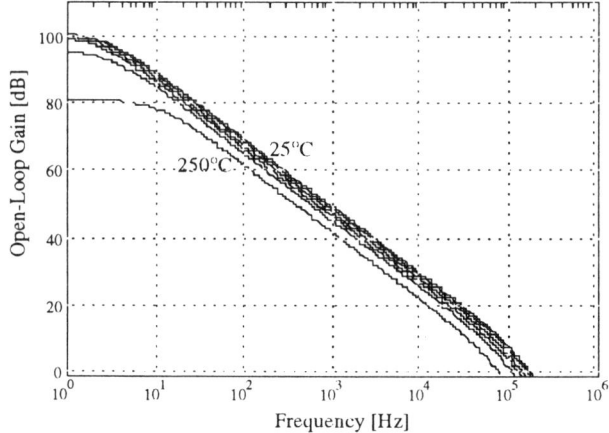

FIGURE 15: Measured Bode diagram of the Miller operational amplifier presented in the previous Figure. V_{DD}=3V and I_{DD}=3µA (1µA in each branch of the circuit). Measurement was carried out at 6 different temperatures (25, 50, 100, 150, 200 and 250°C).

3.3. ESD PROTECTIONS

Bulk circuits utilize large-area diodes, field oxide MOSFETs, SCRs or bipolar transistors as input protections against electrostatic discharge (ESD) damage. Because SOI devices are made in a thin silicon film, the use of diodes, bipolar devices and SCRs is ruled out in SOI circuits. Field transistors are ruled out as well. Because of thermal insulation provided by the buried oxide there is concern that temperature elevation in SOI devices submitted to an ESD pulse would systematically damage SOI CMOS circuits.

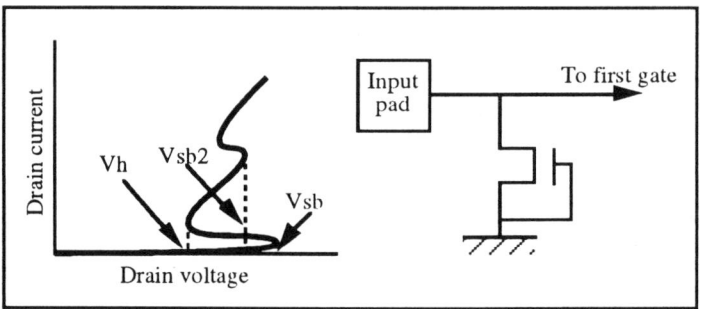

FIGURE 16: Double snapback characteristic in a thin-film FD SOI nMOSFET (left) and ESD input protection using an n-channel SOI MOSFET operating in the snapback mode.

It has been recently demonstrated that n-channel SOI MOSFETs present a double snapback characteristic due to parasitic bipolar action.[61] Typical holding voltage (V_h), snapback voltage (V_{sb}), and second snapback voltage (V_{sb2}) values are 4, 6 and 8 volts for a 1 µm-long device, and 5, 8.5, and 9 volts for a 2 µm-long device, respectively (Figure 16).[62]

One can rely on this snapback mechanism to realize ESD input protections. Indeed, the transistor presented in Figure 16 allows one to use an input voltage as large as V_{sb}, but it clamps the input voltage to the first gate of the circuit to V_h as soon as the input voltage becomes larger than V_{sb}, thereby protecting the circuit from large excursions of input voltage. This type of protection functions for positive as well as for negative ESD pulses. Such protections have been tested under the MIL STD 883C/3015.7 HBM (human body model) ESD stress standard and have show protection from pulses of 4,000 volts. The larger the magnitude of the pulse, the wider the protection transistor. It is found that snapback transistors provide bidirectional protection (positive and negative pulses) of 12.5 volts per micrometer of device (e.g. a 100 µm-wide device offers protection up to 1250 volts). This type of level of protection is comparable to what is usually obtained in bulk.

4. Recent developments

4.1. MICROWAVE SOI MOSFETs

If a high-resistivity (5,000...10,000 Ω•cm or higher) p-type substrate is used the capacitance between the SOI devices and the substrate can be drastically reduced. Using such a substrate reasonable microwave devices can be realized and gains of 13.4 and 17.5 dB have been reported at 2 GHz for gate lengths of 1 µm and 0.25 µm, respectively. These results suggest that microwave SIMOX circuits could provide an economical alternative to GaAs circuits (Table 3).

TABLE 3: Performances of SOI microwave MOSFETs

Device	L	f_T	f_{max}	Noise fig. @ 2GHz	Company	Ref.
MOSFET	1 µm	14 GHz	21 GHz	3 dB	Westinghouse	63
MOSFET	0.25 µm	23.6 GHz	32 GHz	1.5 dB	Westinghouse	64

4.2. SOI BIPOLAR TRANSISTORS

Lateral SOI bipolar devices polysilicon emitters can be realized, also at the expense of process complexity.[65] It is, however, worthwhile to mention that the increase of process complexity necessary to upgrade a CMOS SOI process to (C)BiCMOS is much more modest than for a bulk process. Some very efficient bipolar devices can even be obtained without making any change to the CMOS process. These devices are called "hybrid bipolar MOS" transistors and can be obtained by connecting the otherwise floating body of an SOI MOSFET to its gate (Figure 17).[66,67,68] Such devices can provide common-emitter current gains of 10,000 for L= 0.3 µm.[69] Table 4 presents the performances of some recent SOI bipolar transistors.

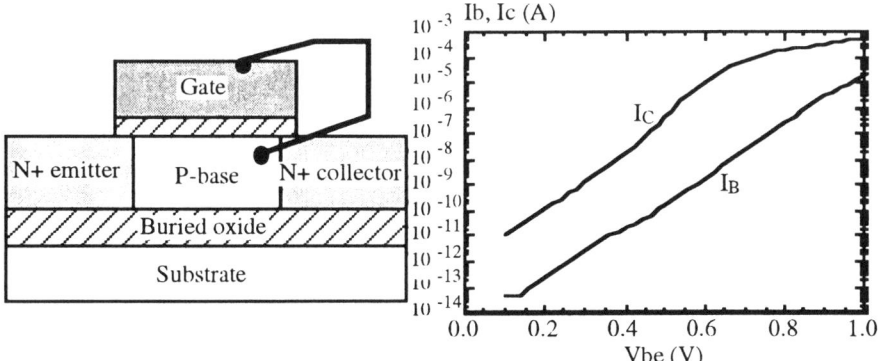

FIGURE 17: Cross section of hybrid bipolar-MOS fully depleted SOI NPN device (left) and measured Gummel plot (right). W/Leff = 20µm/0.6µm, t_{si} = 80 nm, t_{ox} = 30 nm. β_F = 5,000 @ I_C = 10 µA.

TABLE 4: Performances of SOI bipolar transistors

Type of bipolar	L(μm)	β	BV_{CE0}	f_T	Company	Ref.
Lateral	0.2	120	2.5 V	4.5 GHz	UCB	[70]
Hybrid	0.3	10000	2.5 V	-	UCB	[71]
Lateral	-	90	3 V	15.4 GHz	Philips	[72]
Lateral	-	80	> 3 V	10 GHz	Motorola	[73]
Lateral	-	30	2.8 V	20 GHz	IBM	[74]
Vertical	-	50	8 V	12.4 GHz	SIM	[75]
Vertical	-	110	12 V	4.5 GHz	Analog Dev	[76]

It is also worthwhile to mention that it is possible to realize SOI vertical bipolar transistors which are virtually free of Early effect. In these devices, the current flowing from emitter to collector through the base is collected not by the collector-base junction, but by an inversion layer created using the back gate. Such a device has been proposed as early as 1986 [77], and has recently been improved in such a way that Early voltages as high as 30 kV and common-emitter current gains as high as 800 have been obtained. Under these conditions, a current gain x Early voltage product in excess of 10^6 has been demonstrated.[78] Such a device can be used to fabricate high-gain analog amplifiers.

5. Conclusion

SOI CMOS technology appears to offer tremendous potential for low-voltage, low-power circuit applications. Most of the problems related to SOI devices (low drain breakdown voltage, self-heating, and others) tend to fade away as the supply voltage is reduced to 2 V or below. The advent of microwave SOI MOSFETs and novel SOI bipolar structures open the door to mixed microwave-CMOS, BiCMOS and high-performance applications.

References

1. K. Izumi, M. Doken, and H. Ariyoshi (1978) *Electronics Letters*, **14**, 593
2. J.P. Colinge and T.I. Kamins (1987) *Proc. IEEE SOS/SOI Technology Workshop*, 69
3. J.P. Colinge, J. Kang, W. McFarland, C. Stout, and R. Walker (1988) *Proc. IEEE SOS/SOI Technology Workshop*, 68
4. T. W. Houston, H. Lu, P. Mei, T.G.W. Blake, L.R. Hite, R. Sundaresan, M. Matloubian, W.E. Bailey, J. Liu, A. Peterson, and G. Pollack (1989) *Proc. IEEE SOS/SOI Technology Workshop*, 137
5. A.J. Auberton-Hervé, B. Giffard, and M. Bruel (1989) *Proceedings IEEE SOS/SOI Technology Workshop*, 169
6. G.K. Celler, A. Kamgar, H.I. Cong, R.L. Field, S.J. Hillenius, W.S. Lindenberger, L.E. Trimble, and J.C. Sturm (1989) *Proc. IEEE SOS/SOI Technology Workshop*, 139
7. H. Miki, Y. Omura, T. Ohmameuda, M. Kumon, K. Asada, K. Izumi, T. Asai, and T. Sugano (1989) *Techn. Digest of IEEE Internat. Electron Device Meeting (IEDM)*, 906
8. W.E. Bayley, H. Lu, T.G.W. Blake, L.R. Hite, P. Mei, D. Hurta, T.W. Houston, G.P. Pollack (1991) *Proceedings of the IEEE International SOI Conference*, 134
9. L.K. Wang, J. Seliskar, T. Bucelot, A. Edenfeld, N. Haddad, (1991) *Techn. Digest of IEEE Internat. Electron Device Meeting (IEDM)*, 679
10. A.K. Agrawal, M.C. Driver, M.H. Hanes, H.M. Hobgood, P.G. McMullin, H.C. Nathanson, T.W. O'Keeffe, T.J. Smith, J.R. Szedon, R.N. Thomas (1991) *Techn. Digest of IEDM*, 687
11. M.H. Hanes, A.K. Agrawal, T.W. O'Keeffe, H.M. Hobgood, J.R. Szedon, T.J. Smith, R.R. Siergiej, P.G. McMullin, H.C. Nathanson, M.C. Driver, R.N. Thomas (1993) *IEEE Electron Device Letters*, **14**, 219
12. H. Lu, E. Yee, L. Hite, T. Houston, Y.D. Sheu, R. Rajgopal, C.C. CHen, J.M. Hwang, G. Pollack (1993) *Proceedings of IEEE International SOI Conference*, 182
13. GG. Shahidi, T.H. Ning, T.I Chappell, J.H. Comfort, B.A. Chappell, R. Franch, C.J. Anderson, P.W. Cook, S.E. Schuster, M.G. Rosenfield, M.R. Polcari, R.H. Dennard, B. Davari (1993) *Techn. Digest of IEEE Internat. Electron Device Meeting*, 813
14. T. Iwamatsu, Y. Yamaguchi, Y. Inoue, T. Nishimura and N. Tsubouchi, (1993) *Techn. Digest of IEEE Internat. Electron Device Meeting*, 475
15. J.P. Colinge (1991) *Silicon-on-Insulator Technology: Materials to VLSI*, Kluwer Academic Publishers, 124-129
16. D.J. Wouters, J.P. Colinge, and H.E. Maes (1990) *IEEE Trans. Electron Devices*, **37**, 2022
17. J.C. Sturm and K. Tokunaga (1989) *Electronics Letters*, **25**, 1233

18   J.C. Sturm, K. Tokunaga, and J.P. Colinge (1988) *IEEE Electron Device Letters*, **9**, 460
19   J.G. Fossum and S. Krishnan (1993) *IEEE Transactions on Electron Devices*, **39**, 457
20   J.P. Colinge (1991) *Silicon-on-insulator technology: materials to VLSI*, Kluwer Academic Publishers, 132
21   R.H. Yan, A. Ourmazd and K.F. Lee (1992) *IEEE Transactions on Electron Devices*, **39**, 1704
22   H.O. Joachim, Y. Yamaguchi, K. Ishikawa, T. Nishimura and K. Tsukamoto, (1992) *Extended Abstracts of the Solid-State Devices and Materials Conference (SSDM)*, 158
23   J.Y. Choi and J.G. Fossum (1990) *Proceedings IEEE SOS/SOI Technology Conference*, 21
24   C.E.D. Chen, M. Matloubian, R. Sundaresan, B.Y. Mao, C.C. Wei, and G.P. Pollack (1988) *IEEE Electron Device Letters*, **9**, 636
25   R. Sundaresan and C.E.D. Chen (1990) *Proceedings of the fourth international Symposium on Silicon-on-Insulator Technology and Devices*, Ed. by D.N. Schmidt **90-6**, The Electrochemical Society, 437
26   G.A Armstrong and D.W. French (1992) *Solid-State Electronics*, **35**, 1761
27   P.H. Woerlee, AH. Van Ommen, H. Lifka, C.A.H. Juffermans, L. Plaja and F.M. Klaasen (1990) *Technical Digest of IEDM*, 821
28   J.P. Colinge (1991) *Proceedings of the IEEE International SOI Conference*, 126
29   P. Smeys and J.P. Colinge (1993) *Solid-State Electronics*, **36**, 569
30   T.N. Buti, S. Ogura, N. Rovedo, K. Tobimatsu, and C.F. Codella (1989) *Technical Digest of IEDM*, 617
31   T. Hori (1989) *Technical Digest of IEDM*, 777
32   J.P. Colinge, K. Hashimoto, T. Kamins, S.Y. Chiang, E.D. Liu, S. Peng, and P. Rissman (1986) *IEEE Electron Device Letters*, **7**, 279
33   J.G. Fossum, J.Y. Choi, and R. Sundaresan (1990) *IEEE Trans. on Electron Devices*, **37**, 724
34   J.P. Colinge (1987) *IEEE Trans. on Electron Devices*, **34**, 2173
35   L.T. Su, H. Fang, J.E. Chung and D.A. Antoniadis (1992) *Technical Digest of IEDM*, 349
36   G. Reimbolt and A.J. Auberton-Hervé (1993) *IEEE Trans. on Electron Devices*, **14**, 364
37   P.H. Woerlee, C. Juffermans, H. Lifka, W. Manders, F. M. Oude Lansink, G.M. Paulzen, P. Sheridan and A. Walker (1990) *Technical Digest of IEDM*, 583
38   S.M. Guwaldi, A. Zaleski, D.E Ioannou, S. Cristoloveanu, G.J. Campisi and H.L. Hughes (1992) *Proceedings of the Fifth International Symposium on Silicon-on-Insulator Technology and Devices*, Ed. by. W.E. Bayley, Proc. vol. **92-13**, The Electrochemical Society, 157
39   B. Zhang, A. Yoshino and T.P. Ma, *Proceedings of the Fifth International*

	Symposium on Silicon-on-Insulator Technology and Devices (1992) Ed. by. W.E. Bayley, Proc. vol. **92-13**, The Electrochemical Society, 163
40	T Ouisse, S. Cristoloveanu and G. Borel (1990) *Proceedings of IEEE SOS/SOI Technology Conference*, 38
41	P.H. Woerlee, C. Juffermans, H. Lifka, W. Manders, F. M. Oude Lansink, G.M. Paulzen, P. Sheridan and A. Walker (1990) *Technical Digest of IEDM*, 583
42	D. Yachou and J. Gautier (1994) *Proceedings of ESSDERC*, 787
43	L.J. Mc Daid et al., (1989) *Electronics Letters*, **25**, 827-828
44	L.T. Su, K.E. Goodson, D.A. Antoniadis, M.I. Flik, and J.E. Chung (1992) *Technical Digest of IEDM*, 357
45	M. Berger and Z. Chai (1991) *IEEE Transactions on Electron Devices*, **38**, 871
46	O. Le Neel et al. (1990) *Electronics Letters*, **26**, 74-76
47	D. Yachou, J. Gautier and C. Raynaud (1993) *Proceedings of the IEEE International SOI Conference*, 148
48	J.P. Colinge (1991) *Silicon-on-Insulator Technology: Materials to VLSI*, Kluwer Academic Publishers, 177-188
49	D. Flandre and J.P. Colinge (1994) "High-temperature characteristics of CMOS devices and circuits on Silicon-on-Insulator (SOI) substrates", *Proceedings of IX SBMICRO 94*, Rio de Janeiro, 777-786
50	A.J. Auberton-Hervé (1990) *Proceedings of the fourth international Symposium on Silicon-on-Insulator Technology and Devices*, Ed. by D.N. Schmidt, Vol. **90-6**, The Electrochemical Society, 455
51	M. Fujishima, K. Asada, Y. Omura and K. Izumi, (1993) *IEEE Journal Solid-State Circuits*, **28**, 510
52	*ibidem*
53	Y. Kado, M. Suzuki, K. Koike, Y. Omura and K. Izumi, (1993) *IEEE Journal Solid-State Circuits*, **28**, 513
54	Y. Kado, T. Ohno, M. Harada, K. Deguchi and T. Tsuchiya (1993) *Techn. Digest of IEDM*, 243
55	GG. Shahidi, T.H. Ning, T.I Chappell, J.H. Comfort, B.A. Chappell, R. Franch, C.J. Anderson, P.W. Cook, S.E. Schuster, M.G. Rosenfield, M.R. Polcari, R.H. Dennard, B. Davari (1993) *Techn. Digest of IEEE Internat. Electron Device Meeting*, 813
56	T. Eimori, T. Oashi, H. Kimura, Y. Yamaguchi, T. Iwamatsu, T. Tsuruda, K. Suma, H. Hidaka, Y. Inoue, T. Nishimura, S. Satoh and M. Miyoshi (1993) *Techn. Digest of IEDM*, 45
57	E.A. Vittoz (1994) *Tech. Digest of Papers, ISSCC*, 14
58	E.A. Vittoz (1993) *Proceedings of 23rd ESSDERC*, 927
59	D. Flandre, B. Gentinne, J.P. Eggermont, and P.G.A. Jespers (1994) *Proceedings of the IEEE International SOI Conference*, 99
60	Y. Omura and K. Izumi (1991) *IEEE Electron Device Letters*, **12**, 655
61	K. Verhaege, G. Groeseneken, J.-P. Colinge and H.E. Maes (1993) *IEEE Electron Device Letters*, **14-7**, 326

62 K. Verhaege, G. Groeseneken, J.-P. Colinge and H.E. Maes (1993) *Proceedings of the 15th annual EOS-ESD Symposium*, Florida, 215

63 A.K. Agrawal et al. (1991) "MICROX'-An advanced silicon technology for microwave circuits up to X-band", *Techn. Digest of IEDM*, 687

64 M.H. Hanes et al. (1993) "MICROX'-An all-silicon technology for monolithic microwave integrated circuits", *IEEE Electron Device Letters*, **14**, 219

65 R. Dekker, W.T.A. v.d. Einden and H.G.R. Maas (1993) "An ultra low power lateral bipolar polysilicon emitter technology on SOI", *Techn. Digest of IEDM*, 75

66 JP Colinge (1987) "An SOI voltage-controlled bipolar-MOS device", *IEEE Transaction on Electron Devices*, **34**, 845-849

67 JP Colinge (1987) "Voltage controlled bipolar-MOS ring oscillator", *Electronics Letters*, **23**, 1023-1025

68 "P^+-P-P^+ Pseudo-Bipolar Lateral SOI transistor", J.P. Colinge, D. Flandre and D. De Ceuster (1994) *Electronics Letters*, **30**, 1543

69 S.A. Parke, C. Hu and P.K. Ko (1993) "Bipolar-FET Hybrid-mode operation of quarter-micrometer SOI MOSFET", *IEEE Electron Device Letters*, **14**, 234

70 S.A. Parke, C. Hu and P.K. Ko (1993) "A high-performance lateral bipolar transistor fabricated on SIMOX", *EEE Electron Device Letters*, **14**, 33

71 S.A. Parke, C. Hu and P.K. Ko (1993) *IEEE Electron Device Letters*, **14**, 234

72 R. Dekker, W.T.A. v.d. Einden and H.G.R. Maas (1993) "An ultra low power lateral bipolar polysilicon emitter technology on SOI", *Techn. Digest of IEDM*, 75

73 W.M. Huang, K. Klein, M. Grimaldi, M. Racanelli, S. Ramaswami, J. Tsao, J. Foerstner and B.Y. Hwang (1993) "TFSOI BiCMOS technology for low power applications", *Techn. Digest of IEDM*, 449

74 G.G. Shahidi, D.D. Tang, B. Davari, Y. Taur, P. McFarland, K. Jenkins, D. Danner, M. Rodriguez, A. Megdaniss, E. Petrillo, M. Polcari and T.H. Ning (1991) "A novel high-performance lateral bipolar on SOI", *Techn. Digest of IEDM*, 663

75 U. Magnusson, H. Norström, W. Kaplan, S. Zhang, M. Jargelius and D. Sigurd, (1993) "High frequency bipolar transistor on SIMOX", *Proceedings of the 23rd ESSDERC*, 683

76 S. Feindt, J.J.J. Hajjar, M. Smrtic and J. Lapham (1993) "A complementary bipolar process on bonded wafers", *in Semiconductor Wafer Bonding: Science, Technology and Applications*, Proc. Vol. **93-29**, the Electrochemical Society, 189

77 J.C. Sturm and J.F. Gibbons (1986) *in Semiconductor-On-Insulator and Thin Film Transistor Technology*, Chiang, Geis and Pfeiffer Eds., (North-Holland), MRS Symposium Proceedings, **53**, 395

78 K. Yallup, S. Edwards and O. Creighton (1994) *Proceedings of the 24th ESSDERC*, 565

1.2 µm CMOS/SOI ON POROUS SILICON

V.P. BONDARENKO, Y.V. BOGATIREV, L.N.DOLGYI,
A.M. DOROFEEV, A.K.PANFILENKO, S.V. SHVEDOV,
G.N.TROYANOVA, N.N.VOROZOV, V.A.YAKOVTCEVA
Belarusian State University of Informatics and Radioelectronics
P.Brovka 6, 220027 Minsk, BELARUS

1. Introduction

CMOS devices fabricated on Silicon-On-Insulator (SOI) substrates offer several advantages over bulk silicon due to their low power consumption, and inherent immunity to single event and transient upsets as a result of their small active volumes [1]. There are several different technologies to create SOI structures. Thermal oxidation of porous silicon (PS) is a promising technique for the fabrication of high-quality SOI: in contrast to other SOI methods, the Si islands formed are undamaged monocrystalline silicon [2-7].

The principal advantage of this technique is the self-limiting nature of the anodization process, which allows the thickness and uniformity of the SOI layer to be accurately controlled. As a result, thin-film SOI/MOS transistors were fabricated and characterized from the point of view of electrical parameters [4-6] and total dose radiation hardness [7].

In this paper, CMOS SOI technology, electrical parameters of such devices, and total dose gamma-radiation characteristics are reported.

2. Silicon-On-Insulator Structure Formation

The process of SOI fabrication [8] was kept as similar as possible with the existing 1.2 µm CMOS technology. Standard 100 mm n-Si(100) wafers of 4.5 Ohm cm resistivity were used as initial substrates. Then n^+-layers were formed by Sb^+ ion implantation on the face and back surfaces and 1220 ^{0}C furnace anneal. Epitaxial 0.6 µm n-Si layer was then grown on the face surface by dichlorsilane process at 975 ^{0}C. Next 20 nm of amorphous Si was deposited onto epi-layer with subsequent deposition of 0.24 µm Si_3N_4 to produce a mask. Standard RIE process was used to pattern the mask and to define the device islands of feature sizes of 6, 14, 20, and 40 µm.

The anodization process was the only non-standard step and performed in a two-terminal cell using HF/iso-propanol electrolyte. Porous silicon density and structure are known to be complex functions of current density, electrolyte composition, and Si doping level [4, 9]. Therefore, the anodization process (5-7 min) of inhomogeneously doped silicon was controlled potentiostatically. As a result, silicon islands were separated from the substrate by porous silicon layer having a thickness of approximately 1.5 μm.

Porous silicon was oxidized in a three-step process. First, its structure was stabilized [10] and prevented from sintering [11, 12] by low-temperature oxidation at 300 ^{0} C for 1 h in dry oxygen. Porous silicon was then fully oxidized using a steam ambient at 850 ^{0} C and 20 atm pressure. Low temperature and high steam pressure were used to avoid the thermal overheat and outdiffusion from the oxidized porous silicon. This step was also necessary to convert the residual dopant tail of the n^+-Si buried layer into oxide. Finally, the oxidized porous silicon was densified [13] by high-temperature anneal at 1200 ^{0} C in dry oxygen to produce a high-quality thermal layer (Figure 1).

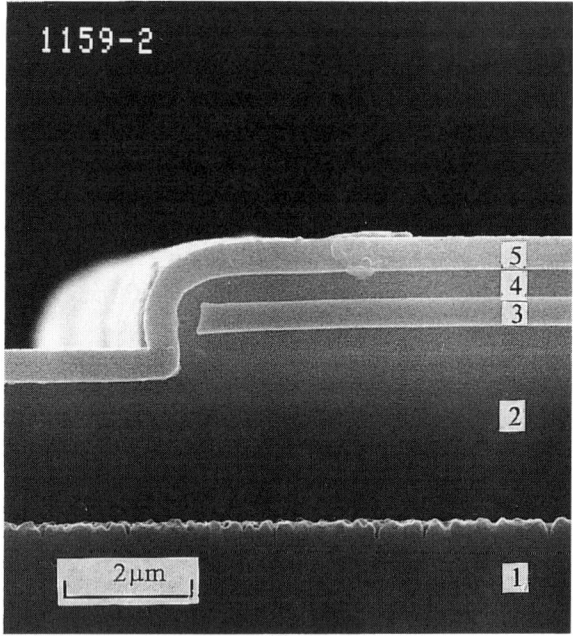

Figure 1. Cross-section SEM photograph of SOI structure formed by PS oxidation

1 - silicon substrate; 2 - oxidized porous silicon; 3 - silicon island;
4 - SiO_2 layer; 5 - poly-Si (for the contrast).

The resulting isolation oxide had a resistivity of 10^{16} Ohm cm, a fixed charge density of $7 \cdot 10^{10}$ cm^{-2}, and was compatible with subsequent device processing. The isolated Si islands had thickness of 0.3 μm, maximal width of 40 μm, background doping of $3 \cdot 10^{15}$ cm^{-3}, and defect density less than 100 cm^{-2} [8].

3. Device Fabrication

P- and N-channel MOS transistors were fabricated in SOI structures using standard 1.2 μm CMOS technology. The gate oxide (30 nm) was formed by pyrogenic oxidation at 850 ^{0} C. Poly-Si gate was doped with phosphorus and had a thickness of 0.45 μm. The rad-hard variant included additional bottom and guard B$^+$ ion implants to control the back threshold voltage and to prevent current leakage along N-channel transistor bottom and sidewalls.

The 23-stage ring oscillators were fabricated and packaged to estimate the stage delay in SOI/CMOS gates under different supply voltage (3...5.5 V), temperature (77...400K) and gamma-irradiation (up to 10 Mrad(Si)) conditions.

4. Radiation

Silicon chips with P- and N-channel SOI/MOS transistors were irradiated in ^{60}Co Gamma Cell under "floating" bias. The dose rate was about 400 rad(Si)/s. Electrical measurements were made less than 1 hour after irradiation. Room-temperature anneal for 100 h after 1 Mrad(Si) total dose was used in several cases. Ring oscillators were irradiated in the same Gamma Cell under 5 V supply voltage and different substrate bias conditions: 0 V; -5 V; and "floating".

5. Experimental Results

5.1. ELECTRICAL PARAMETERS OF SOI/MOS TRANSISTORS

No "kink" was seen in the output characteristics of SOI/MOS transistors. The front channel electron and hole mobilities (550 and 220 cm^2 /V s, respectively) were similar to those obtained on bulk CMOS with similar doping, which supports the idea of low crystallographic defect density and low stress level in SOI structures. N-channel transistors had threshold voltage $V_{th} = 1$ V, subthreshold slope of 80 mV/dec, and leakage current J= 20 pA/μm of channel width. P-channel transistors had the values of -1 V, 90 mV/dec, and 2 pA/μm, respectively.

The small maximal width of isolated silicon island is known to be as one of the problems of porous silicon SOI technology. We have overcome this problem by partitioning devices: powerful output SOI/MOS transistors have been designed in several silicon islands of 26x40 μm individual size, connected in parallel by metallization. As a result, these devices had drain current of 80 mA at $V_{dd} = 5$ V.

5.2. DEGRADATION OF SOI/CMOS PARAMETERS UNDER GAMMA-IRRADIATION

Gamma-irradiation resulted in a certain degradation of device parameters. In the case of P-channel SOI transistors the front gate threshold voltage shift ΔV_{th} was about -0.3 and -0.7 V at 1 and 10 Mrad(Si), respectively, and leakage current decreased down to 0.2 pA/μm at 0 V gate voltage. The threshold voltage decrease is similar to other bulk and SOI transistors [7] and is due to the radiation-induced positive charge in the front-gate oxide.

For the rad-hard N-channel transistors the corresponding values were -0.1 and -0.2 V, and leakage current increased up to 10 and 50 nA/μm ($V_g = 0$) at 1 and 10 Mrad(Si), respectively. Smaller values of threshold voltage shift ΔV_{th} in this case are explained by a partial compensation of interface trap and oxide trapped charge contributions to the shift. Prevent of dramatic leakage current increase was due to the formation of the channel stopped structure: the additional

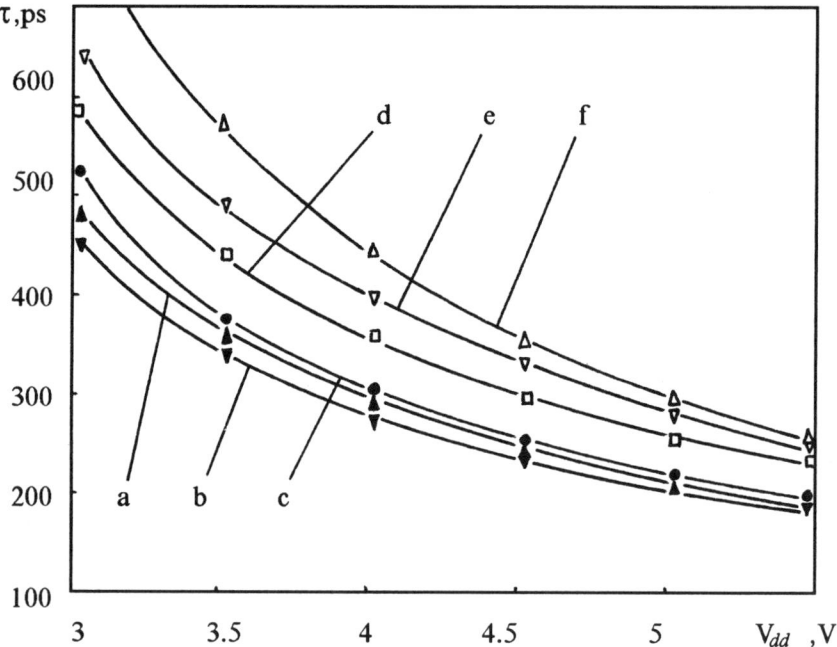

Figure 2. Stage delay vs supply voltage. SOI/CMOS 23-stage ring oscillator. Gamma-irradiation at $V_{dd} = 5$ V, Vsub = 0 V.

(a) before irradiation; (b) 10 min after 1 Mrad(Si); (c) 168 h after 1 Mrad(Si);
(d) 10 min after 10 Mrad(Si); (e) 66 h after 10 Mrad(Si); (f) 168 h after 10 Mrad(Si).

bottom and guard B^+ ion implants were used to control the back threshold voltage and to prevent current leakage along the N-channel transistor bottom and sidewalls.

Room-temperature anneal for 100 h resulted in sufficient recovery of device parameters. Analysis of back-gate subthreshold I-V characteristics has shown the sufficient effect of interface traps to the radiation-induced shift ΔV_{th}.

SOI/CMOS ring oscillators have shown 40% higher speed than reference bulk devices. For $V_{dd} = 5$ V the stage delay values of 150-170, 200-220, and 250-280 ps were measured at 77, 300, and 400 K, respectively. Both supply voltage decrease up to 3 V and gamma-irradiation up to 10 Mrad(Si) resulted in sufficient stage delay increase in passive and active (Figure 2) irradiation regimes, but SOI/CMOS ring oscillators continued stable operating under such environment.

Moreover, irradiation of devices in active regime (under 5 V power supply) resulted in smaller degradation of stage delay in comparison with irradiation in passive regime. This effect can be explained by the switched gate bias of operating ring oscillator. Changing gate bias during gamma-exposure is known [14] to alter MOS radiation response due to partial neutralization of the holes trapped near Si/SiO_2 interface.

The SOI technology developed was used for fabrication of high-speed logic integrated circuits of 74AC series. These SOI/CMOS devices were free of "latch-up" effect and had 40% higher circuit speed in comparison with the same bulk CMOS circuits.

6. Conclusion

The presented SOI technology based on oxidized porous silicon is compatible with standard CMOS processing and does not need expensive, precision installation. Regimes of SOI formation have been adjusted to provide SOI defect density at the level of initial epitaxial structures. The technology can provide the wide range (0.1...10μm) of SOI thickness. SOI/CMOS ring oscillators had 40% higher circuit speed in comparison with the same bulk CMOS devices, and continued stable operating under supply voltage 3...5.5 V, environment temperature 77...400 K and gamma-irradiation up to 10 Mrad(Si).

7. References

1. Colinge, J.P. (1987) Some properties of thin-film SOI MOSFET's, *IEEE Circuits and Devices* 3, 16-20.
2. Tsao, S.S. (1987) Porous silicon techniques for SOI structures, *IEEE Circuits and Devices* 3, 3-7.
3. Holmstrom, R.P., and Chi, J.Y. (1983) Complete dielectric isolation by highly selective and selfstopping formation of oxidized porous silicon, *Applied Physics Letters* 42, 386-388.
4. Zorinsky, E.J., Spratt, D.B., and Virkus, R.L. (1986) The "ISLANDS" method a manufacturable porous silicon SOI technology, *International Electron Device Meeting*, Technical Digest, 431-434.

5. Barla, K., Bomchil, G., Herino R., Monroy, A., and Gris, Y. (1986) Characteristics of SOI CMOS circuits made in N/N+/N anodized porous silicon structures, *Electronics Letters* 22, 1291-1293.
6. Thomas, N.J., Davis, J.R., Keen, J.M., Castledine, J.G., Brumhead, D., Goulding, M., Alderman, J., Farr, J.P.G., Earwaker, L.G., L'Ecuyer, J.,Stirland, I.M., and Cole, J.M. (1989) High-performance thin-film silicon-on-insulator CMOS transistors in porous anodized silicon, *IEEE Electron Devices Letters* 10, 129-131.
7. Matloubian, M., Zorinsky, E.J., and Spratt, D.B. (1988) Total dose radiation characteristics of SOI MOSFETS fabricated using ISLANDS technology, *IEEE Transactions on Nuclear Science* 35, 1650-1652.
8. Bondarenko, V.P., Yakovtceva, V.A., Dolgyi, L.N., Dorofeev, A.M. Vorozov, N.N., and Troyanova, G.N. (1994) SOI structures based on oxidized porous silicon, to be published in *Mikroelectronika* 23 (in Russian).
9. Labunov, V.A., Bondarenko, V.P., Parkhutik, V.P., and Vorozov, N.N. (1981) Investigation of inhomogeneously doped silicon anodization in hydrofluoric acid solution, *Proceedings of the 15th Electrical/Electronics Insulation Conference*, IEEE Publ.No.81CH1717-8, 332-335.
10. Herino, R., Perio, A., Barla, K., and Bomchil, G. (1984) Microstructure of porous silicon and its evolution with temperature, *Materials Letters 2*, 519-521.
11. Labunov, V.A., Bondarenko, V.P., Glinenko, L.K., Dorofeev, A.M., and Tabulina, L.A. (1986) Heat treatment effect on porous silicon, *Thin Solid Films* 137, 123-134.
12. Labunov, V.A., Bondarenko, V.P., Borisenko, V.E., and Dorofeev, A.M. (1987) High temperature treatment of porous silicon, *Physica Status Solidi (a)* 102, 193-198.
13. Yon, J.J, Barla, K., Herino, R., and Bomchil, G. (1987) The kinetics and mechanism of oxide layer formation from porous silicon formed on p-Si substrates, *J. Applied Physics* 62, 1042-1048.
14. Fleetwood, D.M., Winokur, P.S., and Riewe, L.C. (1990) Predicting switched- bias response from steady-state irradiations, *IEEE Transactions on Nuclear Science* 37, 1806-1817.

SOI PRESSURE SENSORS BASED ON LASER RECRYSTALLIZED POLYSILICON

V.A.Voronin, I.I.Maryamova, A.A.Druzhinin, E.N.Lavitska, Y.M.Pankov
"Lviv Polytechnika" State University
Kotlarevsky st. 1, Lviv, 290013, Ukraine

Introduction

In recent years the progress of microelectronic sensors depends on SOI-structures. The use of recrystallized polysilicon layers on insulating structures makes it possible to create low-cost high temperature mechanical sensors, particularly, piezoresistive sensors [1,2], having some advantages in comparison with sensors based on monocrystalline silicon. Among these advantages one could note optimal characteristics in the wide operating temperature range -60...+350°C, extended technological possibilities to create intelligent sensors etc.

The developed method of microzone liquid-phase laser recrystallization of poly-Si layers gives the possibility to obtain high quality SOI-structures which are suitable to create on their basis pressure sensors [3].

Piezoresistive properties of polysilicon layers

The laser-recrystallized polysilicon layers were investigated to create miniature pressure sensors on their basis. Recrystallization of poly-Si layers was carried out by CW IAG-laser irradiation by means of wafer scanning of laser beam. At the region of crystallization front the poly-Si grains turn into elongated (500 µm x 30 µm) monocrystalline (100) areas with preferential <100> azimutal crystallographic orientation. The structure of recrystallized layers is determined by laser treatment conditions and by other technological especialities.

The physical model of the piezoresistance in polysilicon has been developed taking into account contribution in this effect the potential barriers at the grain boundaries (GB) and the structure of poly-Si layers. The carriers' transfer through the potential barrier is supposed to be due to thermoionic emission combined with diffusion [4]. For boron-doped p-type poly-Si the software was created and for the wide temperature and concentration ranges the characteristics of the potential barrier at the GB have been calculated as well as the resistivity and piezoresistance. The temperature dependence of the barrier height was taken into account according to $V_b = V_{bo} (1+ \gamma T)$. Supposing the voltage applied to be small one could consider the grain boundaries and the crystallites as a set of linear resistors.

Therefore,

$$G_l \sim \frac{\rho_g G_{lg}}{\rho_g + 2w \rho_b /(L-2w)} + \frac{\rho_b G_{lb}}{\rho_b + (L-2w) \rho_g /2w},$$

where: ρ_g, ρ_b - resistivities of the grain and the grain boundaries; G_{lg}, G_{lb} - longitudinal gauge factors of the grain and the grain boundary; w - width of the depleted region nearby the grain boundary; L - grain size. The barrier contribution in the resistivity and piezoresistance has been estimated. This contribution was shown to depend substantially on average grain size L (Fig.1). In fine-grained non-recrystallized poly-Si (L = 120 nm) the potential barrier's influence is significant for almost all studied temperatures and dopant levels; in large-grained laser-recrystallized poly-Si (L = 100 μm) this influence should be neglected for the high temperatures and concentrations.

For experimental study of the piezoresistive properties of laser-recrystallized poly-Si layers on insulating substrates the samples were fabricated with the photolithography-formed strain gauges. The relative change of resistance versus strain dependences were measured for both n- and p-type poly-Si strain gauges with different impurity concentrations. The experimentally obtained averaged longitudinal strain gauge factors for laser-recrystallized poly-Si layers are presented in the Table 1. One can see from the table that the longitudinal gauge factor G_l increases with carrier density decrease. Calculated values of the main elastoresistance coefficients are presented in the table also.

TABLE 1. Piezoresistive properties of laser-recrystallized polysilicon layers

Conductivity, carrier concentration, cm^{-3}	Gauge factor G_l	Elastoresistance coefficients m_{ij}
p-type, 10^{17}	55	$m_{44} = 27.5$
p-type, 10^{18}	42	$m_{44} = 21$
p-type, 10^{20}	26	$m_{44} = 13$
n-type, 10^{21}	-22	$m_{11} = -55$

According to our experimental studies the laser recrystallization causes significant (approx. 1.5 times) increasing of the longitudinal gauge factor for poly-Si layers as compared to non-recrystallized ones. For example, in non-recrystallized boron-doped polysilicon layers (N = 5x10^{18} cm^{-3}) G_l =22.8 whereas after the laser recrystallization G_l = 37.5. At the same time resistances of poly-Si piezoresistors after the laser recrystallization strongly decrease as compared to non-recrystallized samples, sometimes this decrease achieves an order of value. The temperature coefficient of boron-doped poly-Si for high concentrations is approximately an order less that in monocrystalline boron-doped silicon and is about 0.01 %/deg. These facts confirm the perspectivity of the laser-recrystallized poly-Si layers to create the piezoresistive sensors.

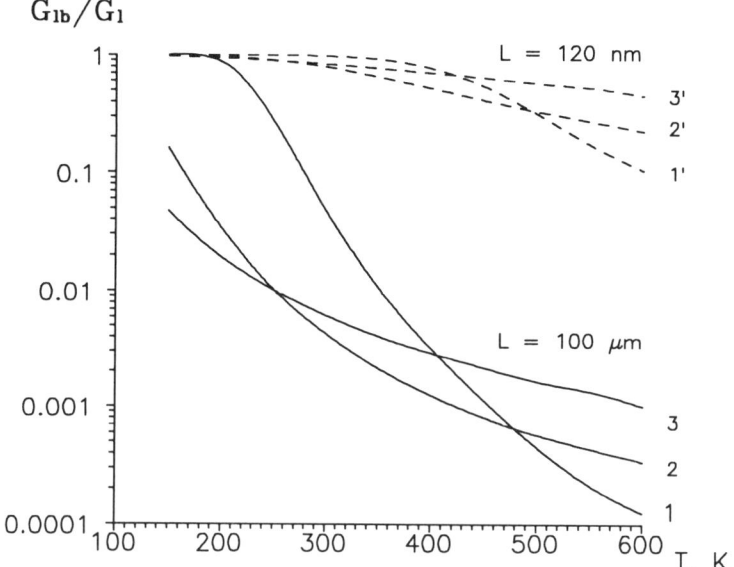

Fig.1. Temperature dependences of the barrier contribution into the piezoresistance of poly−Si for different concentrations: 1,1' − 1×10^{18} cm^{-3}; 2,2' − 4×10^{18} cm^{-3}; 3,3' − 1.6×10^{19} cm^{-3}.

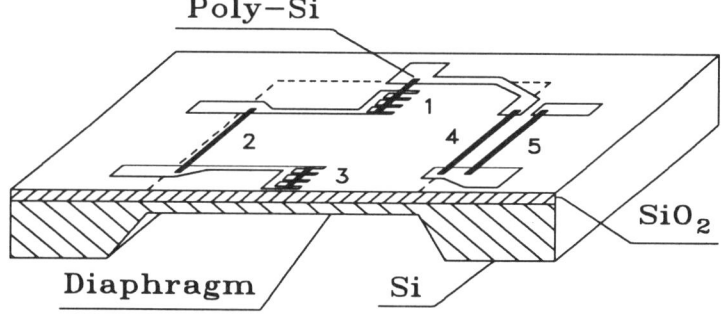

Fig.2. Simplified design of pressure sensor's chip with poly−Si strain gauges.

Piezoresistive pressure sensors

To choose and optimize the parameters, geometrical dimensions and topology of piezoresistive polysilicon pressure sensors the mathematical model was created that would be applied in computer software. The topology and simplified design of the pressure sensor's chip are shown in Fig.2. The 2 mm x 2 mm diaphragm was fabricated by anisotropic etching of Si substrate in KOH water solution. The pressure range was adjusted by the diaphragm's thickness. Taking into account the anisotropic nature of the etching process the diaphragm edges and the scribing strips were oriented in [110] direction. The longitudinal axes of poly-Si strain gauges 1-4 connected into a fully active Wheatstone bridge were aligned in [110] direction corresponding to the laser scanning direction; poly-Si resistor 5 was used for thermal compensation.

The techniques developed to fabricate these sensors were based on technological processes for IC. At the same time this technology includes some specific operations, namely: diphragm fabrication, two-side pattern alignment, chip-to-header bond fabrication. All these operations are important ones because of their significant influence on sensor's performances.

The steps of piezoresistive pressure sensors fabrication are: preparation of initial SOI-structures; laser recrystallization of poly-Si; oxidation; photolithography, including two-side one; forming of the strain gauge pattern; fabrication of the diaphragm; slice scribing to obtain the sensors chips; chip-to-header bond fabrication by means of field-assisted glass-metal sealing. Polysilicon doping by boron was carried out by the ion implantation.

The output voltage characteristics of the sensor with boron concentration 5×10^{18} cm^{-3} are shown in Fig.3 for temperatures between +19 and +153°C. The temperature coefficients of resistance and of sensitivity (output) can be adjusted through appropriate selection of the dopant level. One could see it clearly in Fig.4, where the temperature dependences of output voltage in relative units are presented for two dopant levels: 5×10^{18} cm^{-3} and 1×10^{20} cm^{-3}. The main performances of the developed piezoresistive sensors are presented in the Table 2.

The batch technology of the designed piezoresistive SOI pressure sensors has been developed in cooperation with the "Rodon" concern (Ivan Frankivsk city, Ukraine). These sensors are applied in different branches of industry, science and medicine. The real applications are: machine-building, aerospace investigations, test apparatus etc. The device for arterial pressure measurement is developed on the basis of these sensors also.

Capacitive pressure sensors

Besides that, we're carrying out the investigations to develop the capacitive pressure sensors on the basis of SOI-structures. In these sensors approx. 1 µm thick polysilicon layer was used as a moving capacitor plate that was a diaphragm simultaneously. The controlled gap between the capacitor plates was fabricated by isotropic etching of SiO_2 layer through etch channels. Sensitivity of designed sensor is 10-17 pF/bar. The experimental samples of capacitive sensors are in development for the pressure ranges 0-0.4 bar. This sensors would be applied in medicine, for arterial pressure devices fabrication, in particular.

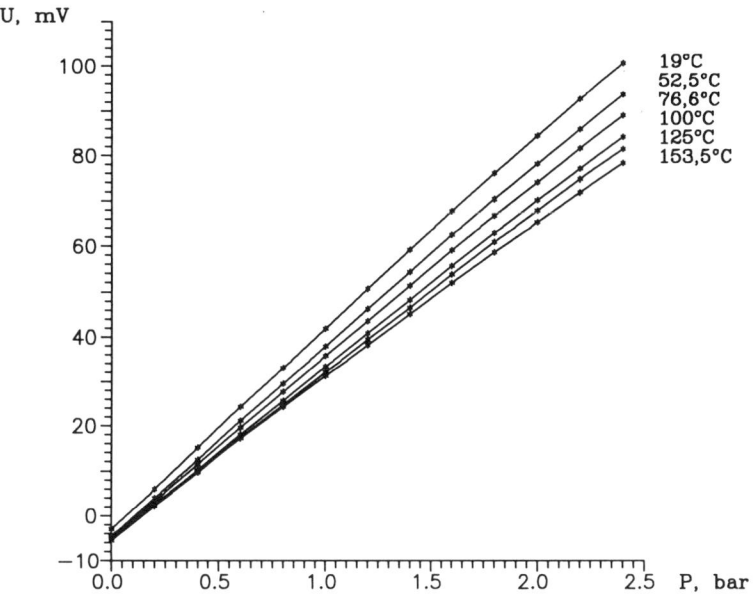

Fig.3. Output voltage vs pressure at different temperatures.

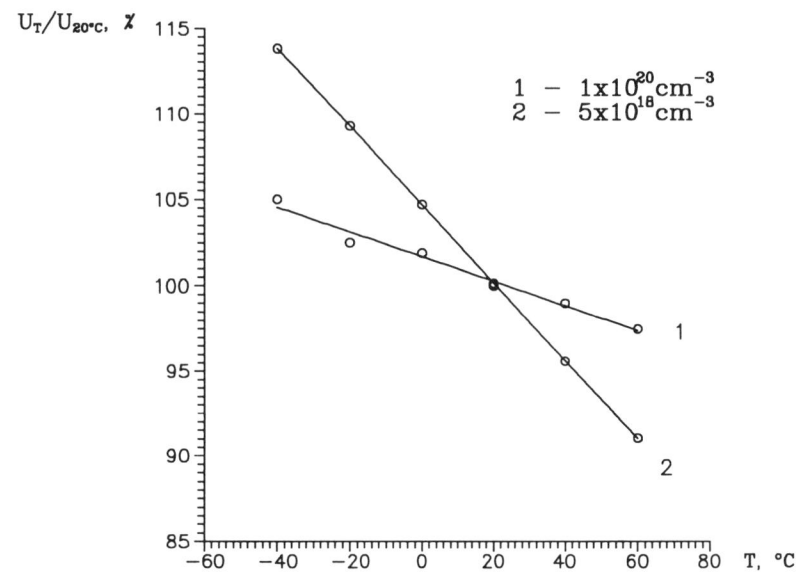

Fig.4. Temperature dependences of sensors output for different inpurity concentrations.

TABLE 2. Technical performances of piezoresistive pressure sensors

Performance	Value
Pressure ranges, bar	0 - 0.4; 0 - 1; 0 - 5; 0 - 10
Maximum overloading, %	200
Ecxitation: - voltage, V	5 - 15
- current, mA	1 - 5
Pressure sensitivity, mV/(V bar)	5 - 15
Temperature coefficient of the offset voltage, %F.S./ °C	0.02 - 0.03
Temperature coefficient of sensitivity, %/ °C	-(0.04 - 0.05)
Temporal drift at 20 °C, %F.S./day	0.03
Operating temperature ranges, ° C	-60...+60; 20...300
Resonant frequency, kHz	100 - 400
Chip's size, mm	5 x 5 x 0.5

Conclusions

The advantages of pressure sensors based on poly-Si SOI piezoresistors are high-temperature operating range, possibility to select the temperature coefficients by the choice of doping concentrations for polysilicon layers and reproducibility of the process of sensors fabrication.

It has been shown that the SOI-structures with laser-recrystallized poly-Si can be successfully used to fabricate piezoresistive pressure sensors with good metrological properties. The laser recrystallization provides the increasing of sensor's sensitivity to pressure and improves its stability.

References

1. Mosser V., Suski J., Goss J. and Obermeier E. (1991) Piezoresistive pressure sensors based on polycrystalline silicon, *Sensors and Actuators* **28A**, 113-132.
2. Obermeier E. and Kopystinsky P. (1992) Polysilicon as a material for microelectronic applications, *Sensors and Actuators* **30A**, 149-155.
3. Voronin V.A., Druzhinin A.A., Maryamova I.I. et al. (1992) Laser- recrystallized polysilicon layers in sensors, *Sensors and Actuators* **30A**, 143-147.
4. French P.J. and Evans A.G. (1989) Piezoresistance in polysilicon and its application to strain gauges, *Solid State Electronics* **32**, 1-10.

INDEX

3-D IC, 222, 217, 246, 247, 265
a-Si film, 93, 146
accomodation volume, 134
accumulation-mode devices, 9, 110, 144, 171, 204, 220, 247
accumulated irradiation dose, 218
AMLCD (active matrix flat panel displays), 183
amorphization, 41
amorphous phase, 95, 134, 184, 226, 275
analog switch, 264
anisotropic etching, 284
annealing, 13, 18, 41, 67, 73, 79, 93, 129, 133, 159, 164, 184, 222, 242, 275
anodization, 22, 41, 275
APCVD, 184
Auger electron spectroscopy, 158
avalanche, 200
bandgap, 18, 56
BESOI (Bonded and Etchback Silicon-On-Insulator), 3, 27, 87, 133
bias-temperature stress, 217
bipolar transistor (BT), 211, 222
body effect, 255
breakdown characteristics, 12, 202, 213
buried channel, 247
buried oxide (BOX), 3, 28, 56, 67, 79, 110, 133, 157, 169, 201, 217, 243, 256
capacitance techniques, 117, 169
capacitive pressure sensor, 284
capture cross section, 145, 244
carbidization, 17
carrier density, 105, 109, 172, 191, 251
channel coupling, 199
channel doping, 199
channel length, 200
channeling phenomenon, 44, 134
charge injection phenomenon, 199, 218

charge pumping (CP) technique, 109, 120, 172, 207
chemical etch, 4, 138, 164
chemical vapour deposition (CVD), 43, 55, 88, 220
chemical/electrochemical deposition of metal, 17
chemomechanical polishing, 88
chip density, 199
cluster, 134, 159
CMOS, 169, 211, 217, 247, 255, 275
coaggregation, 79
concentration profile, 243
conductance techniques, 117
conducting defect, 138
confinement effects, 150
contamination, 242
crystal properties, 15
crystalline quality, 4, 6, 129, 175, 277
crystallinity, 243
crystallites growth, 101
crystallization, 183
current crowding effect, 213
current-based techniques, 109
defect composition, 217
defect density ,5, 41, 65, 89, 277
defect engineering, 27, 50
defect structure, 133
defects formation, 101
degradation, 199
densification effect, 138
depletion, 6, 126, 171, 204, 216, 247
diaphragm, 284
diffusion, 42, 223, 281
diffusivity, 134, 189
dislocations, 23, 29, 50, 222, 246
DLTS (deep level transient spectroscopy) measurements, 191
doped layer, 4
doping levels, 9, 112, 170, 244, 276, 281
doping profile, 202, 211
drain breakdown voltage, 259
DRAM (dynamic random access memory), 7, 46

electrical characterization, 109, 169, 219, 277
electrical conduction, 141
electrochemical anodic treatment, 15, 20
electrochemical etching, 87, 164
electrochemical oxidation, 17
electroluminescence (EL), 21, 147
electromagnetic scanning, 241
electromagnetic separation, 241
electron cyclotron resonance (ECR), 241
electron-spin resonance (ESR), 140
electrophysical properties, 157
encapsulated layer, 33
epitaxy, 18, 28, 40, 56, 101, 137, 175
epiwafer, 88
ESD (electrostatic discharge), 267
ETCH-STOP BESOI, 51
excimer laser, 93, 186
Fermi level, 56, 191, 251
field effect mobility, 191
fixed charges, 112
flat band, 112, 171
floating-body effect, 9, 195, 200, 260, 277
four-point probe, 113
full internal insulation, 217
fully-depleted devices, 8, 111, 133, 176, 199, 247, 256
gamma radiation, 228, 275
gate-all-around (GAA), 222
generation, 171, 193
generation-recombination noise, 247
gettering, 49, 134
glass-layer fusion, 87
grain boundaries, 102, 281
graphoepitaxy, 27
Gummel number, 211, 268
Hall coefficient, 115
Hall-effect, 105
heteroepitaxially grown, 40
heterogeneous reactions, 15
HI-FIPOS (full isolation by porous oxidized silicon with hydrogen implantation), 41
high speed, 199, 212, 279

HIII (high intensity ion implantation), 67
homo-and heteroepitaxial layers, 16
hot carrier instability effect, 183
hot carrier, 199
hot-carrier degradation, 195, 259
hybrid bipolar MOS transistors, 268
impact ionization, 10, 199, 259
implantation, 3, 41, 67, 73, 79, 114, 133, 158, 163, 184, 213, 241, 275, 284
impurity concentration, 282
instability, 194
interface states, 42, 112, 184, 200, 218, 244, 247, 278
internal stress, 74
interstitials, 158
inversion, 110, 171, 208, 216
Ion Cluster Beam technique, 50
ion beam processing, 39, 67, 79, 134, 224, 241
ionizing irradiation, 217
IR-spectroscopy, 157, 163
irradiation behaviour, 147
isotropic etching, 284
junction leakage, 41, 128, 177
kink effect, 11, 46, 195, 258, 277
L-IBEC (lateral-ion beam induced epitaxial recrystallization), 42
laser recrystallization, 101, 184, 281
lateral epitaxy, 28, 43, 101
LDD (lightly doped drain), 214, 260
LDS (lightly doped source), 260
leakage current, 7, 62, 67, 79, 144, 183, 231, 246, 277
lifetime, 41, 62, 103, 116, 175, 199, 241, 260
light emitting diode (LED), 19
liquid sublayer, 27
LOCOS, 220
low dose SIMOX, 3
low frequency noise, 123
low power, 5, 39, 199, 212, 255, 275
low voltage applications, 9, 255
low-frequency noise, 247
LPCVD (low pressure chemical vapour deposition), 101, 184
mainstream applications, 14

MBE (molecular beam epitaxy) -SOI, 49, 56
mechanical grinding and polishing, 4
melt undercooling, 96
melting, 93, 102, 188
mesa isolation, 62
metallization, 277
micropower circuits, 264
microprecipitates, 79
microwave device, 255
microwave ion source, 241
MISFET (metal insulator semiconductor field effect transistor), 83
mobility of carriers, 111, 171, 183, 228, 245, 277
monocrystalline structure, 15, 213, 275, 281
MOSFET (metal oxide semiconductor field effect transistor), 9, 55, 119, 169, 190, 199,
negative output resistance, 261
nitride layer, 67
nitridization, 17
noncrystalline silicon dioxide, 135
nucleation centers, 102, 185
nucleation temperature, 99
numerical simulation, 205
operational amplifier, 265
optical lithography, 46
optical characteristics, 16, 23, 166
optical waveguides, 21
optimization methods, 217, 246
optoelectronic, 15, 50
oxide charge density, 118, 206
oxidized porous silicon (OPS), 21
OXSEF (oxygen-doped silicon epitaxial films), 56
packaging density, 217
packing density, 39, 46, 221
parasitic capacitance, 170
parasitic lateral bipolar, 259
partially-depleted devices, 8, 128, 199, 247
passivating effect, 16, 57, 105
PECVD (plasma enhanced chemical vapour deposition), 93, 184
pedestal-like contact, 213

photo-induced current transient spectroscopy (PICTS), 117
photo-injection studies, 144
photo-magneto-electric (PME) effect, 116
photoconductivity, 116
photodetector, 19
piezoresistive sensors, 281
planar defect, 40
plasma etch, 5, 49
plasma treatment, 229
poly-Si, 27, 93, 101, 169, 183, 213, 268, 281
polycrystalline film, 44, 56, 83, 94
porous Silicon, 4, 15, 41, 275
precipitates, 73, 134, 159, 167
pressure sensor, 281
pseudo-MOS transistor, 109
pulse, 93
pulsed radiation interference, 217
quantum efficiency, 20
radiation defect, 80, 166
radiation hardness, 217, 275
radiation stability, 222
radiative recombination, 18
Random Telegraph Signal (RTS), 124
reactive ion etching, 49
recombination, 121, 141, 171
recrystallization, 28, 93
redistribution, 79
refraction index, 21
relaxation time, 115
residual crystalline defects, 126
RF (radio-frequency) magnetron sputtering, 88
RHEED (reflection high energy electron diffraction) patterns, 89
RIE (reactive ion etch) process, 275
rocking curves, 89
Rutherford Backscattering (RBS), 68, 242
saturation current, 46, 202, 256
scattering, 112, 143
Schottky barrier, 59, 112
Secco etch, 29
second ion mass spectrometry (SIMS), 49, 57, 75, 243
seed, 101

seeding, 95
self-aligned techniques, 213
self-heating effect, 9, 201, 261
semi-insulating layer, 55
SIMNI (separation by implantation of nitrogen), 46, 67
SIMON (separation by implantation of oxygen and nitrogen), 46, 80
SIMOX (separation by implantation of oxygen), 3, 27, 46, 56, 67, 73, 79, 114, 133, 207, 213, 241, 247, 255
sintering, 16, 276
SIPOS (semi-insulating polycrystalline silicon), 56
SIS (semiconductor-insulator-semiconductor), 117
snapback, 267
solid phase crystallization (SPC), 185
SOS (silicon-on-sapphire), 115, 169
SPE-SOS (solid phase epitaxy silicon-on-sapphire), 40
spreading resistance, 83, 114, 212
sputtering, 52, 158
SRAM, 7, 255
SSIC (seed selection through ion channeling), 44
stacking faults, 222
stimulating factors, 73
strain gauge, 284
structural properties, 157, 212, 217, 246
subboundaries, 28, 169, 222
subthreshold slope, 10, 46, 55, 191, 257, 277
TESC bipolar transistor, 211
TFT (thin-film transistor), 183
TH-SPE (thermal induced solid phase epitaxy), 42
thermal oxidation, 16
thermal processes, 16
thermal sintering, 16
thermoionic emission, 281
threshold energy, 188
threshold voltage, 6, 45, 83, 111, 191, 202, 219, 248, 265, 277
topology, 5, 284
transconductance, 111, 171, 212, 219, 258
transient technique, 126, 177
transport measurements, 112
trapping, 141, 183, 206, 220
ULSI (ultra large scale integration), 3, 39
van der Pauw measurements, 114
very large internal surface, 15
VLSI (very large scale integration), 39, 67, 169, 222
waveguiding properties, 19
wet chemical etching, 43
Wheatstone bridge, 284
Zerbst method, 109, 126, 177
ZMR(zone-melting recrystallization)-SOI, 27, 56, 169, 222